TUBE GUITAR AMPLIFIERS
VOLUME 1
HOW TUBES & AMPS WORK

THD - Total harmonic distortion

IGOR S. POPOVICH

DISCLAIMER & COPYRIGHT NOTICE

The information contained in this book is to be taken in the context of a general overview, not specific advice. You should not act on the information contained herein without seeking professional advice. Neither the author nor the publisher (or any other person involved in the publication, distribution, or sale of this book) accepts any responsibility for the consequences that may arise from readers acting in accordance with the material given in the book. Professional advice about each particular case/instance should be sought.

Our choice of designs, parts, models, and brands was based on their availability and educational value. We were not influenced or induced by anybody in our selection, and their use does not mean we endorse or recommend them. You should satisfy yourself that a particular device, component or method is suitable for your intended purpose.

Some circuit diagrams of commercial equipment discussed here were not published by the manufacturers but were posted online by others, and we cannot confirm their accuracy or authenticity. They are used here for review and discussion purposes as "fair dealing," permitted by international copyright laws.

Designs marked with a copyright symbol are the intellectual property of their copyright holders and should not be used without their permission. They are discussed here from an educational perspective only.

INDEMNITY NOTICE

The consequences of any modifications to and deviations from the featured designs are at your own risk, and no responsibility will be accepted by us.

Tube amplifiers involve lethal voltages, high temperatures, and other safety hazards. By purchasing and reading this book, you agree to indemnify its author, publisher, and retailer against any claims of any nature and for any reason. If unsure about any aspect of testing, repairing, or constructing a tube amplifier, engage the services of a suitably qualified and authorized expert.

Published by Career Professionals Australia

Second (revised) edition, 2022

Bulk purchases

This book may be purchased in larger quantities for educational, business or promotional use. Please e-mail us at sales@careerprofessionals.com.au

National Library of Australia Cataloguing-in-Publication Data:

Popovich, Igor S.

TUBE GUITAR AMPLIFIERS, VOLUME 1: HOW TUBES & AMPS WORK

ISBN: 978-0-9806223-5-5
1. Electrical engineering 2. Electronics
I Igor S. Popovich II Title III Index

621.3

CONTENTS OF VOLUME 1

CONTENTS OF VOLUME 1, continued

INTRODUCTION

- About us and this book
- Various types of frames used
- Abbreviations and symbols used

ABOUT US AND THIS BOOK

Welcome to your book, and thank you also for investing your hard-earned money in it. If you borrowed it, thank you for deciding to invest your valuable time in reading it. To understand the story behind this book, please allow me to introduce myself and also my father and technical mentor, Dr. Slobodan (Bob) Popovich. That is why I use "we" in this book. I may be the guy who put it on paper, but even I'm still learning from the Master.

Dr. Bob, long retired from his academic career, was a physics and electronics professor by day and a tube guru by night. His obsession with tubes started early. He went on to build hi-fi and tube guitar amplifiers for local pop and rock bands in our native Yugoslavia.

I started with the theory first, completed my university degree in electronics in 1986, and then learned the practicalities in various engineering positions in Australia. Unfortunately, tubes were out of the curriculum by the early eighties, so I had to educate myself all over again, with Bob's help, of course.

Since those early days, we have been analyzing, experimenting, and building tube gear, which has proved to be a challenging but also satisfying and rewarding hobby.

At a certain stage, we felt we should share our knowledge with a wider audience, so I spent two years writing my book "Audiophile Vacuum Tube Amplifiers." Since there was so much material I couldn't possibly fit it all in one book, the book was published in three large (A4-sized) volumes (the same size as this book).

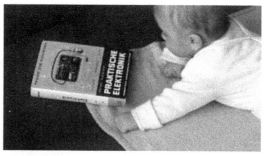

LEFT: Dr. Bob Popovich as a budding student of physics in 1958, repairing a tube amplifier as a means of financing his university studies.

ABOVE: Be careful what kind of book you give your kids to "read", it may have a profound effect on their life. Here I am in 1963, "reading" my first book called "Practical Electronics", in German, no less.

Why this book?

There are a dozen or so vintage books on tube guitar amps and another dozen modern ones. Most are valuable to a guitar player, DIY constructor, or amp fixer and should be read by every serious student of the subject.

Of course, no book is ideal. No book covers everything or explains all aspects of the topic equally well. Although this book is no different in that sense, every author hopes that he brings something new to the table, something worthwhile and valuable. We hope that you will find such helpful info and advice inside these pages.

Just as with my audiophile or hi-fi books, these are not two separate books but one book in two Volumes, so, ideally, both volumes should be read together.

How to read this book

One way to read a technical book like this one is to go to a section or topic that interests you immediately and then keep jumping back-and-forth to the related issues and chapters. This will, I suspect, be the way the more experienced designers and constructors will approach this book.

A more systematic approach is sequential, starting from the beginning and reading in order. This is what I would recommend. Although it seems more time-consuming (since you will read about many issues you may already know a lot about), paradoxically, this approach is often faster. You will not miss anything, and you will not waste time flipping forward and backward trying to clarify an issue that you've overlooked and perhaps not fully understood.

The best way to gain insight into the quality of the design and construction of an amplifier or any other piece of gear, and the only way to experience their sonics, is to have it in your hands. Unfortunately, most sought-after amps are costly, so such an approach is not practical or affordable.

If you own an amp, you could measure it on a test bench or even reverse-engineer its circuit. Some more-or-less accurate circuit diagrams posted on the web by others were obtained through such a method. We have only done it with amps and preamps we purchased and thus felt that we had the right to do whatever we wanted with them.

The next best thing is to have an accurate circuit diagram, which means one published by the manufacturer. Although in some cases DC voltages are not marked, and AC (signal) levels are seldom given, in this book, we will teach you how to determine or at least estimate the operating points, stage gains, and other aspects of such circuits.

Magazine reviews are also helpful. Even a few snippets of information and some results of measurements performed by those reviewers may be enough to identify the circuit topology used and other design aspects.

Learning from the best commercial designers and builders of tube audio is only a means to an end, not the end in itself. Such knowledge should be applied to your own designs and creations, not merely by copying but by experimenting, evaluating, and applying it in a synergistic way to other circuits, situations and designs. Ultimately, if possible, such learning should be crowned by improving on the ideas and designs of others.

WARNING & DISCLAIMER

While most circuit diagrams in this book have been drawn from the actual amplifiers on our test bench, some were obtained from the manufacturers or various Internet sources. Since we cannot guarantee their accuracy and because designs change over time, you should not rely on these diagrams when buying, repairing, or modifying amplifiers. Always verify beforehand the design, components, and wiring of a particular amplifier in an "as-built" state and its actual DC voltages and AC signals in various points.

How we selected the gear for case studies in this book

Buying half-a-dozen boutique amps (at $2,000-3,000 a pop) and another half-a-dozen sought-after vintage examples (at $1,000-2,000 apiece) for case studies in this book would have been a costly proposition. Plus, from the readers' perspective, if you had that kind of money, why would you bother learning about and modifying such expensive amps? You could just buy them and play happily ever after.

Unless you plan to keep them for yourself, modifying highly sought-after (expensive) amps is not a good idea since it almost always significantly reduces their resale value. If you are a collector or fix them for resale, you should leave them as they were designed and built all those years ago. Even the replacement components (resistor and capacitor values, types, and even brands) should be close to the original ones! Thus, we limited ourselves to the more affordable amps, those we could get for a few hundred bucks. Apart from learning how these amps work (the theoretical aspect), the main premise of this book is to get a cheap or "affordable" platform, a faulty, under-appreciated, or unknown amp (whose price hasn't been pumped up into the stratosphere), fix it and improve it, thus making it into a much more valuable and desirable instrument to keep or sell.

Another idea that will be explored in these pages is converting affordable and good-looking solid-state amps with decent cabinets and speakers (if combo) into tube amps.

We received no inducements or equipment samples from any manufacturer; we felt that would compromise our integrity and hinder honest appraisals of their amps. Apart from a couple of exemptions (loaned from friends), we owned most amps featured in this book.

For that reason, you will not find amps from some manufacturers, while others will have two or more of their models featured. Again, we see no problem with that since it reflects the fact that brands such as Fender were trailblazers and de facto standards in the vintage era and are still the most common brands on the used amp market.

Getting in touch with us

If you've liked the book and benefited from it, the best way to repay a favor is to recommend it to your friends and to write a glowing online review. Also, if you spot an error or an omission or have any constructive criticism of the book, I'd like to hear from you to fix it together.

If you want to contribute ideas or projects for the next edition, please let me know if you have ideas on how to make the next edition better.

My e-mail is **igorpop@careerprofessionals.com.au**

I sincerely hope that this book will answer, if not all, then at least some of your questions about the design, building, fixing, and modifying of tube guitar amps and related gear. I wish you every success on your journey through the fascinating world of tube guitar amps!

Igor S. Popovich

VARIOUS TYPES OF FRAMES USED

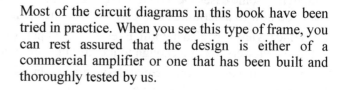

MEASURED RESULTS:

- BW: 15Hz - 35 kHz (-3dB, at $10V_{RMS}$ into 8Ω)
- V_{MAX}: $11V_{RMS}$
- $P_{MAX} = 15W$

Most of the circuit diagrams in this book have been tried in practice. When you see this type of frame, you can rest assured that the design is either of a commercial amplifier or one that has been built and thoroughly tested by us.

RULE-OF-THUMB

Load impedance for triodes:
$Z_{AOPT} \approx 3r_I - 4r_I$
r_I = internal resistance of a tube

These Rules-of-Thumb are simple shortcuts for amplifier builders who don't want to bother with high-level maths, models, and similar highbrow concepts. Easy to memorize, they approximate and summarize much more complex formulas, methods, and concepts.

Although each detailed circuit diagram in this book could be a DIY project by itself, small projects are framed and marked with this soldering-iron symbol.

DIY PROJECT

TUBE PROFILE

Each of the tubes discussed or featured in the designs in this book will have its basic parameters and operational data summarized in a box such as this one.

Commercial designers and manufacturers wish to protect their practical knowledge and "insider secrets." Framed boxes of this kind will emphasize lesser-known practical tips and tricks. Although a magician's hat and a magic wand are used as symbols, there is nothing magical about these tips & tricks; solid scientific and engineering principles underpin all.

TRADE TRICKS

MANUFACTURER'S SPECIFICATION

A list of amp's features and technical parameters published by the amplifier's manufacturer or retailer

SPECS

A WARNING OR A VERY IMPORTANT POINT!

Some issues, myths and warnings are so important that they warrant being emphasized in a frame of this kind.

IMPORTANT FORMULA

The calculator symbol indicates an important or often-used formula.

TRANSFORMERS

Both power & output transformers are critical for proper operation and the tone mojo of tube guitar amps. A few easy-to-measure parameters of commercial transformers are given in this type of a frame.

LOUDSPEAKER PROFILE

Just as with tubes, these frames outline the most critical parameters of specific speaker drivers used in combo guitar amps and speaker cabinets.

DESIGN PROBLEM OR CALCULATION

Intuitive feelings and practical experimentations are fine design and troubleshooting methods, but sometimes one cannot proceed without performing some basic calculations first. It will save you time, hassle, and money!

INVESTIGATION or a CLOSER LOOK

The best defense against false claims, myths and misconceptions is to investigate issues deeper and closer, just any good detective would do, hence the magnifying glass!

ABBREVIATIONS AND SYMBOLS USED

AC	Alternating current	MAX	Maximum	GND	Ground terminal
DC	Direct current	MIN	Minimum	COM	Common terminal
RMS	"Root-Mean-Square", effective value of an AC signal	BNC	Bayonet Neill–Concelman, video & test equipment connector for coaxial cable	ESR, ESL	Equivalent series resistance and inductance (capacitor)
THD	Total harmonic distortion	SQ	Special quality (tube)	AWG	American Wire Gauge
EM	Electromagnetic	PCB	Printed Circuit Board	SPL	Sound Pressure Level
BW	Bandwidth	PP	Push-pull (amplifier)	CF	Cathode follower
MF	Metallized film (capacitor)	OT	Output transformer	SPL	Sound pressure level
CW	Clockwise (rotation of a potentiometer)	CCW	Counterclockwise (rotation of a potentiometer)	DCR	Direct Current Resistance
PP	Peak-to-peak (AC signal)	PPP	Parallel push-pull	TUT	Tube under test
LOG	Logarithmic scale or taper (potentiometer)	NTC, PTC	Negative and positive temperature coefficient (of a resistor or other component)	QF	Quality factor (of a transformer)
LIN	Linear scale or taper (potentiometer)	SET, PSET	Single-ended triode, parallel SET	EMF, CEMF	Electro-magnetive force, counter EMF
TR, VR	Turns or voltage ratio (of a transformer)	E/S	Electrostatic (field, interference or shield)	RCA	Unbalanced audio connector
IR	Impedance ratio (of a transformer)	SLO BLO	Slow blowing fuse, delay fuse	NP	Non-Polarized (electrolytic capacitor)
RC	Resistive-Capacitive coupling between stages or a power supply filtering stage	IC	Internal connection (pin of a tube) or Integrated Circuit	MMF	Magnetomotive Force (of a speaker magnet or transformer)
LC	choke-capacitor power supply filtering stage	GOSS	Grain-oriented silicon steel (transformer lamination material)	FET, JFET, MOSFET	Field effect transistor, Junction FET, Metal Oxide FET
CG	Control grid (of a tube)	SG	Screen grid (of a tube)	SP	Suppressor grid (of a tube)
TPV	Turns-per-volt (of a transformer)	CT	Center tap (of a transformer)	DF	Damping factor
CRC	Capacitor-resistor-capacitor filter	CLC	Capacitor-inductor-capacitor filter	TX	Transfer curve (of a tube)
AF	Audio frequency	RF	Radio frequency	RFI	Radio frequency interference
TC	Tone control	TS	Tone stack	LCR	Inductance-capacitance-resistance meter or circuit
NOS	New Old Stock	NFB	Negative feedback	SB	Standby mode
IEC	International Electrotechnical Commission	DIL	Dual-In-Line (integrated circuit)	UL	Ultralinear (output stage)
IS	Interstage transformer	ZD	Zener diode	DIY	Do it yourself
SS	Solid state (semiconductor)	SW	Switch	TP	Test point
EQ	Equalization	VOL	Volume	HV	High voltage
SMD	Surface-Mounted Device	FX	Effects loop	LV	Low voltage

Currents, voltages and other markings on circuit diagrams

250V DC voltage in the marked node (quiescent state, no signal)

A=55 Voltage amplification of the adjacent stage

1V
~ AC signal voltage in the adjacent node (RMS or effective value)

5mA DC current through the adjacent branch (quiescent state, no signal)

✗ WRONG - how not to do it!

✓ RIGHT- how to do it!

Symbols

PHONE (1/4') JACKS

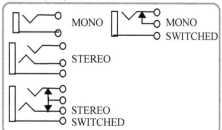

MAGNETIC COMPONENTS

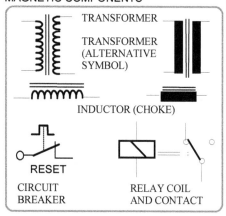

CAPACITORS

AC AND DC SOURCES

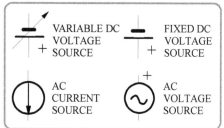

MISCELLANEOUS SYMBOLS

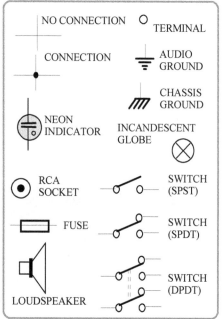

RESISTORS

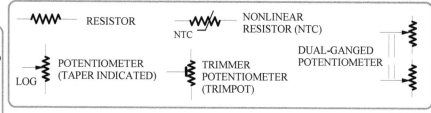

SEMICONDUCTORS

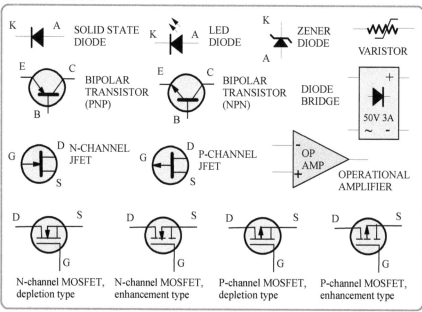

ELECTRON TUBES

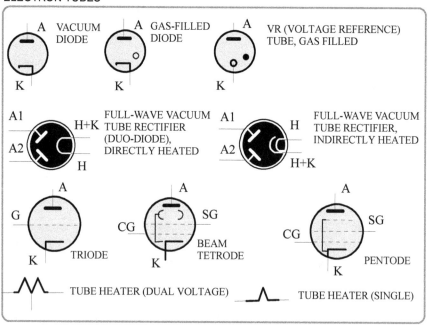

MATHEMATICAL & LOGICAL SYMBOLS

$\approx$ APPROXIMATE $\parallel$ PARALLEL CONNECTION $\triangleq$ EQUIVALENT

TEST INSTRUMENTS

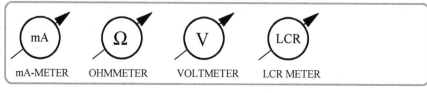

BASIC ELECTRONIC CIRCUIT THEORY

1

- FUNDAMENTAL LAWS OF ELECTRICAL ENGINEERING
- CIRCUIT ANALYSIS TOOLS AND METHODS
- AC VOLTAGES AND CURRENTS
- RESISTIVE, CAPACITIVE AND INDUCTIVE IMPEDANCE
- THE BLACK BOX CONCEPT: THEVENIN'S AND NORTON'S THEOREMS

Before medical students study specialist subjects, they need to pass fundamental courses such as anatomy. By the same token, amplifier builders and fixers must master the basic circuit concepts. AC and DC voltages cause currents to flow in circuits, and such voltages and currents draw (consume) power from power supplies and produce useful or signal power.

Passive electronic components (resistors, capacitors, and inductors) modify voltage and current signals, make them dependent on the frequency of AC signals passing through them, and thus act as filters and tone-shaping circuits.

FUNDAMENTAL LAWS OF ELECTRICAL ENGINEERING

Ohm's Law

Three fundamental electrical engineering laws underpin all others. The first is Ohm's law, named after Georg Simon Ohm, German physicist, and mathematician (1789–1854).

The simplest possible electrical circuit consists of one voltage source and one resistor. Without the load (resistor) connected, the battery terminals are open, and no current flows - there is no "circuit." At such a low voltage (12 volts), the air around the battery is an insulator. Once there is a conductive path across the terminals (nodes) A and G, no matter how large or small such a resistance may be, the circuit is closed, and a current will flow.

Resistors "resist" the flow of current and limit it in proportion to their resistance—the higher the resistance, the lower the current. A resistor is a physical component, a device, while resistance is its physical property.

A DC source could be a 1.5V AA battery or a 12 Volt car battery. Even though we use the battery symbol in circuit diagrams, this DC voltage source does not have to be of chemical nature (a battery); it could be a stand-alone fixed or adjustable mains-powered DC power supply on a test bench or a DC power supply inside a tube amplifier.

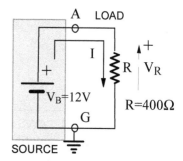

$$I = V_B/R = 12V/400\Omega = 0.03A = 30mA$$

In voltage terms, DC or "Direct Current" means that one terminal is always positive, and the other one is always negative, with respect to each other. In electrical engineering, positive and negative are relative designations.

One terminal is chosen as a referent point, in this case, point G. Often (but not always), this point is grounded or "earthed", physically connected to the neutral or earth terminal of a mains supply and the metal chassis of an amp, so a symbol G, GND or E is used. Likewise, COM or REF may be used to indicate that this point is a common or referent point. Of course, point A could be grounded instead and be considered a referent point. Then point G would be at -12V with respect to ground. Voltage polarities are relative, depending on the referent point.

Ohm's Law describes the relationship between three parameters in a circuit, voltage, resistance, and current. Since there are three parameters, the same Law can be written in three different ways: V=R*I, R=V/I or I=V/R.

These can also be expressed verbally in three ways. The voltage drop on a resistor is a product of its resistance R and the current I flowing through it. The resistance of a resistor is the ratio of the voltage across the resistor and the current flowing through it. The current through a resistor is the voltage drop across the resistor divided by the resistance of the load.

In this case the voltage drop V_R on the resistor equals the battery voltage ($V_B=V_R$), because there are no other resistors in the circuit. Notice the current flowing out of the positive end of the battery and into the positive end of the load (the resistor). Resistors don't have positive or negative ends; the + next to the resistor indicates the polarity of the voltage drop across it! This is how you can tell the difference between "sources" and "loads" in a circuit. Currents come out of the + poles (or terminals) of sources and enter into the + poles of loads.

Resistors as loads

The term "load" is significant in circuit analysis. The battery is the source of electrical energy, the source of electrons, negatively charged particles whose flow is called "electrical current." The resistor is the load, meaning it uses that energy in some way. Some "loads" dissipate energy by converting it into heat. An electrical heater is nothing but a powerful wire-wound resistor in open air or in an oil-filled tank. Other loads convert electrical energy into other forms, such as mechanical energy, as electric motors and loudspeakers do (air pressure changes called "sound").

In amplifiers, resistors are used to produce DC voltage drops (to reduce voltage levels) or to convert the flow of alternating current into AC voltage signals. The unit for voltage (both AC and DC) is called Volt, in honor of Italian physicist Alessandro Volta (1745–1827).

Notice different ways resistors can be "marked" in a circuit. The symbol for resistance unit, called Ohm, is a Greek capital letter Ω (omega). There are larger units, such as kiloohm ($1k\Omega = 1,000\ \Omega$) and MΩ (megaohm, 1 million ohms or $10^6\ \Omega$), and smaller units such as milliohm ($0.001\ \Omega$).

In amplifier diagrams we usually write 1k2 instead of 1.2kΩ. Apart from avoiding awkward Greek letters (apologies to our Greek readers), this has another advantage. Circuits get copied by hand or photocopied and the . (dot) between 1 and 2 in 1.2 kΩ can easily get lost or omitted by mistake. Suddenly, 1.2 kΩ becomes 12 kΩ, which can have serious consequences. The finished amplifier may not work properly or even go up in smoke. Using 1k2 naming convention prevents this from happening!

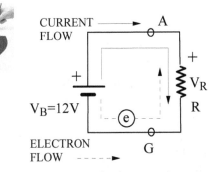

CURRENT FLOW AND ELECTRON FLOW

In modern literature, the positive flow of current is the opposite of the electron flow. Electrons are negatively charged particles, so in this circuit, they will come out of the battery's negative terminal and flow upwards through resistor R towards the battery's positive pole. Opposites attract in love and electronics!

In some older and American books, you may find the direction of electron flow as positive current flow, but there is no need to get confused - the positive current flow is nothing but a convention; you can use either standard, as long as you do it consistently.

The "official" direction of current flow is the opposite of electron flow

Kirchoff's 2nd or Voltage Law

The two laws were named after German physicist Gustav Robert Kirchoff (1824–1887). We'll illustrate them with a simple DC circuit with one source and a load, two resistors in series, R_1 and R_2. The same current flows through the two resistors and "develops" the voltage drops V_{R1} and V_{R2}. Kirchoff's Second Law says: The sum of voltage drops in a closed-loop equals the sum of voltage sources.

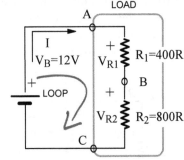

Here we introduce the notion of a "loop," which means going around a closed circuit and writing the voltage equation of such a loop. We can do it in two ways. We can put all voltage sources on one side of the equation and all voltage drops on the other, so we get $V_B = V_{R1} + V_{R2}$.

Alternatively, we can put all voltages on the same side of the equation. We can start at any point and go in any direction. Let's start in point A and go clockwise. The arrow will hit the + side of V_{R1} first, then the + side of V_{R2} and finally the - side of the battery V_B, so we write $V_{R1} + V_{R2} - V_B = 0$. Why zero? Because that's what 2nd KL says, that the sum of voltages in a closed loop is always zero. When you look at the first equation, $V_B = V_{R1} + V_{R2}$, it is exactly the same equation, just written in a different way.

The 2nd KL: The sum of voltages in a closed loop is zero.

Let's calculate the values for the current I and the two voltage drops V_{R1} and V_{R2}! Ohm's law says that $V_B = V_{R1} + V_{R2}$ and since $V_{R1} = IR_1$ and $V_{R2} = IR_2$ we have $V_B = IR_1 + IR_2 = I(R_1 + R_2)$

Finally $I = V_B/(R_1+R_2) = 12/(400+800) = 12/1,200 = 0.01$ A [Ampere] or 10mA.

Ampere is the international unit (abbreviated SI from French: Le Système International D'unités) for electric current, named after André-Marie Ampère (1775–1836), French mathematician and physicist. A smaller unit called mA (milliAmpere) is often used in electronics, 1mA= 0.001A or 10^{-3}A.

Now that we know the current flowing in this circuit, we can calculate the voltage drops using Ohm's law: $V_{R1} = IR_1 = 0.01*400 = 4.0V$, $V_{R2} = IR_2 = 0.001*800 = 8.0V$.

The two resistors "divided" the source voltage of $12V_{DC}$ (between points A and C) in proportion to their resistances, to 4 and 8 Volts. R_2's resistance is twice the resistance of R_1, so twice as much voltage is "dropped" across R_2 compared to R_1. We can write $V_{AC} = V_{AB} + V_{BC}$ or $V_{BC} = V_{AC}*R_2/(R_1+R_2)$

This type of circuit, called a *voltage divider*, is extremely important and we will come back to it many times in this book. It applies to both AC and DC circuits. Remember this formula for voltage division well!

Kirchoff's 1st or Current Law

While the 2nd KL was about voltages in a loop, the 1st law is about currents entering or exiting a "node." A node is a point where three or more "branches" meet. 1st KL: The sum of currents entering a node equals the sum of currents exiting that node.

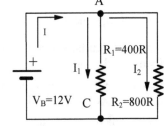

We know that $V_B = R_1 I_1$ so $I_1 = V_B/R_1 = 12/400 = 0.03A$ Likewise, $V_B = R_2 I_2$ so $I_2 = V_B/R_2 = 12/800 = 0.015$ A. Mr. Kirchoff would have us believe that $I = I_1 + I_2 = 0.03$ A + 0.015 A = 0.045 A!

Here we will introduce an important concept called "the equivalent circuit." We have two resistors connected in parallel, and we can calculate or measure their "equivalent" resistance, that of a single resistor that could replace the two of them. In this simple example "the rest of the circuit" means only the battery, but it will illustrate the principle regardless.

The 1st KL: the sum of currents in any node of a network is zero!

The formula for the equivalent or parallel resistance R_P of N resistors connected in parallel is $1/R_P = 1/R_1 + 1/R_2 + ... + 1/R_N$

In this case N=2 (2 resistors) so we have $1/R_P = 1/R_1 + 1/R_2$

Also, $1/R_P = (R_1+R_2)/(R_1*R_2)$ and finally $R_P=R_1*R_2/(R_1+R_2)$. Instead of writing words "in parallel", a sign or symbol "||" is used to denote parallel connection, as in $R_P=R_1||R_2$!

In our case, $R_P=400*800/(400+800) = 320,000/1,200 = 266.67 \ \Omega \approx 267\Omega$. The symbol "$\approx$" means "approximately". Now we can calculate the current supplied by the battery: $I = V_B/R_P = 12/267 = 0.045$ A (45mA), the same result as before.

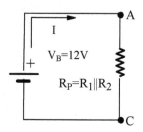

The equivalent circuit between points A and C as far as the battery is concerned

CIRCUIT ANALYSIS TOOLS AND METHODS

Current loop method

Consider this more complex circuit, with two voltage sources and three resistors. Again, it is a DC circuit, but this and all other circuit analysis methods equally apply to alternating current (AC) circuits. To determine voltages and currents in this circuit, the current loop method uses the 2nd Kirchoff's law to write the loop equations for all the loops in the circuit. Here we have three branches, two nodes (A and B), and two independent loops. The orientation of the loops does not matter. In our case, it is clockwise, but you can go anti-clockwise if you prefer. However, once you choose the direction of the first loop, all other loops must be of the same direction.

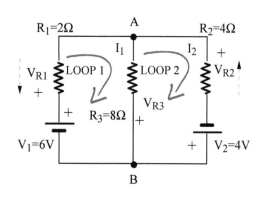

Voltage polarities are written as the arrow "enters" them. Let's follow Loop 1, or current I_1. Starting at B, I_1 flows upward and "hits" the voltage source's - or negative terminal, then it flows through resistor R_1. A current creates a positive potential at the point of entrance and negative potential at its exit, so the voltage drop across R_1 is V_{R1} with its plus as marked. Sometimes a dotted arrow is used as well (as illustrated), but that may be confused with current flow, so a + sign should be sufficient.

I_1 then reaches node A and flows through R_3 downwards into node B and thus completes its loop. Now we can write $-V_1+V_{R1}+V_{R3} =0$. We don't want voltages, so we substitute $V_{R1}=I_1R_1$ and $V_{R3}=(I_1-I_2)R_3$

Notice that the current through R_3 is I_1 flowing clockwise through it, in the "positive" direction, minus current I_2, flowing upwards through it or in the "negative" direction, so the current through R_3 is I_1-I_2!

Finally we have $-V_1+I_1R_1+I_1R_3 -I_2R_3=0$

We write the second loop directly now starting at B. Again, it does not matter at which node you start from, providing you "complete" the loop: $I_2R_3-I_1R_3+I_2R_2-V_1=0$

Notice that the current I2 through the common resistor R_3 is positive in this loop, and I_1 flowing through the same resistor is now negative! We have two equations with two unknowns, I_1 and I_2. Once we solve this system of equations we can calculate any current or voltage in this circuit.

$-V_1+I_1R_1+I_1R_3-I_2R_3=0$ (Eq.1) and $I_2R_3-I_1R_3+I_2R_2-V_2=0$ (Eq.2)

If we substitute resistance values we get $-6+2I_1+8(I_1-I_2)=0$ and $8(I_2-I_1)+4I_2-4+0$ After solving the two equations by substitution, cancellation or matrix method, we get $I_1=1.857A$ and $I_2=1.571A$

Since I_1 is larger than I_2, the resultant current $I_1-I_2 =0.286A$ flows through R_3 downward (in the direction of the larger current, which is I_1). Now we know all currents in the circuit and can calculate any voltage we want.

Node voltage method

In simple circuits, instead of loops, we can write a single equation immediately. This method is one such shortcut. The 2nd KL for node A is $I_3=I_1+I_2$

Since $I_1*R_1 = I_2*R_2 = V_{AB}$ and the 1st KL says that $V_1 = V_{AB}+I_3*R_3$, so we can immediately write $I_3=I_1+I_2$ as $(V_1-V_{AB})/16 = V_{AB}/4 + V_{AB}/8$

Now we have one equation with one unknown, V_{AB}. Multiplying the whole equation by 16, we get $V_1-V_{AB} = 4V_{AB} + 2V_{AB}$ or $V_{AB}= V_1/5 = 2V$!

Now that we know the two voltages, we can calculate all the currents.

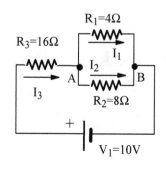

Superposition theorem

The superposition theorem is especially useful in the analysis of tube circuits. In a linear network with two or more voltage or current sources, the current through any element (impedance or branch) is the sum of currents produced by each source separately, providing the internal resistances (or impedances for AC circuits) of these voltage sources are left in the circuit.

This means that we analyze the circuit twice if there are two voltage or signal sources (as in this example), three times with three sources, and so on. Here we have source V_1 with its internal resistance R_1 and source V_2 with its internal resistance R_2. It is the same circuit we analyzed in the section on loop currents, so the results must also be the same! Notice that two currents are flowing in the opposite directions, so $I_3 = I_{3A}-I_{3B} = 0.429-0.1429 = 0.286A$

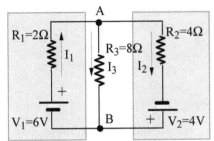

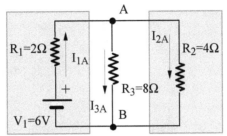

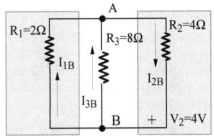

The original circuit to be analyzed, as for the current loop method on the previous page

STEP 1: V_2 removed, analysis with V_1 only

$(V_1-V_{AB})/2 = V_{AB}/8 + V_{AB}/4$

$V_{AB} = 4V_1/7 = 3.428V$

$I_{3A}= V_{AB}/8= 0.429A$

STEP 2: V_1 removed, analysis with V_2 only

$(V_2-V_{BA})/4 = V_{BA}/8 + V_{BA}/2$

$V_{BA} = 2V_2/7 = 1.143V$

$I_{3B}= V_{BA}/8= 0.1429A$

AC VOLTAGES AND CURRENTS

There are two types of physical quantities. Scalars (same root as the word "scale") can be fully described with only their magnitude or numeric value. For instance, temperature and resistance are scalars.

Vectors are described or wholly defined by two parameters, the numeric value, or magnitude, and the direction. For instance, force is a vector. It isn't enough to know how much force you are using to push something away from you; the direction of such force also matters. Vectors are symbolically represented by arrows. The length of the arrow is the vector's magnitude or amplitude, while the angle between such arrow and a referent line is the vector's "phase".

Vectors can be added and subtracted, but not simply by adding their amplitudes. It has to be done in a two-dimensional space. Such addition, multiplication, and other operations on vectors are called vector algebra.

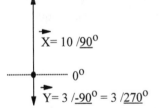

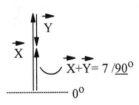

RIGHT: A vector can be represented as a sum of two perpendicular components or its "projections" on X- and Y-axis

$Z=X+Y$

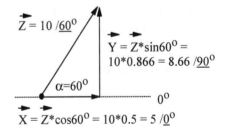

ABOVE: Adding parallel vectors

In the first example, we see two parallel vectors, X and Y, but their directions are opposite. Vector X has a magnitude of 10 units (Volts for voltage or Amperes for current, for example), vector Y's magnitude is three units. Notice how vectors have a little arrow on top of their names to distinguish them from scalars. In our text here, that is awkward to do due to the limitations of the software, so please understand that many physical quantities in this book will be vectors (such as AC voltages and currents), but for convenience reasons, we will write their symbols without their arrows.

In this case, it is easy to add the two vectors. We can move them relative to one another in any way we like, providing we change neither their magnitude (length) nor their angle. Adding a vector of the opposite angle is akin to subtracting its length in the same way that adding a negative number is equivalent to subtracting the same positive number. Rotating vectors or "phasors" were named that way because their phase changes constantly.

Adding vectors that have any angle a between them is more complex; trigonometry rules apply. As a special case, if that angle is 90°, we can use Pythagora's Theorem to calculate their sum (hypotenuse).

To truly understand the origin of sinusoidal waves, such as electrical voltages and currents, we need to study the behavior of a rotating vector Z. Phasor Z has an amplitude of 1, since the radius of the circle is also 1. This is because the maxima and minima of the sine function in mathematics are +1 and -1!

Phasor Z rotates in the counter-clockwise direction, which is the positive direction in mathematics and physics. It starts rotating at the moment in time marked t=0. We are interested in the projections of vector Z on the X- and Y-axis.

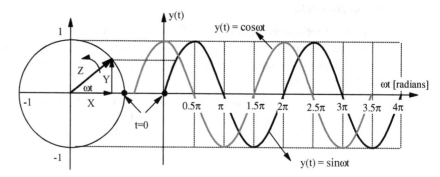

ABOVE: Phasors (rotating vectors) and sinusoidal signals

ANGULAR AND NATURAL FREQUENCY OF A PERIODIC SIGNAL

$$\omega = 2*\pi/T = 2*\pi*f$$

At t=0 moment, its projection on Y-axis is zero, marked with the dot on the waveform. At another point in time, as drawn, its projection on Y-axis is also marked as Y. Again, we see that the two projections are also two perpendicular components of the vector Z. The sum of these two components always equals Z or 1 in this case. If you trace the vertical projection Y in time, you will get a sine wave described by a sine function: $y(t) = \sin\omega t$.

The maximum is reached when the angle $\omega t = 90$ degrees or $\pi/2$ radians, then the projection goes down towards zero, which is reached at $\omega t = 180$ degrees or π radians. Then the same happens in the opposite direction. Eventually, the phasor Z reaches the original position it had at t=0. Since the rotational speed and rotation frequency are constant, the projected waveform repeats itself with the same amplitude and shape. That is why we call such signals periodic - after each period, they repeat themselves.

This period is marked capital T to signify a constant (a scalar) and distinguish it from any moment in time "t". ω (lower case Greek letter "omega") is called angular, or radian frequency and is related to the "natural" frequency by $\omega = 2\pi f$. Since f and T are reciprocal, $\omega = 2\pi/T$!

The same discussion applies to the horizontal projection. At t=0, such a projection is already at its maximum. This wave will look the same as the sine wave or the vertical projection, but it will be shifted 90° or $\pi/2$ radians forward; it will "lead" the sine wave by 90°. This mathematical function is called a cosine function. The sine wave is described as $y(t) = \sin(\omega t)$, and the cosine wave is $x(t) = \cos(\omega t) = \sin(\omega t + p/2)$, so the cosine wave is simply a sine wave shifted in phase by $\pi/2$ radians, 90 degrees or 1/4 of the period.

We will talk about the "leading" and "lagging" phase very soon. These two waves can be represented by two perpendicular phasors (90° between them) of the same amplitude and rotating together with the same frequency. Hence, the phase angle between them never changes.

Lower case symbols are used for instantaneous values of signals, while upper case letters are used for the peak, average and effective values.

To understand these three fundamental terms, we need to move on from the mathematical definition of a sine or cosine wave to electrical signals such as voltages and currents.

When an electrical conductor rotates in a magnetic field, an AC voltage is "induced" at its ends, and if a load is connected to it, an AC current will flow through the load. This is the operating principle of alternators that generate alternating (AC) voltages in power stations.

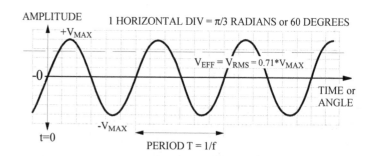

ABOVE: The amplitude, V_{MAX}, effective or RMS value V_{EFF} and period (frequency) of a sinusoidal signal

Generators are the opposite of electric motors. Mechanical energy from a steam turbine or a diesel engine drives the alternator, which converts such mechanical energy into electrical energy. Motors take electrical energy from the power grid and convert it into rotational mechanical energy to drive loads such as pumps, fans, and compressors. Eventually, a "reduced amplitude" version of the alternator's voltage ends up as the "mains voltage" in your house. The frequency, the period, and the waveform stay the same. The mains frequency in most countries is 50Hz, meaning their power generators rotate with a natural frequency of 50 times a second.

In some countries the mains frequency is 60Hz, meaning their power generators rotate faster, 60 times per second. The corresponding angular frequencies are $\omega=2\pi50=314$ radians/second and $\omega = 2\pi60 = 377$ radians/sec. While the amplitude of a sine function was 1, the amplitude of sinusoidal voltages and currents can vary, so in general, we write $v(t) = V_{MAX}\sin(\omega t)$, V_{MAX} is called the amplitude of the sinusoidal voltage and $T=1/f$ is its period.

Power in AC circuits

The effective or Root-Mean-Square (RMS) value of an AC voltage is that value that produces the same heating effect (dissipates the same power) on a resistor as the equivalent DC voltage. The RMS value of a sinusoidal signal is its amplitude divided by $\sqrt2$: $V_{EFF} = V_{MAX}/\sqrt2 = V_{MAX}/1.41 = 0.71V_{MAX}$.

It makes no intuitive sense that an AC voltage or current, which changes direction so many times a second, can produce any active power. However, just because the average value of a sine voltage is zero, does not mean that the power it delivers to a load is also zero!

Here we see one period of sinusoidal voltage $v(t) = V_{MAX}\sin(\omega t)$ and the AC power it dissipates across resistance R connected across it.

Remember, power is a square of the voltage divided by the resistance. Once squared, the power wave has the shape drawn in bold: $p(t)=v^2(t)/R$.

AC power also fluctuates periodically, but since a square of anything cannot be negative, the power on the load always has a positive average. It can be mathematically proven that the average of the power wave is exactly half of its peak value or $P_{AV}=P_{MAX}/2$.

Notice that at the precise moments when the power wave is at that value, the voltage wave is at its RMS value, or $0.71V_{MAX}$. This is not a coincidence, it is a significant relationship, and we'll come back to it many times. Remember that the power on the load halves when voltage drops from its value down to 71% of that value!

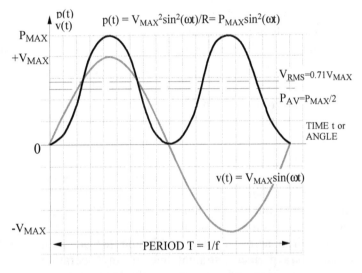

The power of a sine wave (voltage or current signal)

Signals in time- and frequency- domains

A sinusoidal signal can be expressed in two ways. The form we have been using, $v(t) = V_M\sin(\omega t)$, is called a time-domain form. The 18th-century Swiss mathematician Euler defined the complex exponential $e^{j\theta}$ as a point in the complex plane, expressed as $e^{j\theta} = \cos\theta + j\sin\theta$ The "e" in this equation is the base of the natural logarithm, approx. 2.72. "j" is the operator called "i" in mathematics. It is defined as a square root of -1 or $i^2=-1$. This mathematical concept expands the real number system (where anything squared must be positive) to the complex number system.

To avoid confusion, in electrical engineering, which uses i(t) or just "i" to denote AC current, operator "j" is used instead of "i". ω is the angular frequency of the signal. The phase angle θ of the phasor is referenced to the cosine signal, so the two notations are:

TIME-DOMAIN: $v(t) = V_M\sin(\omega t)$

FREQUENCY-DOMAIN: $v(t) = V_M e^{j\theta} = V_M \angle\theta$

The notation $V_M\angle\theta$ is the expression in a polar form, where a voltage phasor is defined by its amplitude V_M and its initial phase angle (in further text just "phase") θ.

Notice that in the definition circle illustrated, the amplitude is $V_M=1$; we are free to scale the phasors any way we like by multiplying (or "amplifying") them with V_M.

The complex plane is used to depict the I-V characteristics or complex impedances of RLC circuits, and amplifier behavior in the frequency domain is studied.

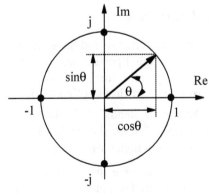

A phasor in a complex plane, with its projections onto the real (Re) and imaginary (Im) axis

RESISTIVE, CAPACITIVE AND INDUCTIVE IMPEDANCE

The equivalent of resistance in DC circuits is the impedance (symbol Z) in AC circuits. While resistance is always a real number, impedance is a complex number, comprised of its real part (resistance) and its imaginary part or reactance: $Z(j\omega) = R(j\omega) + jX(j\omega)$.

Just as conductance (symbol G) is the reciprocal of resistance in DC circuits, the reciprocal of impedance is *admittance*, for which symbol Y is used. It is also a complex number, its real part being *conductance*, while its imaginary part is called *susceptance*: $Y(j\omega) = G(j\Omega) + jB(j\omega)$. Since $Y=1/Z$, its unit is $1/\Omega$, called Siemens [S]. Although Y is inverse of Z, G is not an inverse of R and B is not an inverse of X!

The three basic passive components used in electronics represent three fundamental physical properties, resistance, capacitance, and inductance. As always, to make our journey smooth and gradual, we will start with an ideal resistor, ideal capacitor, and ideal inductor.

Ideal resistors

An ideal resistor has a constant resistance, meaning it does not change with temperature, and, more importantly, does not change with the frequency of the AC signal or with the magnitude of the current flowing through it.

The ideal resistor has no inductance and no capacitance. This means that it behaves in the same way in both DC and AC conditions. The vector diagram shows the current and voltage phasors "in phase" or parallel to one another.

If the current flowing through an ideal resistor is a sine wave of a certain frequency f, the voltage drop such current will produce on the resistor will also be a sine wave, of the same frequency and the same phase.

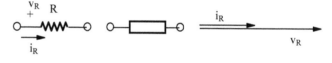

The amplitude of the voltage drop can be determined by Ohms' Law, just as for DC conditions: $V_R = I_R * R$

ABOVE: Ideal resistor - symbols and the vector diagram
BELOW: Voltage and current are in phase (no phase shift)

An ideal resistor is a linear device; there is no amplitude distortion of either current or voltage, there is no phase shift between them, so no phase distortion. The higher the current through an ideal resistor, the higher the voltage drop across it. The V-I graph for an ideal resistor is a perfectly straight line. In any operating point Q, the slope of this curve or the tangent of the angle between that line and the X-axis is constant and equals the ratio of the voltage and the current in that point or the resistance R.

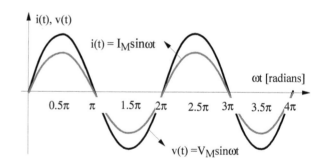

RIGHT: The impedance Z_R of an ideal resistor is pure resistance, independent of frequency. The V-I relationship is linear. Its slope (tan α) is constant and equal to resistance R. The I-V relationship is more commonly used in tube characteristics. Its slope (tan β) is constant and equal to conductance (G=1/R).

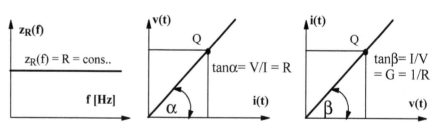

Ideal capacitors

An ideal capacitor would have an infinite resistance for DC current, thus allowing no DC current to pass. The AC current through a perfect capacitor precedes the voltage by 90 degrees. If you cannot remember this fact, you can deduct it logically. Just remember that a discharged capacitor (no voltage across it and no charge held inside it) needs a current to start flowing first to charge it.

So, the current "leads" while the voltage across it develops slowly (raises) or "lags" behind the charging current. The impedance of an ideal capacitor is inversely proportional to frequency f and its capacitance C: $Z_C = 1/(2\pi fC)$. The higher the frequency of the signal, the lower the capacitor's impedance.

This fact will have important repercussions in amplifier circuits, where unwanted or parasitic capacitances will form low pass filters with adjacent resistances and divert high-frequency signals to the ground, thereby limiting amplifiers' frequency range!

IMPEDANCE OF AN IDEAL CAPACITOR
$$Z_C = 1/\omega C = 1/(2\pi fC)$$

A 220µF electrolytic capacitor in a tube amp's power supply will have an impedance (reactance) of Z_C= 1/(220*10⁻⁶*2*3.14*100) = 7.2 Ω at 100 Hz, but only 0.072 Ω at 10 kHz. A 100 times higher frequency means 100 times lower impedance!

Capacitors are used for the following purposes:

1. To block the flow of DC currents, for instance, between the high DC potential anode of an amplifier stage and the grid of the following stage, which is at low DC potential.

2. To bypass AC signals around resistors, for example, in the cathodes of amplification stages, where we don't want such signals to be attenuated by the cathode biasing resistors. The bypass capacitor forms a low-impedance path for audio signals, while the DC current cannot flow through the capacitor, only through the cathode resistor.

3. In various filters, such as tone controls in guitar amps

4. As smoothing capacitors in amplifiers' power supplies, to reduce ripple (AC component on a DC power line) and to provide energy storage.

Ideal inductors

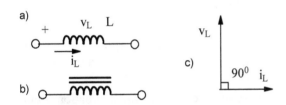

a) Ideal air inductor (no magnetic core)
b) Ideal inductor with magnetic core
c) Vector diagram shows voltage leading current by $\pi/2$ radians or 90°

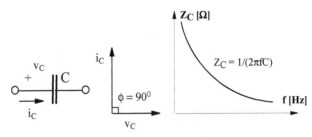

Ideal capacitor and its vector diagram: current leads voltage by $\pi/2$ radians or 90°. The impedance of an ideal capacitor is inversely proportional to frequency.

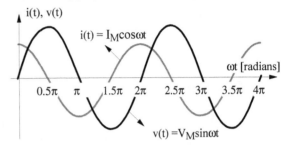

The phase relationship between the AC voltage across and current through an ideal capacitor.

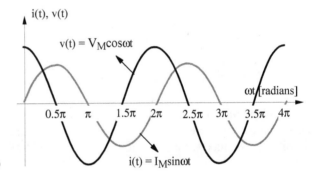

An ideal inductor has no resistance for DC current and no power losses, so there is no energy loss. The voltage on the choke precedes current, or you can say that the current lags voltage by 90 degrees, the opposite of a capacitor!

An inductor is named after a physical phenomenon called "electromagnetic induction, while a "choke" is a popular or pejorative name given to an inductor, due to its effect on alternating currents, which it suppresses or "chokes". In other words, an inductor or choke resists any change in the current that flows through it by inducing an opposite voltage which then tries to push the current in the opposite direction so as to neutralize the current that caused the induction in the first place. That mechanism is called *self-induction*! A choke has a low impedance for DC currents (ideally presents a zero DC resistance) and high reactance to AC currents.

These desirable properties make chokes useful in power supplies (as ripple filters) and in audio circuits as inductive loads.

The reactance of a choke is $X_L = \omega L = 2\pi f L$. Thus, the higher the frequency of an AC signal, the higher the reactance of a choke and the higher the choke's attenuation of the signal!

IMPEDANCE OF AN IDEAL INDUCTOR

$$Z_L = \omega L = 2\pi f L$$

A brief introduction into electromagnetic induction

Let's look at two practical examples, guitar pickups, and loudspeakers. A pickup consists of a permanent magnet (made of Alnico, rare earths, or ceramic material) around which sits a plastic bobbin on which a coil is wound. Six guitar strings are magnetically coupled to the coil through their individual metal pole pieces or slugs. When a guitar string vibrates in the magnetic field, a voltage signal is induced in the coil, between its terminals X and Y. The mechanical force F of the movement, the density of the magnetic field's flux B, and the induced current *i(t)* are related by $F(t)=Bi(t)l$, where *l* is the length of the wire in the coil.

A loudspeaker also has a permanent magnet. When a signal current from an amplifier flows through its voice coil (positioned in its magnetic field), a mechanical force develops, moves its paper cone, and produces air pressure changes and waves that we call sound.

The operation of both of these transducers is based on electromagnetic induction. However, physical force and movement are not necessary for magnetic induction. Chokes and transformers operate without any physical movement. It is enough for the current through a coil to vary in time, as AC current does, to induce an opposing voltage in the coil (inductor).

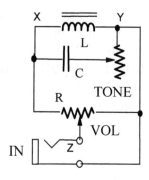

The circuit diagram of guitar electronics is usually very simple. The pickup coil whose inductance is "L" has an RC series combination connected in parallel, a simple 1st order filter that reduces treble frequencies, usually marked "Tone", and a potentiometer marked "Volume" that regulates the amount of signal voltage that will be passed onto the output jack and ultimately to a guitar amp.

Most guitars have more than one pickup, so there is also a switch that select one of them and connects it between points X and Y.

ABOVE: A typical passive circuit inside an electric guitar.

RCL circuits

If a series RCL circuit is connected to an AC signal source, the same current flows through all three components. The voltage on the resistor is in phase with the current, but the voltages across the capacitor and the inductor are shifted in phase. One is lagging the current by 90 degrees (capacitor), the other (inductor) is leading the current.

Since the voltages V_L and V_C are opposing each other, at a certain frequency f_R they will become equal and cancel each other out. Such a frequency is called a *resonant frequency* and the phenomenon is called *series resonance*.

$X_L = X_C$ or $\omega L = 1/\omega C$ Solving the equation for ω, $\omega_R = \sqrt{(1/LC)}$ or $f_R = \sqrt{(1/LC)}/2\pi$.

At the resonant frequency f_R the overall impedance of the circuit is at its minimum, and equals the resistance R. At all other frequencies the resultant impedance is larger than R. Its amplitude can be calculated as $Z = \sqrt{[R^2 + (X_L - X_C)^2]}$ and its phase angle is $\phi = \arctan[(X_L - X_C)/R]$

There is another kind of resonant circuit: the parallel RCL circuit, where the resistor, capacitor, and inductor are connected in parallel, so the same voltage is across each of them, and the currents through them vary with frequency.

A typical example of such a system would be a dynamic loudspeaker that uses crossovers. A similar situation occurs there, the only difference being that at the parallel resonant frequency f_R the equivalent impedance of the whole circuit is at its maximum (R), and the impedance at all other frequencies is lower than R.

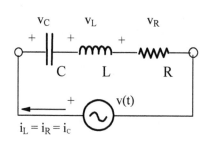

Series RCL or resonant circuit

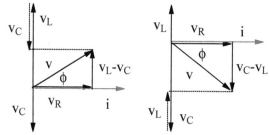

a) predominantly inductive (voltage leading current)
b) predominantly capacitive (voltage lagging current)
c) Resonance (the inductive reactance equals and cancels the capacitive reactance at the resonant frequency), so at resonance the equivalent circuit is a pure resistance

Real inductors

Since inductors are made of many turns of wire on a bobbin (plastic frame), such a large length of relatively thin wire (typically 0.2-0.6 mm diameter) has a measurable resistance. We model real chokes as a series RL circuit, by adding an ideal resistor in series with the choke's inductance. The vector diagram for the real choke (next page) features angles δ and ϕ.

The same current flows through the resistance and the inductance, but there is a phase shift of 90 degrees or $\pi/2$ radians between the voltage drop on the resistance (which is in phase with the current) and the voltage drop on the inductance (which advances 90° ahead of the current). From the right angle voltage triangle we know that $\tan\delta = v_R/v_L$. Since $i = v_R/R = v_L/X_L$ and since $X_L = \omega L$, we get $v_R/R = v_L/\omega L$, $tg\delta = v_R/v_L = R/\omega L$

The quality factor or Q-factor for short, is the inverse of $\tan\delta$: $Q = 1/\tan\delta = \omega L/R$. The lower the tangent delta or the higher the Q-factor, the more the choke approaches the ideal inductor, meaning the lower its ohmic resistance (to DC current) and the lower the power losses.

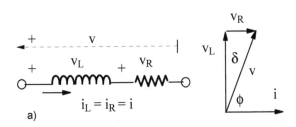

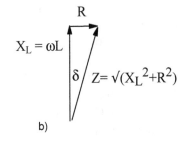

a) Real inductor's voltage and current vector diagram shows voltage leading current by $\phi=\pi/2-\delta$ radians

b) The impedance vector diagram for a real inductor

The losses and a vector diagram of a real capacitor

Real capacitors allow a small leakage DC current to pass through their dielectric. This behavior is modeled by adding a resistor of a very large resistance in parallel to the ideal capacitor. The value of this leakage for a simple plate capacitor (to be discussed soon) would be $R_P= \rho d/A$. ρ (Greek letter "rho") is the specific resistance of the dielectric material, S is the surface area of the plates, and d is the distance between them. Ideally, ρ would be infinite (no conduction at all), meaning the parallel resistor would be an open circuit, and the losses would be zero.

The vector diagram of a real capacitor (with leakage resistance in parallel) features angles δ (delta) and ϕ (phi), the sum of which is 90^0 (degrees). Delta is zero for an ideal capacitor. From the right angle current triangle we know that $\tan\delta = i_R/i_C$. Since $v_C=v_R=Ri_R$ and the same voltage $v_C=v_R=X_Ci_C = 1/(\omega C)i_C$, it follows that $Ri_R = i_C/\omega C$.

If we express $i_R/i_C= 1/(\omega RC)$ and take a tangent of the whole equation we get $\tan\delta = i_R/i_C = \tan[1/(\omega RC)] = D$.

That parameter was named D for "dissipation factor" and is zero for an ideal capacitor. So, the lower the D-factor, the better the capacitor, the lower its dielectric losses. These are thermal losses, so they raise the temperature of the capacitor. This is most noticeable on vintage electrolytic capacitors, which have high losses and are hot to touch. The top of the metal case often even bubbles up, and such a bulge is a sure sign that the capacitor has reached the very end of its life and should be replaced immediately.

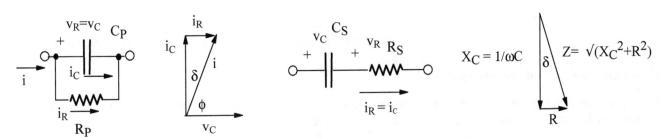

Parallel model of a capacitor and its vector diagram

Series equivalent circuit of a capacitor

The impedance diagram of a real capacitor

Eventually, the build-up of gasses will cause the metal can to rupture, and the capacitor will explode, spewing the gunk (electrolyte) all over the insides of the amplifier! Film capacitors do not use an electrochemical process as electrolytic capacitors do, so there is no gas build-up or the possibility of explosion.

For both models, the impedance diagram of a real capacitor is a right-angle triangle, so Pythagora's Theorem applies again, helping us to calculate the modulus of impedance Z, the loss angle δ, and the phase angle ϕ. The series resistance R_S in the second model is called ESR or Equivalent Series Resistance. It is an important parameter for electrolytic capacitors, which should ideally be zero, but in reality can be quite high, up to 10Ω!

Series- and parallel-connected capacitors

In a parallel connection, each capacitor is connected to the same voltage, so its charge will depend on its capacitance - the higher the capacitance, the higher the stored charge. $Q = Q_1 + Q_2 + ... +Q_N = VC_1 + VC_2 + ... + VC_N = V(C_1 + C_2 + ... + C_N) = VC_{TOT}$.

Parallel connection is used in audio to increase the overall capacitance and to reduce the overall ESR. Assuming identical capacitors and a 6Ω ESR, by connecting four in parallel, the overall ESR will be reduced by the factor of 4, down to $6\Omega/4 = 1.5\Omega$.

In a series connection, each capacitor will carry an equal charge Q. If potential drops across the capacitors are V_1, V_2, ...V_N, then $V_{TOT} = V_1 +V_2 + ...+V_N = Q/C_1 + Q/C_2 + ... + Q/C_N = Q(1/C_1 + 1/C_2 + ... +1/C_N)$ The factor in brackets can be considered an equivalent capacitor, so $1/C_{TOT}=1/C_1 +1/C_2+ ... + 1/C_N$

The series connection is used in high voltage audio power supplies in cases where the DC voltage of the power supply exceeds the voltage rating of available electrolytic capacitors, which is $450V_{DC}$ or $500V_{DC}$ at most. Two $450V_{DC}$ capacitors in series can then operate at voltages of up to $800V_{DC}$ (a margin is always necessary).

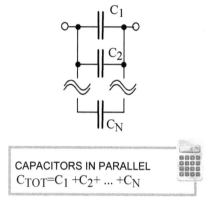

However, since capacitance tolerances of these capacitors are wide, typically +/-20%, in the worst-case scenario, one 470μF capacitor may have a capacitance 20% lower than this nominal value, or 376μF, while the other may have a 20% higher capacitance, or 564μF! Connected in series, they will have different voltage drops across them. The capacitor with a larger capacitance will have a smaller voltage drop across it, while the cap with a lower capacitance will be subjected to a higher voltage drop, which may exceed its rated voltage.

CAPACITORS IN PARALLEL
$C_{TOT}=C_1+C_2+...+C_N$

The rule is $V_1C_1=V_2C_2$, and since in our example $V_1+V_2 = V_{TOT} = 800V$, $C_1=376$μF and $C_2=564$μF we can calculate $V_1= V_{TOT}*C_2/(C_1+C_2)= 0.6*V_{TOT} = 480V$ and $V_1= 800-480 = 320V$!

Indeed, the voltage across C_1 is 480V, above its rated voltage of 450V. In this situation, some capacitors would explode immediately, while others will work fine for a while but will suffer an early failure.

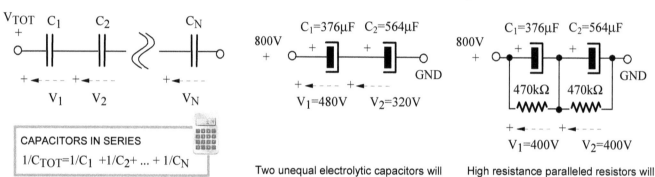

CAPACITORS IN SERIES
$1/C_{TOT}=1/C_1+1/C_2+...+1/C_N$

Two unequal electrolytic capacitors will divide voltages unequally!

High resistance paralleled resistors will equalize the two voltage drops.

Series and parallel-connected resistors

By connecting resistors in series, their resistances add up, while the overall resistance of two or more paralleled resistors is lower than individual inductances. In a parallel circuit, the conductances add up, and since the conductances are inverse of resistances, we get the formula $1/R_{TOT} = 1/R_1 + 1/R_2 + ... + 1/R_N$!

The voltage across individual resistors will be proportional to their values, so the larger the resistance, the larger the voltage drop across it (since the current through the lot is the same). Series resistors form a so-called "voltage divider" network. Such resistive networks are often used in power supplies of tube amplifiers to provide required voltages from the same high voltage power source.

One advantage of paralleled resistors is the division of currents between them. Using a few lower power rating resistors, we can make up an equivalent resistor of a much higher power rating. This handy principle is often used to construct a dummy load for amplifier testing. Say you need an 8Ω load of at least 20Ω rating but have only a bunch of 68Ω, 3Watt resistors. By connecting eight of them in parallel, you will get a total resistance of 68/8 = 8.5Ω (which for practical testing purposes is close enough to the nominal 8Ω) and a total power rating of 8*3 = 24 Watts!

The other advantage is that the overall tolerance is reduced. For instance, if you connect five identical 10% tolerance 10kΩ resistors, the parallel combination will have a resistance of 10/5=2kΩ, but the tolerance of such a network will be reduced by the same factor, 10%/5 = 2%. In such a case, a careful selection of resistor values is usually not required!

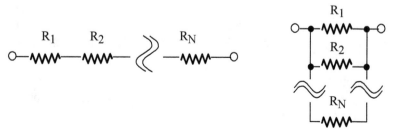

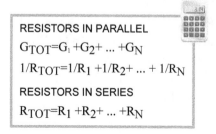

RESISTORS IN PARALLEL
$G_{TOT}=G_1+G_2+...+G_N$
$1/R_{TOT}=1/R_1+1/R_2+...+1/R_N$

RESISTORS IN SERIES
$R_{TOT}=R_1+R_2+...+R_N$

Series and parallel-connected inductors

As with resistances, series-connected inductances add up, while the overall inductance of two or more paralleled inductors is lower than their individual inductances. Parallel connection of chokes is used in filtering circuits of tube amplifiers' power supplies if the current rating of one choke isn't sufficient. For instance, if a total load of the amplifier's high voltage supply is 200mA and 10H, 100 mA chokes are available. Two would be connected in parallel to satisfy the current demand, but the overall inductance would drop to 10H/2 = 5H!

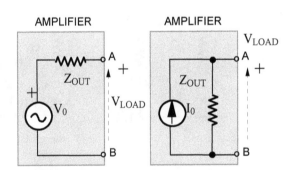

INDUCTORS IN PARALLEL
$1/L_{TOT}=1/L_1 +1/L_2+ ... + 1/L_N$

INDUCTORS IN SERIES
$L_{TOT}=L_1 +L_2+ ... +L_N$

THE BLACK BOX CONCEPT: THEVENIN'S AND NORTON'S THEOREMS

We can evaluate the outward behavior of an electronic device without any knowledge of its internal details, schematics, or construction. By feeding suitable signals at its input and observing or measuring the signals at its output, we can determine what that system or device is (an amplifier, a filter, an oscillator, etc.) and how well it performs its intended function(s). This is the concept of a "black box." For instance, we use this principle when we get an unknown amplifier on our test bench and test its maximum power, distortion levels, frequency range, damping factor, signal-to-noise ratio, crosstalk (if it's a stereo amp), and other parameters of interest.

Two of the most important theoretical theorems in electrical engineering describe any such "black box" behavior. Any linear electronic system or device can be represented as a black box with only two components, a signal source, and an impedance.

Thevenin's equivalent circuit

From the point of output terminals A and B, any linear circuit can be replaced by a series combination of a voltage source equal to the open-circuit voltage at those terminals and an impedance looking back from the two terminals, once all *independent* voltage and current sources have been removed.

Norton's Theorem: From the point of output terminals A and B, any linear circuit can be replaced by a parallel combination of a current source equal to the short-circuit current flowing between these terminals and an impedance that would be measured at those terminals, after the removal of all *independent* voltage and current sources.

The Thevenin's (left) and Norton's (right) equivalent circuit of an audio amplifier

Let's use the simple circuit illustrated on the next page. We are interested in load terminals A and B; the 8Ω resistor connected between them is the load. In the first step, we remove the load, leaving terminals A and B open, and calculate the voltage between them. Since there is no current through the 2Ω resistor between terminals C and A, the voltage at A is the same as the voltage at C. The voltage source V_1 supplies two equal resistors (4Ω) in series, so the voltage between points C and B and thus A and B is half of V_1. So, our Thevenin's voltage $V_0(t)=0.5V_1(t)$.

Whatever the amplitude and frequency of this signal, Thevenin's voltage will have the same frequency and shape (waveform), but its amplitude will be only half of V_1!

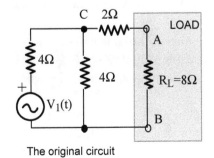

The original circuit

Step 1: Calculating Thevenin's voltage

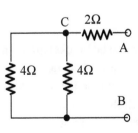

Step 2: Calculating the output impedance

For instance, if $V_1(t)$ is the mains voltage in Australia with an effective or RMS value of 240V, its amplitude V_{1MAX} will be 1.41*240V = 338V, and its frequency will be 50Hz ($\omega=2\pi f=338$), so we can write $V_1(t)=338\sin(314t)$ and Thevenin's voltage will be $V_0(t)=169\sin(314t)$ [V].

In the second step, we remove the ideal voltage source leaving a short circuit where it used to be (since its internal impedance is zero). Now we look "back" to the left, into the circuit from terminals A and B, and calculate the equivalent output impedance. We have two 4Ω resistors in parallel (equaling 2Ω) and then in series with another 2Ω resistor for a total impedance of 4Ω.

A few important points:

1. The impedance of the load has no impact on the calculations.
2. More than one voltage source can be present, but the principle stays the same.
3. Dependent voltage sources are not removed in Step 1 calculations, only the independent sources.
4. The output impedance is usually complex; it has an active (resistive) and reactive (inductive or capacitive) component. We used pure resistance for the sake of simplicity.

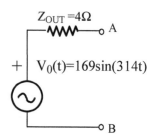

Thevenin's equivalent circuit
for our example

Norton's equivalent circuit

Determining the output impedance of Norton's model is identical to the method for Thevenin's model so that it won't be repeated here. To determine Norton's current, we short-circuit terminals of interest (in this case, A and B) and calculate the current through that branch.

$(V_1-V_{CB})/4 = V_{CB}/4 + V_{CB}/2$ so $V_{CB}=V_1/4$

Since $I_0 = V_{CB}/2$, $I_0 =V_1/8$, and again, assuming that $V_1(t)=338\sin(314t)$, we get $I_0(t)=42.25\sin(314t)$ [A]

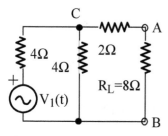

The original circuit

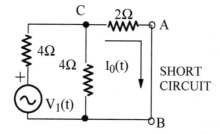

Step 1: Calculating short-circuit current

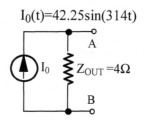

Norton's equivalent circuit

Dependent and independent variables, their units and notations

We have seen that mains voltage can be described by the formula $V(t)=V_{MAX}*\sin(\omega*t) = \sqrt{2}V_{EFF}*\sin(2*\pi*f*t)$ [V]! In this formula, f (the mains frequency) is constant; it is either 50 Hz or 60 Hz (or "cycles per second"), depending on the country you live in. The peak V_{MAX} and the effective value V_{EFF} of the voltage are also nominally constant, although they fluctuate somewhat in real life, just as the frequency does, depending on the quality of the power grid you are connected to. The factors 2π ($\pi=3.14$) and $\sqrt{2}$ are obviously fixed numbers, so the only variable is "t" or time. Thus, time is the independent variable, while the amplitude of the mains AC voltage is the dependent variable.

The [V] at the end of the equation is the unit of the dependent variable. In this case the unit for voltage is a Volt. The square brackets will indicate a unit for the physical parameter described by the formula that precedes it.

Also, since writing the multiplication sign "*" or "x" many times during mathematical analysis can be tedious and slow us down, we will omit those symbols assuming that everyone understands that $V_{MAX}\sin(\omega t)$ means $V_{MAX}*\sin(\omega*t)$!

Furthermore, the symbol "x" for multiplication can be falsely interpreted as some kind of variable called x, so we better not use it. The bracket after sine, cosine, tangent, or cotangent functions is also often omitted, so $V_{MAX}\sin\omega t$ is the ultimately simplified notation, but you must remember that in $\sin\omega t$, the sine function applies to the whole product ωt (the independent variable) and not just the first letter ω!

However, with the root symbol, we should use brackets to indicate what it applies to or what is "under" this symbol; otherwise, errors are almost certain! For instance $\sqrt{2}V_{EFF}$ indicates that only number 2 is under the root, while $\sqrt{(2V_{EFF})}$ would mean that the whole product $2V_{EFF}$ is under the root symbol.

AUDIO AMPLIFIERS

2

- PROPERTIES OF IDEAL AMPLIFIERS
- DECIBELS AND LOGARITHMIC SCALES
- REAL AMPLIFIERS
- DISTORTION IN AUDIO AMPLIFIERS
- WHAT DETERMINES THE SOUND OF AMPLIFIER?
- FLIPPANT, MISLEADING OR SIMPLY "LIBERAL" USE OF LANGUAGE?

To understand real (imperfect) amplifiers, ideal amplifiers should be defined and understood first. This section looks at amplifiers and their behavior from the outside perspective, as "black boxes", without going into their internal construction or operation (yet).

Along the way, we introduce important concepts, such as decibels, logarithmic scales (brush up on your maths skills!), input and output impedances, amplifier bandwidth, various types of distortion, harmonics & timbral spectra, and many others.

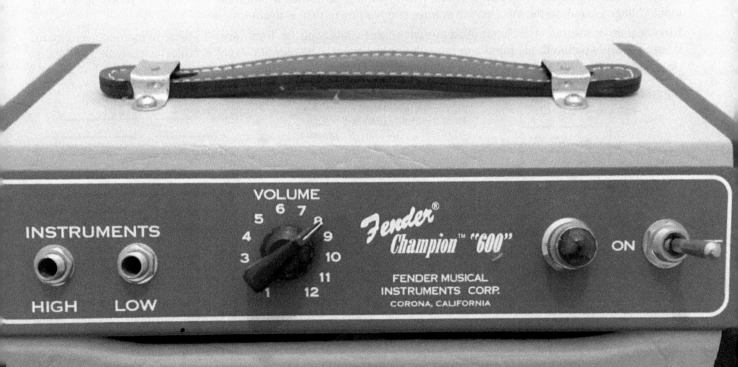

PROPERTIES OF IDEAL AMPLIFIERS

As the very word suggests, an amplifier amplifies some kind of signal from a sensor or transducer, such as a guitar's magnetic pickup, or a music source such as a phono cartridge or a CD player. Before we get into real amplifiers, let's talk about the ideals we are trying to achieve or come as close to as possible

Ideal amplifiers are classified by two parameters, voltage and current, and since these amps have one input and one output, there may be four types of amplifiers: voltage-controlled voltage amps, current-controlled current amps, voltage-controlled current amps, and vurrent-controlled voltage amps.

Voltage-controlled voltage amplifier

The ideal voltage amplifier is a source of AC output or load voltage v_L, which is controlled by and proportional to the input voltage v_{IN}. μ (Greek letter "mju") is the amplification factor, which is constant. The + markings refer to one particular moment in time, since the AC voltage changes polarity f times a second, where f is the frequency of the signal. Sinusoidal waveform will be assumed in the further study, unless noted otherwise, hence the sine waveform inside the voltage source.

The i_L-v_L lines are perfectly straight, vertical and equidistant (equally spaced). This is a linear amplifier without any distortion or power losses. Its input resistance is infinite (there is no input current flowing), so the power input is zero. The output or internal resistance is zero (there are no thermal losses within the amplifier).

The amplifier is oblivious to the resistance of the load. No matter how high or low it may be, its voltage will stay strictly proportional to the input voltage. Whatever voltage comes in, no matter its frequency or shape, it will be multiplied m times and imposed onto the load. Since the input power into the amplifier is zero, its power gain (the ratio of output to input power) is infinitely high. An ideal vacuum triode is one such amplifier.

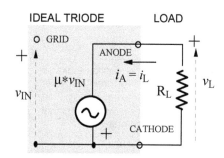

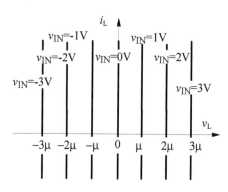

LEFT: The equivalent circuit of an ideal voltage-controlled voltage amplifier. Triode parameters are used, which will be explained soon.

RIGHT: The i_L-v_L characteristic of an ideal voltage-controlled voltage amplifier.

Current-controlled current amplifier

The ideal current amplifier is a source of AC load current i_L, which is controlled by and proportional to the input current i_{IN}, or $i_{OUT} = \alpha i_{IN}$. α (alpha) is the current amplification factor. The i_L-v_L lines are perfectly straight, horizontal and equidistant (equally spaced). This is a linear amplifier without any distortion or power losses. The input voltage is zero, so the input power is zero, and the power gain is again infinite.

Bipolar transistors are current-controlled current sources, although far from ideal. In the common-emitter circuit, whose model is pictured, the input current is the base current. The output current is collector current. The load is connected between the collector and the common terminal, in this case, the emitter.

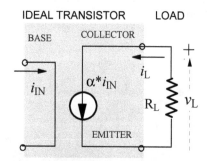

LEFT: The equivalent circuit of an ideal current-controlled current amplifier. Bipolar transistor parameters are used.

RIGHT: The i_L-v_L characteristic of an ideal current-controlled current amplifier. Horizontal "curves" mean the output currents are constant, regardless of the load resistance or load voltage v_L.

Voltage-controlled current amplifier

Voltage controlled current amps supply AC load current that is proportional to the AC signal voltage: $i_L = gm \cdot v_{IN}$. The gm factor is called transconductance or mutual conductance (between the output and input), its unit being mA/V.

Conductance is the inverse of resistance, with a dimension $1/\Omega$. Since $\Omega = V/A$, the unit for conductance has a dimension of current divided by voltage: I/V and in the International System of Units (SI) is called a Siemens, with a symbol S, named after Ernst Werner Siemens (1816–1892), German inventor and industrialist. Siemens AG is a vast multinational industrial concern; tubes made by Siemens in 1950s and 60s are still in circulation.

In American literature, you will come across a unit called micromho. Micro means 10^{-6} or one millionth-part of "mho", which is, believe it or not, ohm spelled backwards. So, 1 Siemens = 1A/V = 1,000 mA/V = 1,000,000 micromhos. European tube testers are calibrated in mA/V or mS, which is 1,000 micromhos. So, if EF86 pentode in a particular operating point has a gm of 1,800 micromhos, that is 1.8 mA/V or 0.0018 S!

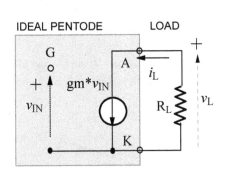

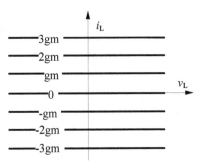

LEFT: The equivalent circuit of an ideal voltage-controlled current amplifier. Pentode parameters are used.

RIGHT: The i_L-v_L characteristic of an ideal voltage-controlled current amplifier. The output current is constant, regardless of the load resistance her voltage.

A pentode is a real-life example of a voltage-controlled current amplifier, albeit not an ideal one, just as a triode is a real-life (meaning "imperfect") voltage-controlled voltage source.

Understanding ideal amplifiers is a vital first step in building long-lasting knowledge and an intuitive feeling for vacuum tube electronics. We will evaluate various tube types and circuit configurations mainly based on how close they come to these ideal amplifiers. There are also current-controlled voltage amplifiers, but we will not talk about them in this book.

DECIBELS AND LOGARITHMIC SCALES

The proper and improper use of decibels

Before we move on to the properties of real or imperfect amplifiers, due to their importance, we must digress a bit and talk about decibels and logarithmic scales. deciBel or "dB" is a unit for a power ratio or gain: $P_G = 10 * \log(P/P_0)$ [dB] where P is the power produced by a system or a device, P_0 is the referent power chosen for a particular purpose, comparison or measurement. Various referent power levels have been used over the decades, and that created confusion among users. We will talk about that in a minute.

The power gain of an amplifier is $P = V^2/R$ we get $P = 10 * \log(P_{OUT}/P_{IN}) = 10\log[(V_{OUT}^2/R_{OUT})/(V_{IN}^2/R_{IN})] = 10\log[(V_{OUT}/V_{IN})^2(R_{OUT}/R_{IN})] = 20\log(V_{OUT}/V_{IN}) + 10\log(R_{IN}/R_{OUT})$ [dB]

If, and only if $R_{OUT} = R_{IN}$, we get $\mathbf{P = 20\log(V_{OUT}/V_{IN})}$

A tube amplifier has an input impedance of $47k\Omega$ and supplies an 8Ω load. Its voltage gain is 15 times (A=15). What is the voltage gain in dB?

These days we simply calculate $P = 20\log(V_{OUT}/V_{IN}) = 20\log15 = 23.5$ dB! That is certainly convenient, but, strictly speaking, it is not correct because this simplification is only correct when $R_{IN} = R_{OUT}$, and in our case, the input and output resistances are very different!

The correct answer is $P = 20\log(V_{OUT}/V_{IN}) + 10\log(R_{IN}/R_{OUT}) = 20\log15 + 10\log(47,000/8) = 23.5 + 37.7 = 61.2$ dB.

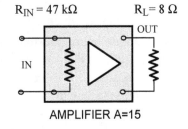

AMPLIFIER A=15

Since the last factor $10\log(R_{IN}/R_{OUT})$ almost always gets omitted (its not even mentioned in nine out of ten books!) we will adopt this simplification too. Although improper from the strict definition point-of-view, such usage of dB works, providing it is used consistently, meaning as long as everyone (mis)uses it in the same way!

Different "types" of decibels and dB scales

If you thought the confusion surrounding the use of dB is over, you were wrong. Humans can and do overcomplicate things on a grand scale, and electronic engineering is a good example of such tendencies. There are two different dB units, dBm and dBV. dBm (also called dBmW, only compounding the confusion) has a communication origin, referenced to the 1 mW of power that a sine source of 0.775 V_{RMS} would dissipate on a 600Ω resistive load. dBV is the unit we have been talking about so far, used for voltage ratios such as amplification and attenuation factors.

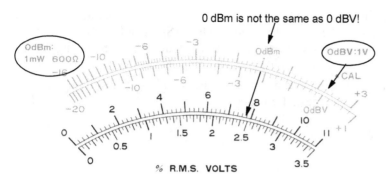

0 dBm is not the same as 0 dBV!

% R.M.S. VOLTS

NF CIRCUIT DESIGN BLOCK CO., LTD.

See how the two dB scales have a different zero on this analog meter face, from a harmonic distortion analyzer: 0 dBm is at 0.775 V_{RMS} (7.75 on the lower scale) and 0 dBV is at 1.0 V_{RMS} (10 on the lower scale)!

The reference power level for dBm is $P_0=0.001$ W, so P [dBm] = $10\log(P/0.001)$. To convert dBV to dBm we need a referent impedance (R), in this case 600Ω. Remember, $P=V^2/R$!

Taking the log of both sides of that equation $\log(P) = 2*\log(V) - \log(R)$ or dBm - 30 = dBV - $10\log(R)$ Finally, dBV= dBm - 30 + $10\log(600)$ = dBm -30 +27.78 so **dBV= dBm-2.218.**

The logarithmic power curve

Let's return to the half-power frequencies, or, as they are more commonly called, -3dB frequencies. To understand the logarithmic nature of the amplifier's power drop at those frequency limits, we need to study the logarithmic power curve, the equation **P = $10\log(P_2/P_1)$ [dB]** in a visual form. The ratio P_2/P_1 is on the horizontal axis (the abscissa) and the decibels of that ratio are on the vertical axis (the ordinate).

When $P_1=P_2$ their ratio is 1, and log1=0 (because $10^0 = 1$) Any number to the power of zero equals one (Point A).

As P_2 becomes larger than P_1, the log curve is in the positive territory, and as P_2/P_1 drops under one, the curve goes into the negative values.

In point E, P_2 is twice as large as P_1, which is +3dB. In point D, P_2 is half as large as P_1, which is -3dB. These are the half-power points.

In point C, $P_2/P_1=10$, and log10=1, so $P_2/P_1=10$dB. Similarly in point B, $P_2/P_1 = 0.1$ so $P_2/P_1=-10$ dB.

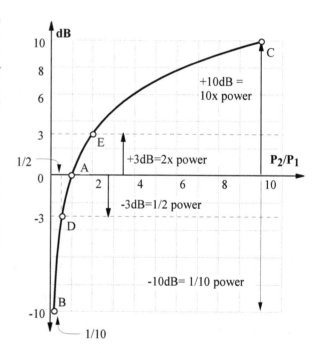

LIN and LOG scales

Four types of X-Y scales are used in electronics. The independent variable (X-axis) and the dependent variable (Y-axis), can be presented using a linear (LIN) or logarithmic (LOG) scale, resulting in four possibilities: LIN-LIN, LIN-LOG, LOG-LIN and LOG-LOG. LOG scale is used when the range of the variable depicted is wide, as in audio frequency range, which is studied from 1Hz to 100,000 Hz, or 5 decades. Each decade is a factor of 10:1.

The example below illustrates the amplifier's gain, where a linear horizontal scale would be impractical since it would be impossible to read the gain from the curve at the frequency extremes, where most of our interest lies.

In some books, you will see the frequency curve of a 1st order system (a simple RC filter or an amplifier stage) with a straight "end" and in others with a curved end. The straight end means the vertical scale is logarithmic (LOG), and it thus has no end; there is no vertical zero "0" marked! A curved end means the vertical scale is linear (LIN) with a definite end at zero level.

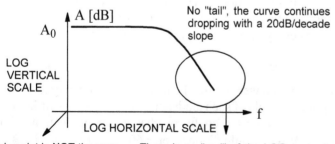

This point is NOT the zero of the vertical LOG scale!   There is no "end" of the LOG scale, it continues into negative values of A (A lower than 1 but larger than 0)

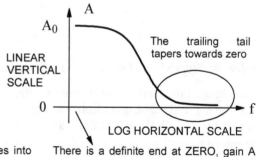

There is a definite end at ZERO, gain A cannot be lower than zero!

REAL AMPLIFIERS

Harmonics and spectra

Before we start analyzing guitar amplifiers, we need to understand the spectra of signals. For instance, a square wave is comprised of an infinite number of sine waves, but only odd harmonics, 1st, 3rd, 5th, 7th, etc., are present; there are no even harmonics. The 1st harmonic is a sine wave of the same frequency as the square wave, the frequency of the 3rd harmonic is three times the frequency of the 1st, and so on.

Notice how the amplitude of the harmonics drops off rapidly in an exponential fashion. Although the illustration only shows spectral components up to the 11th harmonic, the harmonics continue indefinitely; there is the 13th, then 15th, ..., the 127th, ..., the 367th, etc., but because their amplitudes are so infinitesimally small, we usually neglect all other harmonics above the particular value.

For a square wave, the amplitudes of harmonics are $AN = A/N$, so the amplitude of the third harmonic is 1/3 of the fundamental's amplitude, the fifth harmonic has a 1/5 of the 1st harmonic's amplitude, and so on.

This spectral distribution applies to any square wave. For instance, if f_1 was 800 Hz sine wave of the amplitude $A_1=10V$, the third harmonic would have a frequency three times higher or 2,400 kHz and an amplitude three times smaller, or $A_3=A_1/3 = 10/3 = 3.33V$. The fifth harmonic will be $f_5=5f_1=5*800= 4,000$ Hz, its amplitude will be $A_5=A_1/5 = 10/5 = 2.0V$ and so on.

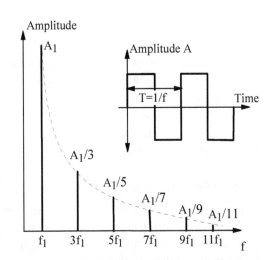

This illustration shows the contribution of the first three harmonics, the 1st, the 3rd, and the 5th. Notice how such a sum already resembles a square wave. As the harmonics are added, the waveform will more and more resemble a square wave.

Although amplifiers are modeled, analyzed, and tested mostly using sine waves of a fixed frequency, there are no such pure tones in nature. Sounds have complex waveforms, comprised of sine waves of various amplitudes and frequencies.

ABOVE: The spectrum of a square wave
BELOW: The fundamental and first two harmonics of a square wave added together.

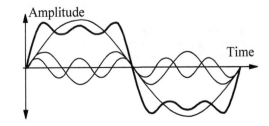

A spectrum of a signal is depicted in an X-Y manner with frequency on the X-scale and the amplitude of various harmonics on the dB vertical scale. Two examples are depicted here, the violin's and the flute's. The first or the lowest frequency is called the "fundamental" or first harmonic. The second harmonic has twice the frequency of the first, the frequency of the third harmonic is triple the fundamental frequency, and so on.

Notice that the amplitudes of some of the harmonics are higher than the fundamental tone. Also, not all harmonics are depicted in the violin's case, only those up to 1,500 Hz.

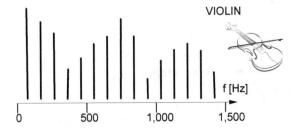

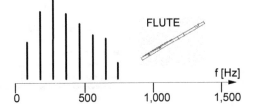

The amplitude and phase characteristics of real amplifiers

An ideal amplifier would have an infinite frequency range, meaning it would amplify all frequencies, from DC (f=0) to infinity. Real amplifiers and preamplifiers behave like bandpass filters. They are unable to amplify very low and very high frequencies as well as they can amplify the midrange band of frequencies.

At a certain low frequency f_L, usually in the region of 10-20 Hz for hi-fi amps and 50-300Hz for guitar amps, the amplification factor drops to 0.707 or 71% of its midrange value A_0 (usually specified at 1kHz). As the frequency is increased, at a certain frequency f_U the amplification factor again drops to 71% of its midrange value. These are -3dB points, also called "half-power" or corner frequencies.

How does output power drop by half if the voltage drops to 71% or $1/\sqrt{2}$ of its mid-band value? It's all a matter of mathematics. The power on the load R is $P = V^2/R$, so the mid-band power is $P_0 = V_0^2/R$, and the half-power is $(V_0*0.707)^2/R = (V_0/\sqrt{2})^2/R = V_0^2/2/R = V_0^2/2R = P_0/2$

The frequency characteristic of a typical amplifying stage is illustrated here using a linear vertical scale, so it is not in dB but in absolute numbers.

The midrange gain is 30 and the phase shift is 180 degrees, meaning the output signal lags behind the input signal by 180°. This amplifying stage inverts the phase of the signal. In popular speak, the input and output signals are "out-of-phase".

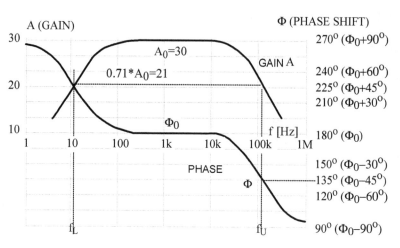

Typical amplitude versus frequency and phase versus frequency characteristics of a real amplifier and the meaning of -3dB points

The phase characteristic of an ideal amplifier would be a straight line - there would be no phase shift between the input and output signals, or such a phase shift would be constant, i.e., the same for all frequencies.

As with the signal's amplitude, the phase relationships also change at frequency extremes. At low frequencies, the phase lag is increased, and at high frequencies, the lag is decreased. Notice that at our half-power or -3dB points, frequencies f_L and f_U, the phase changes $+/-45^o$ from the midrange phase. At f_L the phase is $180^o+45^o = 225^o$, and at f_U the phase is $180-45 = 135^o$.

What frequency range should we aim for in amplifier design?

If you look at the frequency range of musical instruments, illustrated below, none of their fundamental harmonics exceed 4 kHz. Bass guitar fundamental tones go only up to around 220Hz, and a guitar's only up to 900 Hz. So, it may seem that even the shonkiest of amplifiers with the most limited frequency range would be fine for those two instruments. However, that is not so.

Only the fundamental tones are considered here. While the highest notes played on a violin or piano are only around 3kHz and 4kHz, respectively, their harmonics extend well into the ultrasonic range (above 20kHz)! It is the harmonics and their relative amplitudes and interactions that define specific instrument sounds.

The frequency range of various instruments and human voices compared to the piano scale

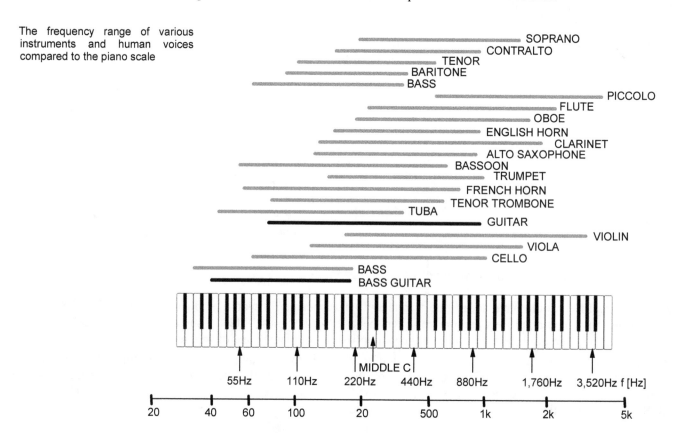

The same applies to the bass and guitar amps. Ultimately, the "wideband" amps generally sound better than those whose frequency ranges are narrower. Although you cannot directly hear those sub-harmonics and high-frequency overtones, you will hear a different tonal presentation or "voicing" with them present and absent.

However, the wideband school of audio design is not universally accepted. Some audio designers choose to limit the upper-frequency extension of their designs, claiming that such amplifiers sound softer and more musical. There is some truth in their claim that significantly extending the bandwidth only gives prominence to higher harmonics. The odd ones (5th, 7th, etc.) are particularly objectionable (harsh sounding and irritating). Plus, it is not just the higher harmonics by themselves, but also the intermodulation distortion products that lie in the 20-60 kHz band, which limited bandwidth amplifiers would cut off or at least attenuate significantly, but wider bandwidth amplifiers would reproduce at full levels!

HF extension needs to be limited if high levels of negative feedback are used, which can become positive at such high frequencies and make amplifiers unstable. So, the truth is that designs that produce low levels of THD and IM distortion and use low levels of negative feedback will probably benefit from wide bandwidth. In contrast, poor designs (distorting and prone to oscillating) will suffer from it.

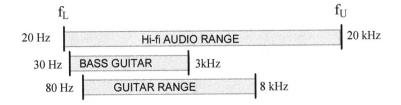

Rules-of-thumb for the minimal target frequency ranges of hi-fi, bass and guitar amplifiers

The rule-of-thumb is usually 10:1, meaning that an amp's upper-frequency limit should extend to the frequency ten times the highest fundamental tone produced by an instrument. If we apply that rule to a bass guitar amp, they should reproduce overtones up to 3 kHz, and the guitar amps' range should extend up to 8kHz or even higher.

Since hi-fi amps have to reproduce the highest of notes (such as piccolo's) and their overtones, even the minimally accepted standard of 20kHz isn't usually enough. Our hi-fi amps typically go up to 35-40k Hz or even higher.

The ten-to-one rule does not apply at the bottom end; no guitar speaker can reproduce frequencies below 40 Hz anyway at any meaningful level, so 30 and 80 Hz low-frequency limit for bass and guitar would be more than adequate. Hi-fi amps should go down to 20 or so Hz.

That is why output transformers in hi-fi amps are so much bigger than those in guitar amps. Smaller transformers' cores saturate at much higher frequencies, in many cases of guitar amps even at 80-100 Hz, which would be unacceptable for a hi-fi amp.

Power bandwidth

The -3dB frequencies of power amplifiers depend significantly on the output power level. It is customary to perform these measurements at two power levels, usually at 1 Watt ($2.83V_{RMS}$ into 8Ω load) and the rated power output. The frequency range at the rated power output will always be narrower than the -3dB range at 1 Watt. In tube amplifiers this is primarily due to the imperfections of the output transformer, whose performance is limited at both low and high frequencies.

In this example, a tube amplifier rated at $P_R=15W$ was tested and the half-power frequencies (also called -3dB frequencies) were found to be $f_{LR}= 19$ Hz and $f_{UR} = 29$ kHz. "R" stands for "rated" power. When the same amp was tested at a lower power output level of 1 Watt, the -3dB frequencies were $f_{L0}=11$ Hz and $f_{U0}=37$kHz, meaning the frequency range was much wider. A tube amplifier's frequency range will be wider at low and narrower at high power levels. That is why quoting a frequency range without specifying the power level and the load impedance is meaningless.

An amplifier's narrower frequency range (or bandwidth) at higher power levels is due to the limited primary impedance and the saturation of the output transformer's magnetic core at low frequencies and parasitic capacitive effects at high frequencies.

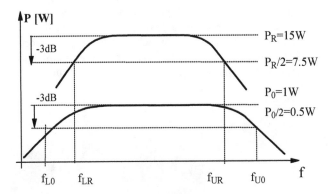

How bandwidth of a certain tube amplifier varies with output power level, at 1W output and the rated 15W output

Guitar amps versus hi-fi or "audiophile" amplifiers

Guitar players, tube guitar amp designers, and DIY builders can gain a great deal of knowledge from books on hi-fi tube amplifiers. The operation of tubes and most of the basic building blocks (components and circuits) are the same, as are the topological and construction principles. However, there are some fundamental differences.

In essence, a hi-fi tube amp is a reproduction device that amplifies the low voltage, low power input signal from the source (turntable, CD player, D/A converter, etc.) into a high current, high power electrical signal needed to drive the electromechanical transducer (loudspeaker). Both should be as "neutral" as possible, meaning they should add as little of their sound coloration to the recorded sound as possible. Real amps and especially loudspeakers are never neutral; they always add their "personality" or voicing to the ultimate result.

The basic job of a guitar amp is also to amplify the weak signal from a guitar pickup and drive the loudspeaker, but a guitar amp is also an instrument, not just a transducer, since its second function is to add its own voice to the overall result. In other words, a guitar amp should produce a tone or a variety of tones that guitar players are looking for. They should distort the incoming guitar signal in specific desirable ways.

In addition to tonal and distortion controls, a guitar amp may include additional effects, the most common being tremolo (amplitude modulation of the signal) and reverb or echo.

In terms of functionality, modern hi-fi amps are very Spartan. Many don't even have volume control (these are called power amplifiers) since the preamplifier performs volume control. Tone controls are a thing of the past in hi-fi, as are presence and loudness controls. The prevalent contemporary approach in hi-fi circles is minimalist in essence, "Less is more!", meaning the fewer switches, potentiometers, cables, etc., the signal goes through, the better/cleaner/more pure the sound.

Guitar amps, being musical instruments, tend to include various controls, of which gain, master volume, tone (bass, middle and treble), "overdrive," "bright," and power attenuation are the most common. Then there are manufacturer-dependent tonal controls named "presence", "resonance", "texture", "contour", "crunch", "fuzz", "Tubescreamer, "Vaporizer" and heaps of others.

TUBE GUITAR AMPLIFIERS

- portable, sound adjustable by positioning and slanting of the amp
- distortion welcome, at least the desired kind
- the amplifier is part of the instrument - its distortion behavior and voicing are of paramount importance
- a wide range of inbuilt effects and controls (tone, presence, gain boost, reverb, tremolo, ...)
- a narrow range of tube types used (the top five being 12AX7, 6L6, 6V6, 6BQ5 and EL34)
- designs are very similar and predictable
- tube guitar amps are relatively affordable (low powered ones priced under $600, high power ones $1,000-2,000)

SOUND SOURCE: GUITAR

AMPLIFIER + SPEAKER (COMBO)

Hi-fi TUBE AMPLIFIERS AND SOUND REPRODUCTION

- stationary, setup critical for optimal results
- neutrality and distortion minimization are the main aims
- minimalist approach, volume control only, no tone controls
- a wide range of preamp and power tube types used
- designs & topologies vary widely (stereo amps, monoblocks, OTL amps, directly-coupled and transformer-coupled amps, etc.)
- audiophile amps are expensive, $5-10,000 or much more

ABOVE: The simplest guitar & amp system

BELOW: The simplest turntable-based hi-fi system

SOUND SOURCE: TURNTABLE OR CD-PLAYER

PHONO STAGE AND/OR LINE-LEVEL PREAMPLIFIER

STEREO AMPLIFIER

A PAIR OF LOUDSPEAKERS

DISTORTION IN AUDIO AMPLIFIERS

While certain types of distortion may be a desirable feature of guitar amplifiers, it is generally avoided in high-fidelity ones. I say "generally" because not all distortion is bad or unpleasant sounding, and also because specific measures that reduce distortion may also negatively impact the other aspects of sound quality. A typical example is a negative feedback.

Harmonic distortion (THD)

Assuming a simple sine wave signal at the input of a real amplifier, the output will also be a sine wave but slightly distorted. Such a distortion cannot usually be detected visually, by observing the waveform on an oscilloscope, for instance, except in the cases of severe distortion. Nevertheless, should such a signal be brought to the input of a distortion meter or a spectrum analyzer, the presence of new harmonics will be detected.

Real amplifiers generate harmonics due to their nonlinear input-output or "transfer" characteristics. We will study these in great detail to understand the various measures that can be taken to "straighten" or "linearize" the operation of nonlinear devices such as vacuum tubes. By the way, for the transistor fanatics in our midst, bipolar transistors are even less linear than vacuum tubes. Pentodes are less linear than triodes, so the most linear of all amplifying devices is a triode!

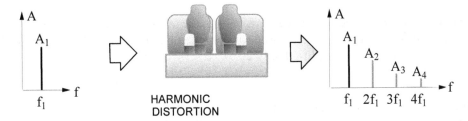

Usually, the second harmonic (twice the signal's frequency) has the highest amplitude, followed by the third, the fourth, and so on. The relative amplitudes of various harmonics depend on the amplifying device (triodes distort differently from pentodes and beam tubes), the power level at which the measurement is taken, the design of the circuit (single-ended or push-pull, and many other factors). More on different spectral "signatures" very soon.

Intermodulation (IM) distortion

To explain this type of distortion, let's look at the way it's measured. Two pure sine wave signals are mixed and fed into an amplifier. One is of a lower frequency f_1 (usually the mains frequency, 50 or 60Hz), the other of fifty times higher frequency f_2. Assuming f_1=50Hz, f_2=2,500Hz. Various commercial analyzers use different frequencies but the principle behind their operation is the same. The amplitude of the lower frequency signal is adjusted so that it's four times higher than the amplitude of the higher frequency signal (A_1=4A_2).

An ideal amplifier would amplify both signals equally (say 20 times), but a real amplifier will also generate two unwanted signals (or "side-bands) of the higher frequency signal. One will be of f_2-f_1 frequency (or 2,500-50 = 2,450Hz in our case), the other will have a frequency f_2+f_1 (or 2,500+50 = 2,550Hz)!

In severe cases of distortion, second side-bands will be also be generated, f_2-2f_1 and f_2+2f_1. The situation is analogous to AM (amplitude modulation) radio, although frequencies in question here aren't in the radio band but in the audio range.

The output signal will be an amplitude-modulated carrier. The f_2 signal is the high frequency (HF) carrier, while the low frequency (LF) signal (f_1) modulates the LF signal's amplitude.

IM distortion is more unpleasant to the human ear than harmonic distortion. Thus, its reduction should be even higher on the priority list of an amplifier designer than the reduction of harmonic distortion.

Going back to the wide bandwidth debate, a poor quality amplifier may generate distortion tones at, say, 28 kHz and 31 kHz. Due to IM distortion, these will then produce sidebands, one of which, 31-28 = 3 kHz, will fall in the audible range, dead smack in the midrange where the human ear is the most sensitive!

Attenuation distortion

The illustrated situation shows a complex audio signal at the amplifier's input (thicker trace), comprising two sine waves, a fundamental, and its 3rd harmonic (thinner traces). If the third harmonic lies in the frequency region where the amplification factor A of the amplifier starts dropping (around and above the f_U frequency), the 3rd harmonic will be amplified less than the fundamental tone, and the waveform of their sum, the output signal, will differ from the waveform of the input signal. This distortion is called attenuation distortion because higher harmonics are attenuated (or not amplified as much) compared to lower harmonics.

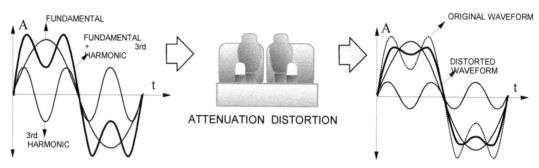

For instance, assuming the fundamental tone has frequency f_1=3kHz, the third harmonic will be at f_3=3*5=15kHz. A guitar amplifier with A=20 and f_U=9 kHz (numbers chosen for the sake of simplicity) will amplify the fundamental 20 times, but the 3rd harmonic will only be amplified only 20*0.71 = 14.14 times!

Delay distortion

The phase angle between the input and output signal stays constant through the midrange frequencies but changes significantly at frequency extremes. Just as in the last example, the illustration shows a complex input audio signal comprising a fundamental and its 3rd harmonic. Again, if the third harmonic lies in the frequency region where the phase angle changes from its midrange value, the 3rd harmonic will be shifted in phase by the angle θ.

Their sum, the output signal, will differ from the waveform of the input signal. This kind of distortion is called delay distortion. Looking at how different the shape of the resultant output voltage is from the input waveform, you'd think that this kind of distortion would be the most serious and malign of all. The truth is that our ears do not object to this kind of distortion. In fact, they don't even detect it! Exactly why that happens (or rather why doesn't it happen) is beyond the scope of this book, but boy, aren't we glad, one less issue to worry about.

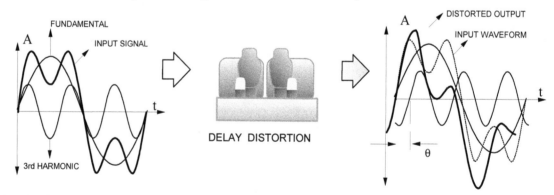

Amplifier behavior in time- and frequency domains

Now that we understand the concept of harmonics in a two-dimensional or time domain, we can study it in a 3D space and introduce the frequency domain. The amplitude-time graph (next page) shows a distorted periodic waveform that can be broken down into two components, the fundamental harmonic and the second harmonic, of twice the fundamental frequency ($2f_1$). There is also a phase shift a (alpha) between the fundamental and the second harmonic. This is the "time domain" in which we see these three signals' amplitude and phase relationships. Test instrument that displays signal waveforms in the time domain is called an oscilloscope.

If we depict these harmonics in a three-dimensional space and add frequency as the 3rd dimension, it would look like this. The projection on the A-f plane will result in the spectrum, as illustrated. This is the frequency domain. We don't see the waveforms anymore (we know they are sinusoidal signals, anyway), but we see the frequencies, the absolute and relative amplitudes of all harmonics, something we don't see in the time domain. Test instrument that displays amplitudes of signal harmonics in the frequency domain is called a spectrum analyzer.

The two depictions illustrate different aspects of the same signal. In this case, we have only the fundamental (1st harmonic) and the second harmonic, but generally, higher-order harmonics will be present as well.

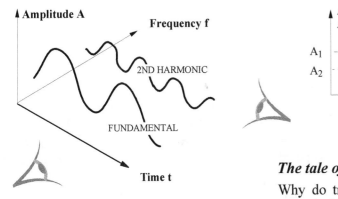

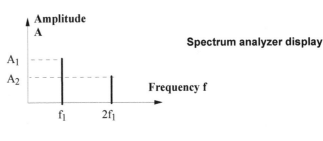

Spectrum analyzer display

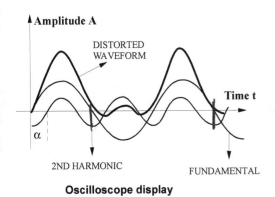

Oscilloscope display

ABOVE: The relationship between the time and frequency domains in a 3D space

The tale of two spectra: triodes versus pentodes

Why do triode amplifiers usually sound better than those using pentodes? The spectra of the 2A3 single-ended amplifier at 1/2 Watt and 3 Watt power outputs show a dominant second harmonic. At lower power levels, that is the only harmonic measurable. The third and fourth only appear when the amp approaches its maximum power of 3.5W. The harsh-sounding 3rd harmonic is "masked" by the harmonically pleasing 2nd harmonic.

The pentode's sonic signature is very different. Even at low power levels, the third and the fifth harmonics are present, and higher-order ones are measurable. At higher power levels, those odd harmonics increase rapidly (from -53 dB to -25 dB, in the case of the 3rd harmonic), which makes most pentode amps sound shrill and harsh compared to triodes.

RIGHT: Harmonic distortion spectra of two low powered SE amplifiers, 6F6 pentode and 2A3 triode, for two power levels.

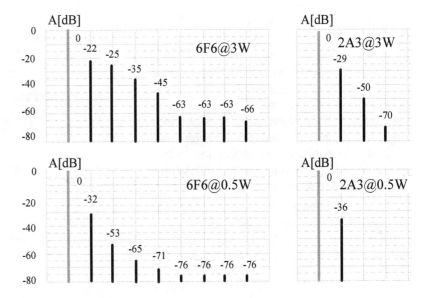

Calculating distortion figures from the harmonic distortion spectrum

How do we determine harmonic distortion in % from such spectral figures in dB? The fundamental formula is dB = 20log(x), so "x" would be our harmonic distortion coefficient H (but not in % yet!) Remember, if the base of the logarithm is not specified, it is assumed to be 10, so log(x) means $\log_{10}(x)$. Therefore $x = 10^{(db/20)}$!

Let's say the 2nd harmonic H_2 is 40 dB below the zero level of the 1st harmonic H_1. We have $H_2 = 10^{(db/20)} = 10^{(-40/20)} = 10^{(-2)} = 0.01$ To convert that figure into percentages, multiply it by 100 and get $H_2 = 1.0$ %

Remember, -20 dB means 10% distortion, -40dB equals 1% distortion, -60dB corresponds to 0.1% distortion, or, in other words, for every 20 dB drop the distortion reduces by the factor of 10!

What if, for some reason, the amplitude of the 1st harmonic is not 0dB but some other figure, such as -4 dB? Well, this is where the beauty of decibels comes into play. Since the dB is a relative unit, to get a relative difference between the two harmonics, simply subtract one figure from the other!

Let's say the 2nd harmonic is at -40 dB while the 1st harmonic H_1 is at -4 dB. Since all harmonics must be referenced to the fundamental's level we have $H_2 = H_2 - H_1 = -40 - (-4) = -40 + 4 = -36$ dB!

Now $H_2 = 10^{(db/20)} * 100$ [%] $= 10^{(-36/20)} * 100$ [%] $= 1.585$ %

Harmonic Distortion

Calculating THD from individual harmonics

If the amplitudes of the individual harmonics are known (in Volts), the overall THD (Total harmonic Distortion) figure can be calculated as THD [%] = $\sqrt{(H_2^2+H_3^2+ ...+H_N^2)}/H_1$ *100%

THD
THD [%] = $\sqrt{(H_2^2+H_3^2+ ...+H_N^2)}/H_1$ *100%

Let's calculate THD figure for the SE triode amplifier at the output level of 3 Watts (spectrum on the previous page). The fundamental H_1 is at 0dB, the 2nd harmonic H_2 is at -29dB, the 3rd H_3 at -50 and the 4th H_4 at -70 dB.

Since -29dB=20log(H_2/H_1) we get H_2/H_1=0.03548. Since -50 = 20log(H_3/H_1) we get H_3/H_1=0.003162, and H_4/H_1=0.00031623. Our harmonics are already expressed as percentages of H_1, so the square root of H_1^2 in the formula cancels out H_1 in the denominator and we have

THD= $\sqrt{(H_2^2+H_3^2+...+H_N^2)}$ if H_2 to H_N are in % of H_1. In our case THD [%] = $\sqrt{(0.03548^2+ 0.003162^2+ 0.00031623^2)}$ = $\sqrt{(3.548^2+0.3162^2+0.031623^2)}$ = 3.562 %

WHAT DETERMINES THE SOUND OF AN AMPLIFIER?

Why do tube and solid-state amplifiers sound different?

If you are reading this book, you most likely don't need to be convinced of the sonic beauty of tube guitar amplifiers.

Although they have numerical scales, the graphs on the right qualitatively illustrate two of the many reasons why tube and SS amps sound different. So, don't get stuck on the figures; we are talking principles here.

First, notice that SS amplifiers' distortion at low power levels is relatively high, and then it comes down with an increase in output levels, only to jump suddenly once the clipping levels are reached. Distortion of a typical tube amplifier is low at low power levels and gradually creeps up with increasing output levels. Clipping is gradual and soft unless high levels of feedback are used, in which case it resembles solid-state clipping.

Perhaps that explains why tube amps sound better at low power levels and why their distortion at high power levels is more pleasant and tonally rich than the distortion of transistor amps. When SS amps reach clipping levels, one would need to be totally deaf not to notice such unpleasant harshness.

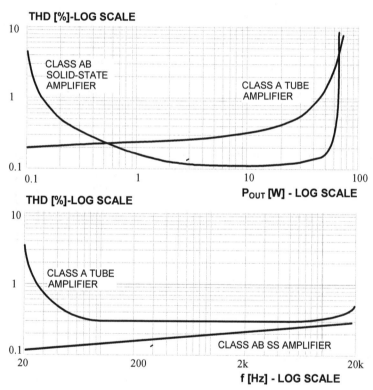

ABOVE: THD versus output power characteristics of a typical Class AB solid state and Class A tube amplifier (above) and THD versus frequency characteristics of a typical Class AB solid state and Class A tube amplifier

The second graph illustrates that THD (Total Harmonic Distortion) in SS amps increases steadily with increasing frequencies. Tube amps distort more at very low frequencies, especially at high power levels, due to the output transformers' core saturation. Still, then the harmonic distortion is more or less constant through the mid-band, increasing only in the upper-frequency band above 10kHz.

Many solid-state guitar amplifiers use integrated circuits, mainly at the lower end of the power range, around the 10-15 Watt output capability. They sound even harsher than SS amps using discrete transistors since their clipping is very abrupt and very symmetrical, containing loads of shrill odd harmonics.

Why do amplifiers sound different at different volume levels?

Volume control potentiometers in amplifiers do not use linear taper (the relationship between output resistance of a potentiometer and the angle of rotation) since human ears are not linear sensors but react to sound pressure in a logarithmic manner. That is why the logarithmic taper is commonly called an audio taper.

Human hearing is also frequency-dependent, and since our ears are most sensitive in the frequency range between 500 Hz and 5kHz, amplifier and speaker designers should strive to minimize the distortion of their amplifiers and speakers in the midrange.

As signal frequencies drop into the bass region, higher SPL levels are needed to achieve the same perception of loudness. Bass needs to be boosted at lower volumes to restore the original balance, so many vintage amplifiers had a "loudness" switch or potentiometer. This simple filter did not (could not!) boost low frequencies but attenuated the high frequencies instead. Even some guitar amps had loudness-compensated volume controls.

The same conclusion applies to higher frequencies, although to a lesser extent. The nature of our hearing correlates quite well with the frequency-dependent nature of amplification, so one of the most common methods used in audio engineering is to study the three frequency bands separately.

Our modeling of vacuum tube amplifiers and audio transformers will also be done that way, with separate models for the low-, mid-, and high-frequency ranges. The limits of these frequency bands aren't strictly defined; the low-frequency range extends up to 100 or 200 Hz, the high frequencies are above, say, 7kHz, and the midrange is the range in between.

Another issue is that as loudness changes, the frequency balance of an amplifier's response or its "voicing" changes as well; a distortion that sounds pleasant and cool at a lower volume may sound shrill and irritating at higher volume levels.

A few famous guitar players even used a variable transformer ("Variac") to power their amps through. Instead of simply connecting their amps to 115V mains (in the USA), they would reduce the supply voltage to the amp to 90 or 95 Volts. That would change the operating DC voltages in the amp and change their voicing considerably. Ah, the lengths guitarists go to in search for (musical) nirvana!

Why tubes of the same type sound different?

Most audiophiles and guitar players will agree that different tubes (of the same type) make the same amplifier sound different. Tube "rolling" makes it easy to compare the sound of various tubes by simply swapping them around in an amplifier and listening critically. If you take a few tubes of the same type from multiple manufacturers, say 12AX7 (ECC83), you will notice that anodes or plates are of different shapes and sizes. So are the cathodes and the grids, but these usually cannot be seen from the outside.

Some anodes are twice as large as the others (JJ has the smallest anode here, Ei the largest). Some anodes are "boxed," others flat, some are ribbed, some smooth; even more importantly, they are made of very different materials, some shiny silver in color (nickel?), such as Ei, some matte gray (JJ and RCA), others black (Arcturus).

Other parts of these tubes will also differ. The material used for cathodes, heaters, grids, and other electrodes, the chemical composition of the glass will vary, the thickness and construction geometry of electrodes will be different, and so on. These variations are not apparent, tubes would need to be destroyed and taken apart for further inspection, and only chemical analysis would detect metallurgical differences.It is indeed a miracle, or rather a long stretch of the imagination, to call all these very different tubes "12AX7".

Sure, they all have a similar amplification factor m and the other two basic tube parameters (transconductance and internal impedance) but may not behave in the same manner in other respects. No wonder they sound different in the same amplifier!

Other parts will also impart their character onto the final sound of an amplifier. Passive components (resistors, capacitors, inductors and transformers) are a significant factor in the final voice of an amplifier, but not the only one. Hookup wire, shielded cables, tube sockets, even the construction method (PCB or point-to-point) all impact the sound. Even fuses and power cords, far removed from the signal chain, contribute sonically!

In the early days of my audio journey, the engineer in me dismissed such claims as ridiculous, until I realized that neat models and simple measurements are one thing, but the messy reality is something else.

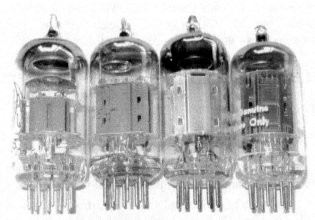

ABOVE: All marked 12AX7, but are they all the same tube? L-R: JJ ECC83S (Slovakia), RCA (USA), Ei (Serbia), Arcturus brand (unknown origin)

Why amplifiers that measure well don't necessarily sound better?

The first issue is the lack of standard load, which would emulate a typical loudspeaker. What is a typical loudspeaker is the ultimate question, and the answer is that there is no such thing! Thus, we test amplifiers using a resistive dummy load, whose impedance is independent of frequency and is pure resistance.

Dynamic loudspeakers used with guitar amps are complex inductive loads whose impedance is highly frequency dependent. They also generate a counter electromotive force (EMF), which is fed back into the output stage of an amplifier and interacts with it in the most unpredictable ways.

Furthermore, most tests (frequency range, input impedance, output impedance, and damping factor, harmonic distortion, maximum power) use only a single test frequency of fixed amplitude. However, music signals are complex waveforms comprised of dozens of harmonics of rapidly changing amplitude and phase.

Another measurement problem is that THD measurements do not distinguish between pleasant and rich-sounding even harmonics and harsh and irritating odd harmonics. It is not so much the amplitudes of the harmonics that determine the sound of an amplifier, but its whole spectral "signature." The spectral analysis gives us the relative amplitudes of all harmonics. Hence, such a test is more meaningful than simple THD measurements. However, despite knowing the number of dB each harmonic is below the fundamental, we still cannot predict how the overall spectral signature would sound to our ears.

We measure things that are easy and convenient to measure, not necessarily those that have the most significant impact on sound fidelity! The most important aspects such as rhythm and pace, the sense of "presence", "musicality," "microdynamics," "responsiveness," or "transparency" cannot be correlated with measurements at all.

The dielectric theory of sound

Audiophiles tend to hate capacitors with passion (especially the electrolytic kind!), claiming they are detrimental to sound quality, and there is a lot of truth in that claim. Even some guitar players and amp designers have noticed a difference in the sound of various coupling caps in guitar amps!

Capacitors are inevitable; they are everywhere; you are sitting inside a huge capacitor right now. The wiring in the roof of your house is one conductive plate, the earth or ground is another, and you and everything around you (including the air) is the dielectric between the two plates.

Even if you remove all the coupling capacitors in an amplifier, there will still be dozens of capacitances left. These are parasitic capacitances between the cables, components (resistors, capacitors, inductors, tube sockets), and chassis. Another issue is tubes' internal capacitances, and there is sweet nothing you can do about them.

All other components, from potentiometers to filtering chokes, also have parasitic capacitances, and all those capacitances will impact the sound of an amplifier. You cannot see them or touch them, they are not discrete capacitors but parasitic (unwanted) capacitances distributed across cables and components.

The sound of a capacitor, or any other component for that matter, is determined by two main factors. One is its conductive parts, the cathode, the grid and the anode of the tube, the conductors in a cable, the metal plates and leads of a capacitor, the winding wire of a transformer. The other factor is the dielectric between those conductors.

The same output transformer will sound different if an impregnated paper is used for insulation between winding sections instead of plastic film such a Mylar7 or Kapton7.

The sound of audio transformers will also depend on the properties of the core or the laminations used. A transformer with GOSS (Grain-Oriented Silicon Steel) core will sound different from an otherwise identical transformer wound on a standard 3% silicon steel core.

FLIPPANT, MISLEADING OR SIMPLY "LIBERAL" USE OF LANGUAGE?

When is a "tube amp" not a "tube amp"?

Bugera BC15 is marketed as a "tube amplifier." The fact that its output stage is solid-state is not even mentioned. We are not singling out this particular brand or model for any specific reason; there are dozens if not hundreds of such "marketing" examples. BC15 was chosen not just because of this questionable practice but also for extensive use of meaningless language in its promotional literature, the second point we are trying to make here.

The first issue begs the fundamental question: Does having a single tube make a hybrid amp into a "tube amp"? The problem is in the definition, what is a tube amp? Manufacturers like to use terms such as "pure Class A" operation of a tube stage or a whole amp, as if there is something as silly as "*impure* Class A"! By the same token, why not use "Pure tube amp" instead for amps without any solid state components in the signal path (or *at all*!) and "tube amp" for anything else? That would make a bit more sense.

Interestingly, some reputable online retailers such as Breakwater, probably to prevent returns and misleading advertising accusations, clearly emphasize that this is a hybrid amp, while most others don't.

Now to the meaningless language and unsubstantiated claims. Under the "features," Bugera makes some interesting claims. First, "hand-selected 12AX7 valve". Hand-selected? Where, in the tube factory or on the amp assembly line? By whom? How? This could mean anything, from "taken by hand from a bulk box of 100 tubes and inserted into the amp" to "tested on a tube tester or some selection rig which measures distortion levels and the breakup characteristics"?

Secondly, "Authentic vintage amp design and classic guitar sound." "Authentic vintage amps (meaning those from the 1950s and 1960s) were not of the hybrid type; they did not mix valves with silicon in the preamp and output stages!

Thirdly, "dedicated 2-band EQ..." and "dedicated master level control ..."? Dedicated to what?

For good measure, so that the boys from Bugera and Behringer (the Music Group of companies) don't feel unfairly singled out, even iconic brands such as Fender and Marshall aren't immune to the use of debased language and marketing fluff. This is verbatim from Marshall promotional literature: "The Marshall MG100HFX amp head packs fully analog tone circuits, providing the distinctive Marshall tone and pure organic flexibility."

"Pure organic flexibility"? Oh please!

The three biggest lies in guitar amp marketing

1. Most amps touted as "all tube" are not all tube!

Strictly speaking, even if only silicon diodes are used as rectifiers in the amp's power supply (no silicon in the signal chain), the amp isn't "all tube" anymore. Likewise, even if there are no semiconductors in the main signal path, but an amp uses digital reverb or a send-return loop with transistors or integrated circuits, it is not "all tube" anymore.

2. The use of printed circuit boards does not make amplifiers more reliable or better sounding. It just makes them cheaper to manufacture.

In fact, the opposite is true. In hi-fi, it has been universally accepted that point-to-point wired amps (without any PCBs) sound better. There is no reason why guitar amps would be different.In trying to justify their use of PCBs, Supro website claims that "Terminal strip construction (à la vintage Supro) is not reliable or allowed by safety regulations". Many guitar amps suffer from overheating components, which in turn damages the PCB and causes secondary failures. Overheating components in hand-wired amps cannot damage terminal strips that easily, and reliability isn't affected as much.

As for its illegality based on safety, we could not verify the accuracy of that claim. There was no reference in Australian Wiring Rules (or any other national standard we could find) that specifically outlaws terminal strips. Indeed, they are still being manufactured and sold worldwide.

3. Most push-pull amps marketed as "Class A" amps do not work in class A most of the time.

Since many guitar players do not understand what class A means, the brazen manufacturers can use this marketing strategy with impunity. Class A refers to one of the two possible audio operating regimes, not amplifiers' quality levels.

Class A means that tubes never go into cutoff (their anode currents never drop to zero). Class B means that one of the pair (or two of a quad, and so on) of tubes will conduct more and more, but the other will stop conducting its part of the signal and enter a cutoff state.

Class AB is not a class by itself; it simply means that at low output power levels, the output stage works at class A, meaning that both tubes are conducting the signal, one more, the other one less. As the output signal increases, the output stage crosses into class B operation at the point when one tube stops conducting at all, so all the signal is passed by the other tube. Thus, when a push-pull amp is marketed as a Class A amp, it is never a complete lie. All push-pull amps work in class A at low power levels. The exact crossing point into class B depends on many design factors.

However, there are some push-pull amps that are purposefully designed to continuously operate in class A and never to cross in class B. However, these are rare, and the main reason is their low power. The true class A push-pull amp will only produce the output power equal to double the power its output tube would produce in a single-ended operation.

Remember, SE amps always operate in class A; operation in any other class would mean that a portion of the output signal was cut off, which would be unacceptable and the amp unbearable to listen to!

The manufacturer touts Supro 1622RT Tremo-Verb as a 25 Watt Class A amplifier using a pair of 6973 (EL84) pentodes in push-pull. According to our tests and tube data sheets (below), it is impossible to get more than 12-13 Watts of class A power from those two tubes, let alone 25 Watts! These figures are for output stages using fixed bias. If cathode biasing is used, the maximum class AB power levels drop from 20 and 24 W to 15W and 17W, respectively!

Typical Operation with Fixed Bias:

6973 (EL84)   *Values are for 2 tubes*

	Class A	Class AB		
Plate Voltage.	250	350	400	volts
Grid–No. 2 Voltage.	250	280	290	volts
Grid–No. 1 (Control–Grid) Voltage*.	−15	−22	−25	volts
Peak AF Grid–No. 1–to–Grid–No. 1 Voltage.	30	44	50	volts
Zero–Signal Plate Current.	92	58	50	ma
Max.–Signal Plate Current.	105	106	107	ma
Zero–Signal Grid–No. 2 Current.	7	3.5	2.5	ma
Max.–Signal Grid–No. 2 Current.	16	14	13.7	ma
Effective Load Resistance (Plate to plate).	8000	7500	8000	ohms
Total Harmonic Distortion.	2	1.5	2	%
Max.–Signal Power Output	12.5	20	24	watts

How does one know that the first column above is for Class A operation? The first giveaway is the low anode supply voltage, 250V, compared to 350-400V in class AB. The second indicator is that the anode current increases very little between idle (92mA) and full power (105mA). The anode current at maximum power in class AB is more than double the idle current (a jump from 50 to 107mA)!

Likewise, Supro Royal Reverb amp is switchable between Class-A operation using cathode biasing (35 Watts claimed) and Class-AB with a fixed bias for 45 Watts output with a tube rectifier or 60 Watts with solid-state rectifiers.

Datasheet below shows 14.5 Watts in Class A with 250V anode supply and -16V bias, and 17.5 Watts with 270V anode supply and -17.5V bias. That is very far from the claimed 35 Watts.

PUSH-PULL CLASS A₁ AMPLIFIER, VALUES FOR TWO TUBES   6L6-GC

Plate Voltage	250	270 Volts
Screen Voltage	250	270 Volts
Grid-Number 1 Voltage	−16	−17.5 Volts
Peak AF Grid-to-Grid Voltage	32	35 Volts
Zero-Signal Plate Current	120	134 Milliamperes
Maximum-Signal Plate Current	140	155 Milliamperes
Zero-Signal Screen Current	10	11 Milliamperes
Maximum-Signal Screen Current	16	17 Milliamperes
Effective Load Resistance, Plate-to-Plate	5000	5000 Ohms
Total Harmonic Distortion	2	2 Percent
Maximum-Signal Power Output	14.5	17.5 Watts

PUSH-PULL CLASS AB₁ AMPLIFIER, VALUES FOR TWO TUBES

Plate Voltage	360	360	450 Volts
Screen Voltage	270	270	400 Volts
Grid-Number 1 Voltage	−22.5	−22.5	−37 Volts
Peak AF Grid-to-Grid Voltage	45	45	70 Volts
Zero-Signal Plate Current	88	88	116 Milliamperes
Maximum-Signal Plate Current	132	140	210 Milliamperes
Zero-Signal Screen Current	5.0	5.0	5.6 Milliamperes
Maximum-Signal Screen Current	15	11	22 Milliamperes
Effective Load Resistance, Plate-to-Plate	6600	3800	5600 Ohms
Total Harmonic Distortion	2	2	1.8 Percent
Maximum-Signal Power Output	26.5	18	55 Watts

6L6-GC data sheet clearly specifies maximum push-pull output power levels obtainable in class A and in Class AB

ELECTRONIC COMPONENTS

3

- FIXED AND VARIABLE RESISTORS
- CAPACITORS
- OTHER COMPONENTS: FUSES AND SWITCHES
- PASSIVE SEMICONDUCTORS: DIODES AND ZENER DIODES

Passive electronic components don't amplify audio signals themselves (in contrast to active components such as tubes or transistors) but perform other crucial functions in circuits. Their physical properties (resistance, capacitance, inductance, etc.), power, voltage, and current ratings, and even the type of technology used (there are many different types of resistors and capacitors) must be carefully considered before using them in your design.

Since most of these choices will affect the operation and sound (voice) of an amplifier, there is no need to use exotic or expensive components to chase that ultimate tone. A simple change of a component type is usually enough.

FIXED AND VARIABLE RESISTORS

A resistor is an electronic component that "resists" the flow of electrical current. The higher its resistance, the less current will flow through the resistor, assuming they are connected across a constant voltage source, a DC battery, or a DC power supply inside an amp.

Of course, resistors resist the flow of signal or AC currents as well, and the same rules apply to both AC and DC currents and voltages as far as resistors are concerned. We will soon see how that is not the case with capacitors, inductors (chokes), and transformers, three other classes of passive components that make up tube amp circuits. By passive, we mean a component that cannot amplify power. Those are called active components and include tubes and transistors.

Carbon composition resistors

Carbon composition (CC) resistors are made of a finely ground mixture of powdered carbon and silica filler, with a small quantity of resin to act as a binding agent, compressed into a slug. CC resistors suffer from aging, their values drifting significantly over time. Their tolerances are the poorest of all resistor technologies. However, since they were the cheapest to produce in the 1950s and 1960s, they were widely used in vintage equipment.

They do sound "warmer" and more "musical" than the film types (especially compared to metal film), but also "grainier." Due to their high noise levels, they should never be used as grid resistors or in the first stage of a guitar amp where the signal levels are low and where the following stages would amplify any noise generated in that stage.

Carbon film, metal film and metal-oxide resistors

Carbon film resistors (also called carbon deposit type) start their life as a ceramic rod onto which a carbon coating is deposited in a furnace or by spraying it or dipping the rod into a carbon solution. Afterward, a spiral groove or track is cut into the coating to increase the length of the resistance path. The pitch and the length of the groove are varied to adjust the resistance to the desired value.

Carbon film resistors have a higher negative temperature coefficient than carbon composition types. The resistive material used in metal film resistors is usually nickel-chromium (NiCr), although other alloys containing gold, platinum, palladium, rhodium, and tantalum nitride are also used.

Metal-oxide film resistors are made by the oxidizing reaction of a vapor or by spraying a tin chloride solution on a heated glass or ceramic rod, so a thin film of tin oxide is fused onto the substrate. Antimony oxide may be added to increase the resistivity of the film. Metal oxide resistors can withstand higher operating temperatures than carbon or metal film resistors and are more stable and reliable.

Wire-wound resistors

Wire-wound resistors are made by winding a certain length of a special resistive wire around a ceramic rod. The whole thing is then dipped or coated in epoxy or enclosed in an aluminium case which can be bolted onto a chassis to act as a heatsink for better cooling.

Since they are constructed in the same way as inductors, the inductance of wire-wound resistors is relatively high compared to other resistor technologies. There are non-inductive wire wound resistors made using special winding methods.

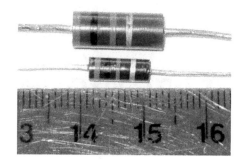

ABOVE: 1/2W (lower) and 1W (upper) carbon composition resistors. The ruler is in cm.

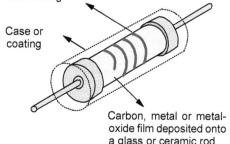

Spiral groove to increase and adjust resistance, incised into the film coating

Case or coating

Carbon, metal or metal-oxide film deposited onto a glass or ceramic rod

ABOVE: The internal construction of film resistors.
BELOW: The cross-sectional view of a wirewound resistor.

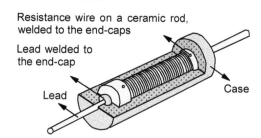

Resistance wire on a ceramic rod, welded to the end-caps

Lead welded to the end-cap

Lead

Case

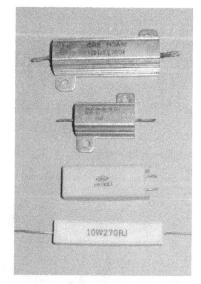

ABOVE: Typical examples of wirewound resistors.

Resistor color chart and standard values

Standard resistor values follow geometric progressions named E12, E24 and E48 series (10%, 5% and 2% tolerance, respectively). The values between 1 and 10 are listed below. All other values are obtained by multiplying those with 10, 100, 1000, etc.

E12 series (10% tolerance): 1.00 1.20 1.50 1.80 2.20 2.70 3.30 3.90 4.70 5.60 6.80 8.20

E24 series (5% tolerance): 1.00 1.10 1.20 1.30 1.50 1.60 1.80 2.00 2.20 2.40 2.70 3.00 3.30 3.60 3.90 4.30 4.70 5.10 5.60 6.20 6.80 7.50 8.20 9.10

Example: Red - Violet - Red - Silver: 27 x 100 = 2k7 (2,700 ohms or 2.7 kilo ohms), tolerance +/-10%. Instead of writing kohms or kΩ all the time, we will simply write "k", assuming that it is obvious we are talking about a resistance of a resistor!

Color	Band 1	Band 2	Band 3	Band 4
	1st digit	2nd digit	Multiplier	Tolerance
Black	0	0	1	
Brown	1	1	10	+/- 1%
Red	2	2	100	+/- 2%
Orange	3	3	1000	
Yellow	4	4	10,000	
Green	5	5	100k	
Blue	6	6	1M	
Violet	7	7	10M	
Gray	8	8	100M	
White	9	9	1G	
Gold			0.1	+/- 5%
Silver			0.01	+/-10%
None				+/- 20%

Say you need an 82k anode resistor and are rummaging through your resistor stash, which is not sorted. Well, 8 is Gray, 2 is Red, the multiplier is 1,000 (to get from 82 to 82,000), which is Orange. So you need Gray-Red-Orange-Silver (from E12 series, with 10% tolerance) or Gray-Red-Orange-Gold (from E24 series, 5% tolerance) or Gray-Red-Orange-Red (from E48 series, 2% tolerance).

Tolerance indicates the acceptable range of values for each resistance. The resistance of a nominally100k resistor from E12 series, with +/-10% tolerance, can range from 100k -10% or 90k to 100k+10% or 110k!

Variable resistors: potentiometers and rheostats

Potentiometers are variable resistors, the ones used in audio are of a single-turn construction (270°), but precision 10-turn pots are also widely used in applications where a more precise control of resistance is required.

Potentiometers that need to be adjusted frequently, such as those that control the volume or tone in amplifiers have a plastic or metal shaft onto which a knob is attached, either a push-on or screw-type.

Potentiometers that are adjusted rarely, only during servicing or calibration, instead of a shaft and a knob have a slot for a screwdriver. These are usually mounted inside an amplifier or an instrument, so the user has no access to it, and are called trimmer-potentiometers or "trimpots".

Film potentiometers are made by depositing a conductive film onto a plastic substrate. The conductive film is usually carbon or "cermet", a special conductive ceramic material. The wiper or sliding arm makes mechanical contact with the conductive film. As such, it is the weakest link since such contacts are abrasive and gauge the conductive layer, eventually making pots "scratchy" or "crackly" or even losing contact altogether.

The two main types of potentiometers are linear taper, usually marked "B," and logarithmic or audio taper, marked "A."

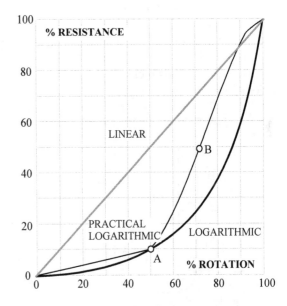

ABOVE: Percentage of resistance versus percentage of rotation curves for a linear potentiometer, nominal (ideal) logarithmic or audio potentiometer, and real (practical) logarithmic potentiometer

The logarithmic potentiometers are used for volume controls because human ears respond to sound pressure in a logarithmic manner.

Manufacturing true log tapers by carbon or plastic deposits is very difficult, so manufacturers usually resort to approximating the log curve with two linear segments. As illustrated, up to 50% rotation, where amplifier's volume controls are set most of the time, the deviation from the real log curve) is quite small, and at higher volumes such error is not noticeable at all.

Just like their fixed-resistance relatives, rheostats or high-power resistors are wire-wound components. The thinner the wire, the finer is the step-adjustment (the jump in resistance between adjacent turns), but the lower the current rating.

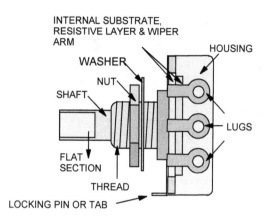

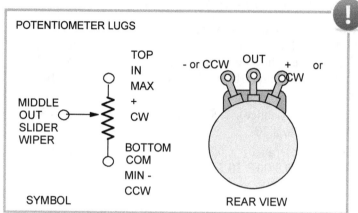

Two kinds of shafts are used for potentiometers. A split shaft (usually ribbed) is usually used with push-on plastic knobs. While pushing the knob down the shaft, the knob squeezes the two split halves together, which, when the knob is fully attached, expand laterally and hold the knob in place.

Fully round shafts or round shafts with a flat section (as illustrated above) are meant to be used with screw-type knobs. The knob is also pushed down onto the shaft but is held in place by one or two lateral screws. Currently produced knobs tend to use primarily screws with a hexagonal head, so tiny Allen-keys of the exact size are required. We prefer knobs that have slit heads, so a small screwdriver is all you need. These are easier to loosen or tighten than knobs with hex grub screws.

CAPACITORS

The simplest capacitor is formed by two conductive plates with a dielectric in-between. The most common dielectric materials used are air (in variable or "tuning" capacitors), mica, oxide films (in electrolytic capacitors), plastic films or paper (in film capacitors) and ceramics. The capacitance of this simple structure can be expressed as $C_P = \varepsilon_0 * k * A / d$ where ε_0 is the absolute dielectric constant (also called "permittivity") of free space, whose value is $\varepsilon_0 = 8{,}85 * 10^{-12}$ Farads/meter, and k is the relative dielectric constant of the dielectric. Typical values are listed in the table below.

DIELECTRIC	k	DIELECTRIC STRENGTH [kV/cm]
Air	1.0006	30
Polystyrene	2.5-2.7	197
Polyester	3.2	650
Polycarbonate	2.57	150
Polypropylene	2	236
Teflon	2	240
Impregnated paper	4-6	500
Mica	4.5-7.5	1,180
Aluminium Oxide	8.4	146
Tantalum Pentoxide	27	1,470
Ceramic (low loss)	6-20	80
Ceramic (high k)	100-1,000	350
Titanium Dioxide	80-120	6,000
Barium Titanate	200-16,000	300

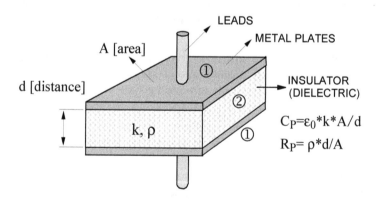

$$C_P = \varepsilon_0 * k * A / d$$
$$R_P = \rho * d / A$$

ABOVE: The simplest capacitor is formed by two conductive plates with a dielectric in-between (which could be air)

The term "capacitance" refers to the amount of electric charge a capacitor can store for each volt of potential difference between its plates: C= Q/V [Coulomb/ Volt]. This unit is called a Farad, in honor of Michael Faraday, famous for his work in electromagnetism (Faraday's Equations) and electrochemistry.

When a discharged capacitor is connected to a DC voltage source (for instance, a battery), a transient current will flow in this simple circuit; electrons will flow towards one electrode and accumulate there. The charge on the other plate will be of equal magnitude but positive, which means it will be depleted of electrons.

An electric field now exists between the two plates, and a certain amount of electric energy is stored as electric "stress" in the dielectric material between the plates. Since the charging current has stopped flowing in such a steady-state, the electric field is stationary and is thus called an electrostatic field.

The energy stored in a capacitor is $E = \frac{1}{2}QV = \frac{1}{2}CV^2$. For instance, in a typical tube amplifier's anode power supply, a 470µF electrolytic capacitor, charged to 400V will store energy of $E = \frac{1}{2} * 470 * 10^{-6} * 400^2 = 37.6$ J (Joule).

A solid state amplifier working at 40V and using 10,000 μF of capacitance in its power supply will have a stored energy of only $E = \frac{1}{2} * 10{,}000 * 10^{-6} * 40^2 = 8$ J.

Film capacitors

Film capacitors are non-polarized and are rated in terms of their capacitance, maximum working DC voltage, and dielectric material. Two main types are film & foil and metalized film capacitors. Film and foil technology uses two thin metal sheets or foils, acting as electrodes, and a plastic film in between them, acting as the dielectric. The three are then rolled together in a tubular shape. The thicker the dielectric, the higher the working voltage of a capacitor, and the bigger its physical size.

Metalized film technology does not use metal sheets or foils; the electrodes are metal layers deposited on both sides of the plastic film instead. Generally, metalized units are much smaller than the film&foil capacitors for the same value and working voltage. Metalized capacitors have a higher voltage rating due to their self-healing property. An internal arc through the metalized paper or plastic film burns the very thin metal layer away, effectively clearing the short circuit.

Film & foil capacitors are made in two technologies. In extended foil capacitors, foils and film are wound offset, with one metal foil protruding over the plastic film on one end and the other foil out the other end.

Once the winding is finished, the foils are crushed, and metal leads are soldered or welded to the crushed ends. This method effectively shorts the capacitor's parasitic inductance without affecting the capacitance.

The tab construction requires that plastic film be wider than the metal foil so metal tabs can be inserted in the roll, and the leads are then soldered onto the tabs. Once the leads are attached, the rolled capacitor is vacuum impregnated with a mineral vax, rigid resin, or synthetic or mineral oil (for paper capacitors).

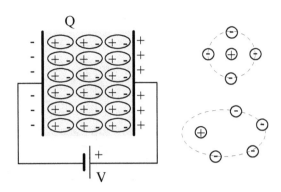

ABOVE: Without an electric field an atom has a positive nucleus and a cloud of negative charge made up of electrons rotating around it. Placing it inside an electric field "polarizes" the charges inside the atom and pulls the negatively charged cloud slightly away from the positive nucleus.

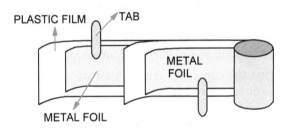

Film & foil capacitor with tab construction (ABOVE) and extended foil construction (BELOW)

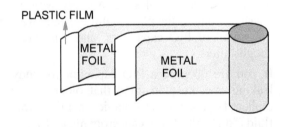

A code imprinted on the case or body of a capacitor usually declares what the dielectric is. The codes are KP - polypropylene film&foil, KS - Polystyrene film&foil, KT - Polyester film&foil, MKC metalized polycarbonate, MKP - metalized polypropylene, MKT - metalized polyester, MKY - metalized low-loss polypropylene, MKL or MKU - metalized lacquer (cellulose acetate).

Modern capacitors have their value or a code printed on their bodies. The code has three digits and one letter. The first two digits are the value, the third digit is the multiplier, and the letter is the tolerance. Some vintage units were marked using a color-coding system similar to that of resistors. The values are given in the tables.

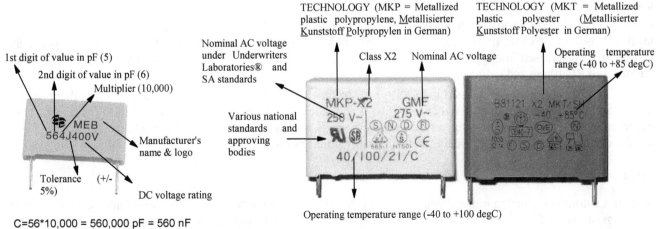

C=56*10,000 = 560,000 pF = 560 nF

ABOVE: Two examples of X2-rated (for mains filtering applications) film capacitors, in metallized polypropylene film and metallized polyester film. These make great coupling capacitors as well.

NUMBER	Multiplier	LETTER	Tolerance
0	1	F	+/- 1%
1	10	G	+/- 2%
2	100	H	+/- 3%
3	1,000	J	+/- 5%
4	10,000	K	+/- 10%
5	100,000	M	+/- 20%
6	NOT USED	P	+100% -0%
7	NOT USED	Y	+50% -20%
8	0.01	Z	+80% -20%
9	0.1		

Color	Band 1 1st digit	Band 2 2nd digit	Band 3 Multiplier	Band 4 Tolerance above 10pF	Band 4 Tolerance below 10pF	Band 5 DC voltage rating
Black	0	0	1	+/-20%	+/- 2pF	
Brown	1	1	10	+/- 1%	+/- 0.1pF	100V
Red	2	2	100	+/- 2%	+/- 0.25pF	250V
Orange	3	3	1000	+/- 3%		
Yellow	4	4	10,000	+/- 4%		400V
Green	5	5	100k	+/- 5%	+/- 0.5pF	
Blue	6	6	1M			600V
Violet	7	7				
Gray	8	8	0.01	+80% -20%		
White	9	9	0.1	+/-10%	+/- 0.1pF	
Gold			0.01	+/- 5%		
Silver			00.1	+/-10%		

The color-coding system for film capacitors.

Of all film capacitors, the polystyrene ones have the best electrical properties, very low temperature coefficient, almost linear, and minimal losses (low dissipation factor of approx. 0.01%). Their maximum temperature is only 85_0C. Polypropylene capacitors are nearly as good, with a slightly higher temperature coefficient and dissipation factor. They can operate up to 105^oC and are cheaper than polystyrene equivalents!

Polyester capacitors are the least stable and have the highest losses but are the cheapest and are widely used in budget guitar amps.

Paper-in-oil capacitors

Paper-in-oil (PIO) is a vintage technology that precedes the use of plastics. Oil loses its insulating properties when subjected to high temperatures, and paper is inferior to plastics in every way, so these capacitors don't last nearly as long as the plastic film types. However, many audiophiles prefer the sound of PIO capacitors, and this type of film capacitor is highly sought-after!

If you are opening a PIO capacitor to study its construction or have a leaking one, keep in mind that older units used PCB (polychlorinated biphenyl), a synthetic organic chemical compound used as a dielectric fluid ("oil"). PCBs are cancerogenic and were banned in the USA in 1979. Notice the printing on the vintage Mitsubishi capacitor (dating back to the early 1980s), "No PCBs."

ABOVE (Top-to-bottom): Aluminum foil PIO capacitor (1μF), and two Japanese PIO caps from 1970s, by Nippon Chemi-con and Mitsubishi (both 68 nF). The Mitsubishi capacitor is rated at $450V_{AC}$, which is roughly 450*2.8 = 1,260 V_{DC}!

Ceramic capacitors

Ceramic capacitors are non-polarized types manufactured in a myriad of shapes, sizes, and technologies. The main ones are rod- (or tubular), disc- and monolithic type. Tubular type can have radial or axial leads and can be molded (resembling carbon composition resistors), enamel-coated, or dipped phenolic. Monolithic ceramic caps can also look like low-wattage resistors (axial); these are glass-sealed types. The other variant is dipped polymer-coated, usually rectangular.

Class I ceramic capacitors have a linear temperature coefficient of capacitance (TCC), expressed in parts-per-million-per-degree-Celsius (ppm/^oC). Below 100Mhz their capacitance is very stable with temperature changes and aging, so they are used in applications where accurate and stable capacitance values are required, such as tuned circuits and RC filters and networks.

The ordinary kind is called a "Class II high-K" type. K refers to their high dielectric constant, enabling them to pack more capacitance into a smaller size than Class I, but their tolerances are wider and stability inferior. Their capacitance varies widely with temperature, as does their dissipation factor (losses) and insulation resistance. These are the small round brown caps widely used in cheaper guitar amps.

Mica capacitors

Mica is a naturally occurring mineral whose layered structure makes it easy to cleave or split into sheets of uniform thickness, down to $5*10^{-6}$ mm! It has a high dielectric constant (6.85) and high insulation voltage (3,000 volts per mil), and it is chemically inert. This makes it an ideal dielectric; mica capacitors do not age or change their properties over time and are very stable under temperature and electric stress.

Alternate layers of metal foil and mica sheets are sandwiched together, or silver is deposited directly onto mica sheets instead of foil. The finished stack is clamped together, and leads are spot welded or soldered onto it. The assembly is then treated (usually with silicone) to prevent moisture absorption, sealed in a molded case, or dipped in epoxy resin. Synthetic mica was also developed, a mixture of aluminium and magnesium oxides, sand, and cryolite, then heated to 1,300 °C and gradually cooled over a few days. It has a similar layered structure to natural mica but is more difficult to cleave.

Mica capacitors are made in values from 1pF to 100nF. The relatively low upper limit is because mica sheets are not flexible and cannot be rolled into tubular structures like film capacitors. Very low leakage currents and dissipation factors, together with small tolerances and high-temperature stability, make mice capacitors invaluable in high precision and high-frequency applications. 1kV voltage ratings are common, with some capacitors rated even up to 10kV!

ABOVE: Dipped mica capacitors

Trimmer capacitors

Disc ceramic trimmer capacitors have a rotor, a thin disc made of Class I or Class II ceramic dielectric. A silver layer is deposited, screened onto the top surface of the rotor. A similar electrode is applied to the stator, which also serves as the outer shell of the capacitor. Turning the rotor changes the effective plate area.

While disc ceramic trimmers go from minimum to maximum capacitance value in half-a-turn, tubular trimmers cover their capacitance range in many turns. They are used for more precise adjustments of small capacitance values (kind of like single-turn and ten-turn potentiometers).

The other trimmer capacitor types are compression mica types, where a thin layer of mica is sandwiched between two spring metal conduction plates. An insulated screw through the center of this assembly changes the compression on the plates and varies the distance between them, thus altering the capacitance. Usually, it is not just the mica sheet acting as dielectric but a layer of air as well.

Mica trimmers are nonlinear and not particularly stable; their value drifts with time. The capacitance values range from 1 pF to 3 nF.

ABOVE: A typical disc ceramic trimmer capacitor.

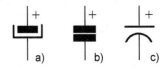

ABOVE: Three common symbols for electrolytic capacitors. The + sign is usually omitted, since it can be determined from the orientation of the symbol. In a) and c) the upper electrode is positive.

Electrolytic capacitors (elcos)

No other electronic component has as many symbols in use as elcos. Symbol c) is mainly used in vintage American literature, while a) and b) are more common in Europe. To add to the confusion, many American books and diagrams also use symbol c) for non-polarized capacitors (film, mica, etc.). We will use symbol a) in this book. The bottom electrode is shaped to symbolize the can into which the anode is immersed.

During manufacturing, a thin oxide film is formed on the anode (positive metal electrode). This film acts as a dielectric. It is surrounded by an electrolyte, into which the negative electrode is also immersed, whose job is only to make contact with the outer surface of the dielectric via the electrolyte. This contact is made by electrolytic conduction, and that is why this class of capacitors is called electrolytic.

The thickness of the oxide film is proportional to the forming DC voltage, but the thinner the film, the higher the capacitance. That is why capacitors rated at 450 or 500V seldom go higher than 680 µF in value!

The oxide film deteriorates without the forming voltage; that is why electrolytic capacitors deteriorate more on the shelf than working inside an amplifier. It is also why NOS capacitors that have been on a shelf for years or even decades should not simply be installed into an amplifier.

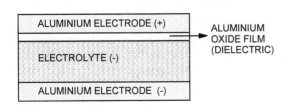

Cross-section through an axial electrolytic capacitor

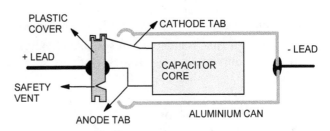

The mechanical construction of an axial electrolytic capacitor

They should be subjected to the reforming process, initially connected to lower DC voltage (say 100V instead of 400V). Then the voltage is gradually increased to their rated voltage and about 10% higher. A healthy capacitor should handle it without any ill effects (500V instead of 450V, for instance).

The anode must stay positive. If the polarity of the outside imposed voltage is reversed, the capacitor will be destroyed and most likely will explode, spewing the electrolyte all over the amplifier. The same happens when their rated voltage is exceeded. It is critical to observe the marked polarity and voltage rating of these capacitors.

The life of an electrolytic capacitor

In their data sheets for long-life 350-450V_{DC} aluminium electrolytic capacitors Vishay specifies the following:

Useful life at 85°C: 10,000 h, Useful life at 40°C, 1.4 x I_R applied: 400,000 h and Shelf life at 0 V, 85°C: 1,000 h

Notice that an increase in operating temperature from 40°C to 85°C reduces life by the factor of 40! (400,000/10,000 = 40). Also, notice that the shelf life is only 1/10 of the operating life (working in equipment), 1,000 hours compared to 10,000 hours. Electrolytic capacitors last longer when energized in equipment than sitting on the shelf! The applied voltage constantly forms the oxide film, which quickly and irreversibly deteriorates on the shelf.

When you buy elcos, ask for capacitors' age and a money-back guarantee. Most sellers have no clue about how long their wares have been sitting on the shelf.

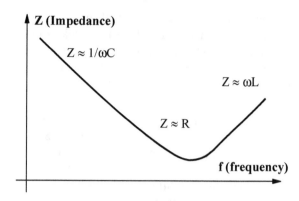

ABOVE: The impedance versus frequency curve for electrolytic capacitors. Above a certain frequency the capacitor turns into an inductor!

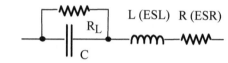

ABOVE: A simplified model of an electrolytic capacitor.

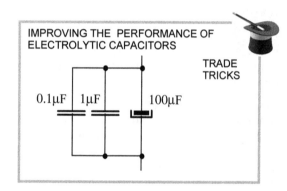

Improving the HF behavior of electrolytic capacitors

Film caps behave in the same fashion across the whole audio frequency spectrum, unlike electrolytic caps, whose capacitance drops and whose inductance and resistance rapidly rise with the increase in the frequency. Elcos lose their capacitance at higher frequencies, where their resistive and inductive components become dominant. A qualitative frequency curve above shows these three regions.

The equivalent circuit of a capacitor shows the leakage resistance RL, which models a resistance that allows a minimal but always present DC leakage current to flow. The other three components are a series LCR circuit. L is the ESI or Equivalent Series Inductance, and R is ESR or the Equivalent Series Resistance. If you still want to use electrolytic caps, bypass all of them with a small film capacitor. Usually, one bypass is enough (1µF film in our example), but if you are fanatical, bypass it again with progressively smaller value film caps.

OTHER COMPONENTS: FUSES AND SWITCHES

Fuses

A fuse is a component designed to act as the weakest link in an electrical circuit and burn out ("blow") when the current through the circuit exceeds its rated value. In its application in audio, it protects the primary winding of the mains transformer and the wiring inside the amplifier, so neither of these burns out before the fuse does.

Thus, it is essential not to replace a fuse of a specific current rating with one of a higher rating. If the current through the primary of a mains transformer is 1A and we replace a 1.5A fuse with a 3A fuse, should a fault develop (a short-circuit on the secondary side), the $150 transformer may burn out before the five-cent fuse!

There are many types and standards of fuses. 3AG (American size) and M205 (European size) are the two miniature types most commonly used in electronics. Both are so-called "cartridge-type" of cylindrical shape. AG stands for "automotive glass", since these sizes (2AG, 3AG, etc.) were originally developed for use in cars.

Paradoxically, cars today use blade-type fuses, which are easier to change in a hurry by simply unplugging them, while 3AG and M205 are enclosed in a fuse holder whose top cap needs to be unscrewed for a fuse to be changed. 3AG fuses are 31.8mm long and 6.35mm in diameter, while metric-sized M205 fuses are 20mm long.

In terms of speed, there are fast and slow-acting (SLO-BLO) fuses, designed to withstand inrush currents that flow when equipment is powered-up, but only for a short time. If the higher current persists the fuse melts. The naming convention of 3AG fuses is best illustrated with an example: 3AGDA1.25 = 1.25A 250V Slow 3AG fuse.

M205 fuses a include letter D for "Delayed action" in their model number: M205DA01.6R = 1.6A 250V Delayed M205 fuse. An example of a fast-acting M205 fuse would be M205002R (2A, 250V Fast M205 fuse)

The current draw of an amplifier should never exceed 75% of the nominal fuse current rating. Thus, a 1 Amp drawing amplifier should use a fuse of at least 1/0.75 = 1.35A rating!

Ambient temperature also plays a role in fuse behavior. Depending on the topology and mechanical design of the chassis and component positioning, internal amplifier temperatures can reach very high levels, especially during hot summer days or long concerts.

A 30°C room temperature and 30°C temperature rise next to a mains transformer means a fuse would be at 60°C. According to the curve, a slow-blow fuse would need to be derated to around 80%. For instance, with a current of 1.5A, we would need a slow-blow fuse rated at 1.5/(0.75 x 0.80) = 2.5A!

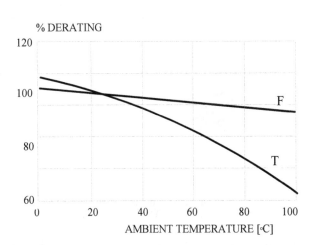

Temperature derating curves for FAST ACTING (F) and SLOW BLOW (T) fuses

Switches

Switches are classified according to the principle of operation and the configuration of contacts. The main types are toggle switches, slider switches, and rotary switches.

Configurations are described in terms of the number of poles and positions ("throws"). A single-pole means a single switch, while a double-pole describes two mechanically linked switches. "Single throw" means one active contact, double-throw means one common contact alternatively connected to two active contacts.

The naming convention is xPyT, where x is the number of poles and y is the number of throws. So, SPST = single pole single throw, SPDT = single pole double throw, DPDT = double pole double throw, 3PDT = triple pole double throw, and so on.

Universal rotary switches have 12 contacts (lugs). 12 is the product of poles and throws (1x12, 2x6, 4x3, 3x4, and 6x2). So it can be an SP12T switch, a single-pole switch with 12 positions, DP6T, two-pole 6-position switch, 3P4T, three-pole 4-position switch, 4P3T, four-pole 3-position switch, and 6PDT, six-pole 2-position switch.

SPDT and DPDT toggle switches (illustrated below) are very common in tube guitar amps. In one position, the central lug (1) is connected to one of the end lugs (say 2), flicking the switch over ("throw") disconnects (1) from (2) and connects (1) and (3). The same happens to the bottom three contacts, which form an entirely separate (electrically speaking) switch or "pole."

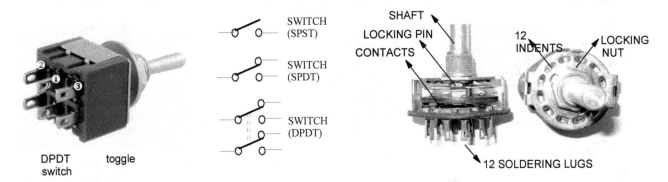

PASSIVE SEMICONDUCTORS: DIODES AND ZENER DIODES

Rectifier diodes

A silicon diode is a PN-junction, which conducts current when forward biased (P-side or anode is positive with respect to the cathode) and does not conduct when reverse-biased (anode is negative with respect to cathode). As we'll see very soon, vacuum tube diodes behave similarly but work on a different physical principle.

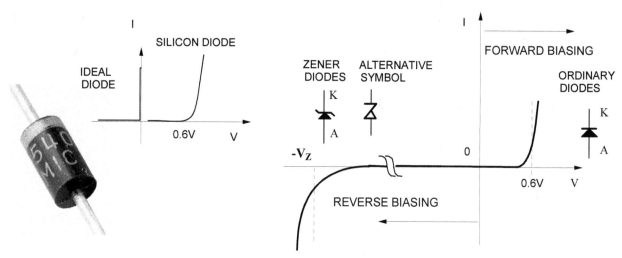

ABOVE: While ordinary silicon diodes work in the forward biasing regime, Zener diodes are reversely biased. The inverted SS symbol indicates that a range (section) has been omitted, that the graph is not to scale, as here with reverse (negative) voltage.

1N4004 is a general-purpose 1A rectifier with a maximum reverse voltage rating of 400V. The VRRM (Peak Repetitive Reverse Voltage) ratings for other diodes in the same series are 50V (1N4001), 100V (1N4002), 200V (1N4003), 400V (1N4004), 600V (1N4005), 800V (1N4006) and 1,000V (1N4007)! As you can see, there is some method to the naming madness, but logically one would think that 4007 meant 4A, 700V, but it does not; it means 1A, 1,000V! 1N5404 (pictured on the previous page) is a general-purpose 400V rectifier diode rated at 3A.

Zener diodes

While ordinary silicon diodes work in the forward biasing regime, Zener diodes are reversely biased (negative DC voltages on the anode). As the reverse bias on the diode increases, the diode does not conduct, and its current is zero. Once the reverse voltage reaches the value of Zener or avalanche voltage $-V_Z$, the Zener diode starts conducting abruptly; its current increases rapidly while the voltage stays at the $-V_Z$ level.

The curves are exaggerated for educational purposes; in reality, the curve around $-V_Z$ is much sharper and more vertical! Zener diodes are available in a range of voltages. Zener diodes of different V_Z voltages can be connected in series if a higher Zener voltage is needed, or diodes with identical V_Z can be paralleled if a higher current capacity is required. More on Zener diodes and their use as voltage regulators in Volume 2 of this book.

Duo-diodes and rectifier bridges

Duo-diodes are used extensively in switch-mode power supplies and DC-DC converters. Their ratings range from low voltage Schottky diodes, such as MBR3045PT (45V, 30A Schottky duo-diode, common cathode) to high reverse voltage super-fast passivated rectifiers such as STPR1660CT. The numbers mean duo-diode (CT=Centre Tap) rated at 16A (first two digits is the current rating), 600V max. reverse voltage (last two digits 60).

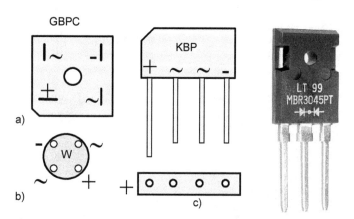

Like the one pictured, most duo-diodes have their cathodes joined. There are also duo-diodes with common anodes, but they are much less common.

Rectifier bridges come in various sizes, ratings, and packaging. The naming is not uniform, so if you are trying to decipher voltage and current rating from the number code, that is not always straightforward. BR810 is probably the easiest example, meaning 1,000V, 8A bridge rectifier. KBP04M is a 400V, 1.5A bridge, while GBPC5006 is a 600V 50A bridge.

Although semiconductor components are cheap, it pays to keep a stash of salvaged diodes and bridges. UPS, battery charges, PC power supplies, and similar equipment with DC-DC converters and switch-mode power supplies are good sources of quality diodes and bridges.

The most common modern types of rectifier bridges:
a) high power bridge, lug-side (bottom) view, GBPC type
b) low to medium power bridge, bottom view, W-type
c) medium power bridge, side and bottom view, KPB type

PHYSICAL FUNDAMENTALS OF VACUUM TUBE OPERATION

4

- ELECTRON EMISSION AND VACUUM DIODE
- VACUUM DIODE IN DC AND AC CIRCUITS
- ELECTRONIC TUBES - NAMING CONVENTIONS, BASES AND PIN NUMBERING
- TRIODE PARAMETERS AND CHARACTERISTICS
- STATIC PARAMETERS OF A 12AX7 (ECC83) DUO-TRIODE

To drive a motor car, you don't need to understand how the internal combustion engine works or know what is under the bonnet. The same can be said about building tube guitar amps.

However, to truly master the art & craft of tube design and construction and to have confidence in your skills, you need some basic knowledge of vacuum tubes or valves. This chapter will equip you for more advanced topics that follow.

ELECTRON EMISSION AND VACUUM DIODE

Thermionic emission

In any material, its atoms or molecules are at rest only at the absolute temperature of -273°C. At any other temperature, they are in a continual state of motion. Higher temperatures make this movement more vigorous and increase the velocity of electrons in the material. This "liberation" of electrons from a metallic electrode by virtue of its increased temperature is known as thermionic emission, and the emitter electrode is called a cathode.

When ions or electrons of sufficient energy bombard a metal electrode, some of that energy is absorbed by surface electrons which may be enough to exceed the work function of the metal and liberate the electrons. This phenomenon is called a secondary emission, meaning the released electrons are secondary electrons while the "primary" electrons provide the energy.

Oxide-coated indirectly-heated cathodes and their advantages over the filamentary type

Early vacuum tubes were directly heated or filamentary type. The heater filament acts as a cathode as well, i.e., it emits electrons itself. Some are still highly sought after, almost exclusively in hi-fi amps: 45, 2A3, 300B, and 211 triodes.

Most vacuum tubes used in guitar amplifiers have oxide-coated cathodes. In indirectly heated tubes, an insulated heater wire is placed inside a cathode, a metal sleeve that gets energy as heat from the heater (conducted through a ceramic insulator), and the cathode then emits electrons. The cathode is made of nickel, nickel with a few percent of cobalt or silicon, platinum or Konal (an alloy made of nickel, cobalt, iron, and titanium), and then coated with a mixture of barium and strontium oxides.

The heater is a tungsten wire loop coated with a heat-resistant insulating material such as aluminium or beryllium oxide, placed inside the cathode sleeve. This coating conducts the heat to the cathode and insulates it electrically from the heater wire. The oxides provide electrons at temperatures as low as 750°C (dull-red heat).

Oxide-coated cathodes have many advantages over filamentary ones, the most important of which is their much longer life of several thousand hours. SQ (Special Quality) tubes are guaranteed for 10,000 hours. At slightly reduced filament voltages, some tubes last 100-200,000 hours!

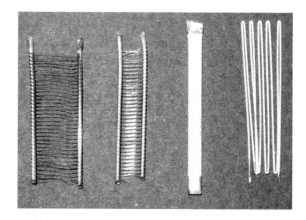

ABOVE: (L-R): The screen grid and the gold control grid with their vertical mechanical support rods. Next is the white oxide-coated cathode sleeve and the folded heater filament pulled out of the cathode sleeve of 6L6 beam tetrode

BELOW: Construction of an indirectly-heated cathode-heater assembly

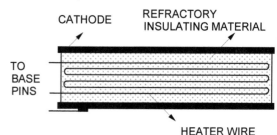

The space charge or "electron cloud"

A thermionic diode is the simplest vacuum tube with only two electrodes, a cathode and an anode. Assuming a filamentary diode, where the heater acts as a cathode (for instance, 5Y3 or 5U4), the filament heats up and emits electrons once the heater voltage is applied.

If there is no positive voltage on the anode, the emitted electrons disperse into the surrounding space and form a cloud or "space charge" around the cathode. The cloud is negatively charged and repels newly-arrived electrons back to the cathode. Soon an equilibrium is established, meaning that as some excited electrons leave the cathode, as many others are forced back to it. The density of the cloud is the greatest near the filament and reduces with the increased distance from it.

When connected to a source of positive voltage VB ("B" for battery) anode will start attracting negative electrons from the cloud. Those closest to the anode will be swept to it first. As electrons reach the anode, others come from the cathode and take their place in the electron cloud. The cloud is not affected by the anode, which is unable to disperse it. The electron cloud reduces the inter-electrode potential from the linear distribution (dotted line) and causes the tube's nonlinear character. This flow of electrons is an electric current through the vacuum and is said to be space-charge limited.

NOTICE: The adopted convention for current flow is the opposite of the electron flow!

Anode characteristic of a vacuum diode

If we connect a variable DC voltage source across a diode (+ to the anode and - to the cathode) and hook up an mA-meter in series to measure the DC current, with an increased voltage between the anode and cathode V_{AK} (or "across the tube"), the current will increase exponentially. The anode draws electrons from the space charge, and the cathode replenishes these electrons. The anode current is limited only by the space charge, and the cathode can produce more electrons than the anode can attract.

However, above a specific voltage V_C, the rise will become slower, the current has reached the saturation level. The anode has depleted all the electrons from the space charge, and the cathode cannot produce enough electrons to reestablish the space cloud. The anode current is said to be temperature limited. Tubes usually work in the space charge region.

VACUUM DIODE IN DC AND AC CIRCUITS

How to graphically analyze non-linear circuits

To illustrate the fundamental properties of a vacuum diode, let's connect it to a source of relatively high DC voltage, 200V in this case. With no load connected, the internal resistance of the diode will limit the current in this circuit, and all the power would be dissipated into heat inside the diode.

However, the DC current through the diode may exceed the maximum value declared by the manufacturer, and the device may be damaged or destroyed, so we need to connect a load in series with the diode. It does not matter where we place it, between the anode and the DC source (as illustrated) or between the cathode and the source.

To enable us to perform the DC analysis of this simple circuit, numerical values have been chosen - 200V for the DC source and load resistance of 25,000 Ω. The basic concept of Ohm's law applies here, but Ohm's law assumes linear resistances, and here we have a nonlinear diode, whose resistance we don't know. Moreover, the diode's resistance changes with the magnitude of the current flowing through it. Likewise, the voltage drop across the diode is not linearly proportional to the current.

There are two ways to analyze this kind of circuit. We can use the I-V curves for the specific diode (provided by its manufacturer) and find the solution graphically. The other option is to simplify things prudently and "linearize" the diode. We will do that soon while performing the AC analysis.

The circuit is a simple voltage divider. Depending on the static resistance of the tube and the load, the DC source's voltage V_B ("B" for battery) will be split in some proportion between the two: $V_B = V_L + V_A$ ("A" for anode-cathode voltage)!

The I-V curve for the diode is given (by its manufacturer), and on it, we draw a "load line" for the resistor R_L. We only need two points to define a line. One is the battery voltage V_B (200V). We mark that point on the voltage or X-axis (point A). The second point can be chosen arbitrarily. Simply select a voltage and calculate the corresponding current.

For instance, if we move 200V to the left of point A (right on the current or vertical-axis), the current in that point B would be $I_B = 200/25,000 = 8$ mA. We mark that value and have our point B. Now we draw the load line for $R_L = 25k\Omega$.

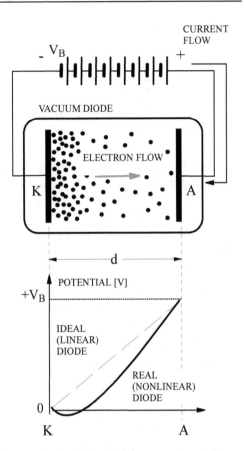

ABOVE: The physical behavior of a vacuum diode
BELOW: The anode characteristic of a vacuum diode for two cathode temperatures (T2 > T1).

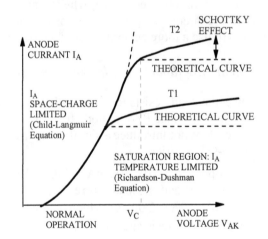

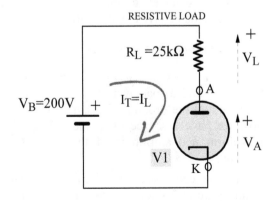

ABOVE: Vacuum diode in a simple series DC circuit with a resistive load

As soon as we've drawn the R_L line, we get the intersection of the two I-V curves. That is our operating or "quiescent" point Q. All we need to do now is draw a horizontal line towards the vertical or current axis and read the value of the current in point Q as $I_Q = 3.3$ mA.

The current through the diode tube and the load is the same current $I_Q = I_L = I_A$, so we can call it simply I.

Similarly, we draw a vertical line down to the voltage axis and read the $V_A=120V$. That is the voltage drop on the diode tube. The rest of the 200V battery voltage is dropped across the load resistor ($V_L=200-120 = 80V$). We have "solved" our circuit; we know all its currents and voltages.

The "static" or DC resistance of the tube R_I in point Q is the voltage drop across it divided by the current $R_I=V_A/I_L = 120V/0.0033A = 36,364\ \Omega$ or approx. 36kΩ!

For each value of the load, the operating point Q will be different. Two additional cases are illustrated, for the load resistor of twice the previous value or 50kΩ and half the previous value or 12.5kΩ (steeper curve).

For increased load resistance, the current decreases, and the operating point moves to the left. Less and less voltage is dropped across the tube, and more and more voltage is applied across the load. As an exercise, calculate the internal DC resistance of this tube in Q1 and Q2.

"Plate resistance" is a term that may be confusing to some. It is not the electrical resistance of the material the anode or "plate" is made of, but the parameter used to model the behavior of a tube as a variable resistance, when viewed from the outside, from the circuit perspective. So, instead of "plate resistance," we will use a more accurate term, "internal resistance."

Linearizing diode's anode characteristics

Since the load resistance is usually much higher than the internal resistance R_I of a diode tube, linearizing their I_A-V_A curves does not introduce a significant error, as illustrated for 6AX5-GT rectifier.

The slope is $R_I=120V/300$ mA = 400 Ω. With a power supply of $400V_{DC}$ and a load current of 200mA the load is $R_L=400V/0.2A = 2,000\ \Omega$, which is five times higher than the 400 Ω internal resistance of the 6AX5 diode.

In audio amplifiers diodes are used mostly as rectifiers, so we will limit ourselves to that application. Let's see how a vacuum diode behaves in alternating current circuits.

Vacuum diode in AC circuits

A vacuum diode V1 is connected across a source of AC voltage v, and a load resistor R_L is connected between its cathode and the ground. The same topology as before, only now we have a source of a sine-voltage. What will be the waveform and amplitude of the AC voltage on the load resistor?

Again, we can solve this circuit graphically in a tedious, point-by-point fashion. For each point on the sine wave of the input voltage, we would find a projected point, reflected of the diode's I-V curve.

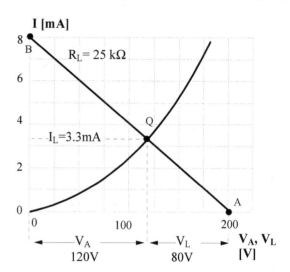

ABOVE: Graphical analysis of the series DC circuit with a vacuum diode

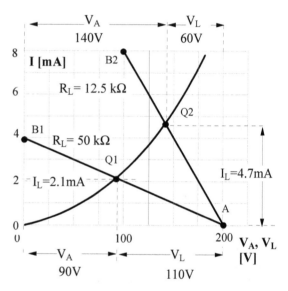

ABOVE: Graphical analysis of the series diode DC circuit for two different load values

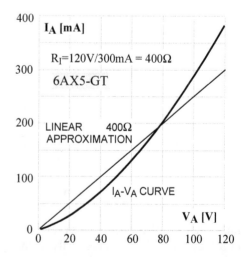

ABOVE: Linearizing a diode curve does not introduce a significant error and can be used for quick approximations in practice. The same applies to triodes (coming soon).

RIGHT: Vacuum diode and its load connected to an AC voltage source.
BELOW: The graphical point-by-point method of constructing the output current's waveform

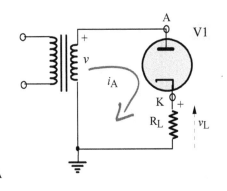

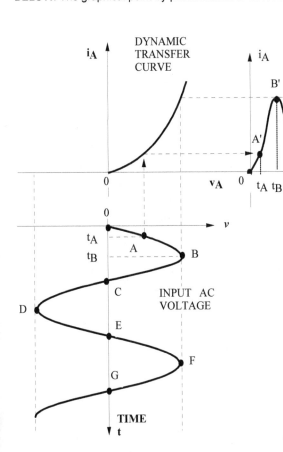

The vacuum diode passes on only the positive halves of the sine voltage. The valve is fully open, and the AC current flows. The diode does not conduct negative for negative peaks because the AC signal makes its anode negative with reference to its cathode. The valve is closed, and no current flows through it.

This unidirectional property of the diode makes it behave as a rectifier because it "rectifies" or "straightens" the input AC signal into a rectified pulsating signal where there are no negative currents. The average value of a sine wave is zero (positive peaks cancel negative peaks), but the rectified signal has a specific positive average value. We can loosely call this rectified signal a DC current or voltage, but it is far from ideal, steady DC voltage. Nevertheless, it is a crucial first step in AC-DC conversion.

Notice a few important things:

1. The waveform thus constructed is the current through the circuit, not the voltage. For a resistive load, the voltage waveform would be proportional to this current.

2. The output signal is distorted due to the curvature of the diode's transfer curve. A linear transfer curve (a straight line) would result in no distortion of the rectified signal.

Dynamic resistance of a vacuum diode

A diode's resistance to AC currents ("impedance") is different from its resistance to DC currents. The dynamic conductance at any point along the I-V curve is the slope of the tangent at that point. Once we draw the tangent, we arbitrarily choose ΔV and ΔI and read their two axis values. Greek capital letter "delta," symbol Δ, stands for "difference." We have $\Delta V = 150-120 = 30V$ and $\Delta I = 4.9-3.3 = 1.6$ mA!

The dynamic internal resistance of the tube is $r_I = \Delta V/\Delta I = 30/0.0016 = 18,750\ \Omega$ or $18.75k\Omega$!

The symbol for dynamic or AC resistance is a lower case letter r. In this case, the AC internal resistance of the diode is about half its DC resistance ($18k\Omega$ versus $36k\Omega$).

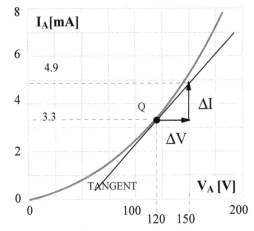

Determining the internal dynamic resistance of a diode from its I-V curve

Ideal and real diode

As we will shortly do with triodes, we distinguish between two idealized concepts, the ideal diode, and the linear diode. The ideal diode has no internal resistance; it is completely linear and behaves in a way best illustrated by its I-V characteristic. For all AC voltages at its input, the ideal diode cuts off their negative half-waves and passes their positive half-waves unchanged. There is no voltage drop V_A across the ideal diode, so its I-V characteristic is a flat line $I_A=0$ for negative voltages and a vertical line $V_A=0$ for all positive signal values.

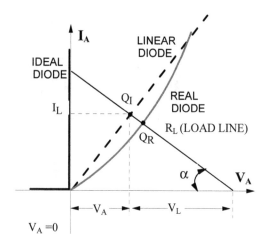

ABOVE: The comparison of I-V characteristics of an ideal diode, linear diode and real diode

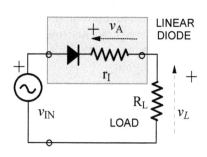

This assumption of no internal resistance and no voltage drop across a diode is unrealistic for design use, so we introduce a concept of a linear diode, which is simply an ideal diode in series with its internal resistance r_I. The shape of the positive waveform is still the same, but its amplitude is reduced by the voltage drop on the diode (V_A). The angle of the load line with the X- or voltage-axis is $\alpha = \Delta I_A/\Delta V_A = 1/R_L$. The current flowing in this circuit is $I_L = V_{IN}/(r_I + R_L)$.

By linearizing circuits, we introduce an error; the real operating point Q_R is not the same as the idealized Q_I. The actual values of the currents and voltages are also different from the linearized results. This doesn't mean that models, simplifications, and approximations don't work and are therefore useless, but that a designer needs to understand their limitations and decide how appropriate such manipulations are for each situation.

Since the difference between the real and linearized diode I-V characteristic isn't significant, the output voltage of a real diode does resemble the output voltage of the linearized diode, but with some distortion of the waveform. This may not be noticeable by an untrained eye and thus would be difficult to depict on such a small drawing.

LEFT: The equivalent model of a linear diode with a load (half-wave rectifier circuit). The "load" is the rest of the power supply filtering circuit and the audio section of the amplifier.

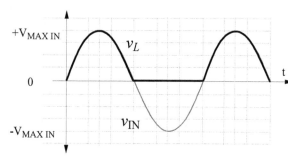

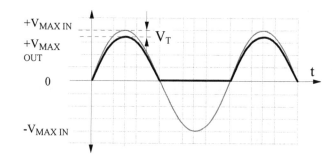

ABOVE: The input and the output voltage (half-wave rectified waveform) of an ideal diode (LEFT) and a linear diode (RIGHT).

The anode (power) dissipation of a tube

This kinetic energy of electrons (due to their speed or motion) traversing from cathode to anode and hitting the anode's surface at such a high speed is converted into heat at the anode. The heat cannot be allowed to accumulate, for that would quickly raise the temperature of the anode's material and eventually melt it away, destroying the tube. So, the heat must be dissipated from the anode into the surrounding vacuum, through the glass bulb, and into the ambient air.

The heat generation rate equals heat dissipation away from the anode for both rectifier diodes and amplifying tubes in steady-state operation. Hence, their temperature remains constant and within allowable limits.

In a quiescent point Q (determined by I_A and V_A), the dissipated heat is equal to electrical power, $P = I_A V_A$. For instance, directly heated duo-diode 5U4-GB in Q1 dissipates $P = 40V \times 0.2A = 8W$.

Indirectly heated 5V4G for the same current (Q2) dissipates $27V \times 0.2A = 5.4W$ and is thus a more efficient rectifier since it wastes less electrical energy into heat. For the same load current of 200mA, the voltage drop on 5U4 is much higher than on 5V4 (40 versus 27 Volts)!

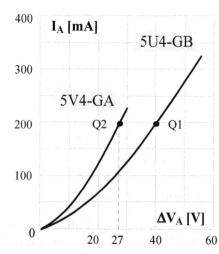

I_A-V_A curves for two vacuum tube rectifiers (duo-diodes), 5V4-GA and 5U4-GB.

Say you have a guitar amp whose audio circuit draws 200mA and uses a 5U4-GN rectifier, and you substitute a 5V4-GA rectifier tube. The DC voltages at its output (pin 8) will now be 40-27 = 13V higher! That will slightly change the operating conditions in your amp (alter its "voicing") and increase its maximum output power a bit.

Although 5V4 is an indirectly heated rectifier, both are pin-compatible, so you can perform this substitution. However, notice that 5U4 can work with anode currents of over 300mA, while 5V4 can only supply up to around 220mA!

A welcome bonus of using indirectly heated rectifiers is an inherent delay. With directly-heated rectifier tubes such as 5U4 or 5Y3, the DC voltage appears much faster than when indirectly heated ones are used, which take time to transfer the heat from their heaters to their cathodes. That delay enables audio tubes in a guitar amp (almost always of indirectly heated variety) to warm up before high DC voltages are applied to their anodes. In effect, indirectly-heated rectifiers provide an inherent or inbuilt "soft-start" feature! Standby mode is thus less important or even unnecessary with such tubes.

ELECTRON TUBES - NAMING CONVENTIONS, BASES AND PIN NUMBERING

Vacuum tubes, as they are called in the United States, have also been called "electronic tubes" or "thermionic valves." The term "valve" is used mainly in the UK and Australian English, and in other languages, like "valvola" in Italian or "válvula" in Spanish. In some countries, the term for the electronic tube is "lamp," such as "lampa" in Serbian and "lampe" in French. Germans call it "Röhre."

While the term electron "tube" probably comes from the tubular shape of the glass bulb, the term valve most likely originated in the mechanical-electrical analogy that explains the operating principle of the triode.

Just as turning the handle on a mechanical valve (water tap) controls the flow of water (or any gas or liquid) through it, changing the voltage (or bias) on the tube's control grid modulates the flow of electrons through it.

Americans use the colloquial term "plate" instead of the proper term "anode." In this book, we will use "vacuum tube" and "anode."

Tube naming "conventions"

European tube makers at least tried to follow some kind of naming system. The same cannot be said for their American competitors. The European system uses four letters followed by 2 or 3 digits. Let's look at a few examples.

For instance, ECC83 means a double triode (CC) with 6.3V heating (E) and using a Noval socket (8). ECC40 is a dual triode (CC) with 6.3V heating (E) and using a RimLock socket (4).

EL34 is a power pentode (L) with 6.5V heater (E) and Octal socket (3). PL519 means a TV output pentode (L) with 300mA heater current (P), "5" indicates a Magnoval socket.

ECF86 is 6.3V (E) voltage amplifier triode (C) and preamplifier pentode (F) using a Noval socket (8).

PY88 is a single diode (Y for "half-wave rectifier"), with 300mA heater (P) using a Noval socket (8).

Special quality tubes where marked SQ and/or named by reversing the order of two of the letters and numbers. For instance, E82CC is the SQ version of ECC82, while E86C is an SQ version of EC86.

The KT prefix (KT66, KT88, etc.) means "Kinkless Tetrode". The number has no meaning, apart from the fact that the higher the number, the higher the anode dissipation of the tube.

The American naming convention is less consistent and not at all self-explanatory. The first number indicates heating voltage, so 12AX7 is a tube with a 12.6V heater, while 6L6 has a 6.3V heater.

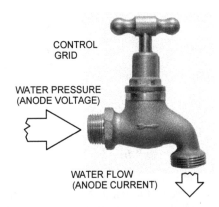

CONTROL GRID

WATER PRESSURE (ANODE VOLTAGE)

WATER FLOW (ANODE CURRENT)

ABOVE: The water tap analogy of a triode explains why vacuum tubes are also called "valves".
BELOW: European tube naming system

1st number		
	3	Octal
	4	Rimlock
	5	Magnoval (large 9-pin)
	8	Noval (small 9-pin)
	9	Miniature 7-pin

1st letter		
A		4.0V heater
B		180mA heater
C		200mA heater
D		Battery-powered tubes, 1.25-1.4V heater
E		6.3V heater voltage
F		13V heater
G		5.0V heater
P		300mA heater current
U		100mA heater current
O, Z		Cold cathode tubes

2nd, 3rd & 4th letter		
	A	Single diode
	B	Duo-diode
	C	Triode (voltage amplifier)
	D	Power triode
	E	Tetrode
	F	Preamplifier pentode
	H	Hexode
	L	Output (power) pentode
	M	Indicator tube
	N	Thyratron
	X, W	Gas-filled tube
	Y	High voltage half-wave rectifier
	Z	High voltage full-wave rectifier

The last digit indicates the number of electrodes + heater, so 2A3 triode has three elements; one electrode is a heater as well (directly-heated cathode). 6L6 has five electrodes + a separate heater, for a total of six elements. 12AX7 has six electrodes (two triodes) and a common heater, thus 7 in total.

Then there are "numerically-named" tubes such as 6922, 5687, and many others, which are usually the "industrial" versions of their consumer-aimed equivalents. These numbers have no meaning, and that is why I titled this section naming "conventions" instead of naming "standards." As different tube manufacturers developed new tubes, they simply named them at their convenience without any regard for standardization.

Not to be outdone, the USA military established their own naming, where numbers are usually preceded by the acronym JAN, which stands for "Joint Army-Navy."

Tube bases and pin numbering

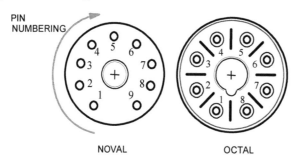

NOVAL   OCTAL

No matter what base a tube uses, the pins are always numbered in a clockwise fashion if viewed from underneath (pins facing you). The same view is of their sockets if viewed from the lug side (where components are soldered).

TRIODE PARAMETERS AND CHARACTERISTICS

Triode "curves" in 3D

Since there are three variables of interest, grid voltage (grid-to-cathode), anode (anode-to-cathode) voltage, and anode current, the triode's behavior is a contour in 3-dimensional space. Dealing with 3D curves on 2D paper is awkward; drawing two variables in two-dimensions is much easier, with the third variable as a parameter.

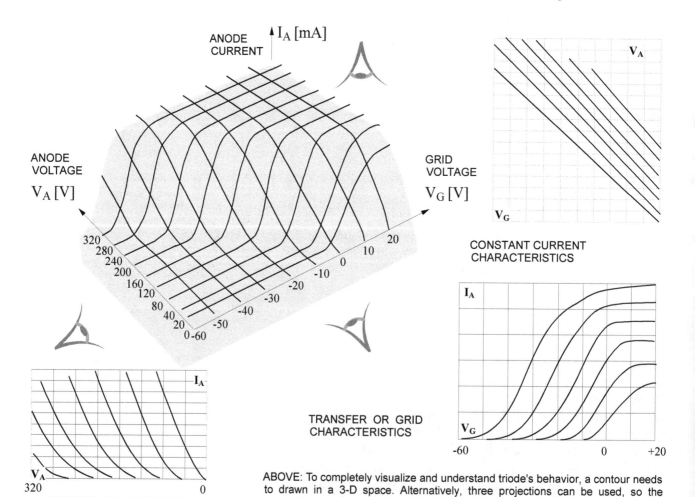

ABOVE: To completely visualize and understand triode's behavior, a contour needs to drawn in a 3-D space. Alternatively, three projections can be used, so the characteristics can be drawn in two dimensions.

There are three possible triode graphs. The I_A vs. V_A graph with V_G as a parameter is also called anode or plate characteristics. The I_A vs. V_G graph with V_A as a parameter depicts one or more transfer curves, and V_A vs. V_G graph with I_A as a parameter is called "constant current characteristics" (since I_A is kept constant, which is the meaning of "parameter")!

Current control by triode's grid, amplification factor and "reachthrough"

We have seen that the current through a vacuum diode is a function of one variable, the anode voltage: $I_A = KV_A^{3/2}$. In triodes, apart from depending on the anode voltage, the anode current is a function of another variable as well, the voltage on the control grid: $I_K = K(V_G + DV_A)^{3/2}$

Since $I_K = I_G + I_A$ (cathode current is a sum of grid and anode currents), and in most circuits (but not all!) the grid current is negligibly small compared to anode current and can be considered zero, this equation is usually written as $I_A = K(V_G + DV_A)^{3/2}$, but that is only valid if $I_G = 0$!

The constant K is called perveance, as in a diode, and the constant D comes from the German noun Durchgrief, meaning "reach through" since the anode's electric field has to reach through the grid to "get" electrons from the cathode. It is a measure of the degree of anode's control over the current, its "effectiveness." The concept is not used in English literature at all; Anglo Saxons, for some reason, exclusively use the inverse concept called voltage amplification factor, for which the Greek letter μ (pronounced mju) is used.

Since $\mu = 1/D$, they write this equation as $I_A = K(V_G + V_A/\mu)^{3/2}$, or, more often, they use the alternative way of expressing anode current, $I_A = A(\mu V_G + V_A)^{3/2}$, where constant A loses its meaning, it is not perveance any more.

For instance, the D-factor for LS50 power pentode is 20% or 0.02, meaning its amplification factor is $1/0.02 = 5$. For 12AX7 triode, with $\mu = 100$, D is $1/100 = 0.01$ or 1%. These differences can be interpreted in the sense that the anode current in 12AX7 is much more sensitive to grid voltage changes (20 times more sensitive, since its μ is 20 times higher!) than for the LS50 pentode.

Using the reach through figures for interpretation, in LS50 with D=20%, 20% of anode current changes are due to the anode voltage influence and 80% as a result of grid voltage changes, while in 12AX7 triode, only 1% of anode current changes come from changes in the anode voltage and 99% result from the grid action, meaning that the grid is 20 times more effective in the 12AX7's case.

STATIC PARAMETERS OF 12AX7 (ECC83) DUO-TRIODE

"Tube profile" frames

We will use "Tube Profile" frames to outline the essential data about the tubes used in various amp designs. Of course, once we publish such a box for the first time, we will not repeat it every time such a tube is used; that would be a waste of space.

The 1st line will be a short description of the tube, in case it says that 12AX7 is a high m duo-triode, meaning the two triodes are identical. In some cases where the two triodes are different, it will say "dissimilar" triodes.

TUBE PROFILE: ECC83 (12AX7)

- High μ duo-triode, Noval socket
- Heater: 6.3V/300mA or 12.6V/150mA
- $V_{AMAX} = 300V_{DC}$, $V_{HKMAX} = 180V_{DC}$
- $P_{AMAX} = 1W$, $I_{KMAX} = 8$ mA
- TYPICAL OPERATION:
- $V_A = 250V$, $V_G = -2.0V$, $I_0 = 1.2$ mA
- gm = 1.6 mA/V, $\mu = 100$, $r_I = 62.5$ kΩ

A socket type is the next piece of information; here it is a Noval socket, Noval meaning "new," since it was developed after the "old" octal or 8-pin socket. A Noval socket is also called a 9-pin miniature socket.

Next, the tube's heater voltage and current draw are specified. 12AX7's center-tapped heater can be connected in two ways. When the two halves are paralleled, the 6.3V AC or DC voltage is brought to pin 9 (CT) and to the heater's ends, pins 4 and 5, connected together. At 6.3V, the heater draws 150mA. If pin nine is left unused, 12AX7 needs 12.6V AC or DC to be connected to pins 4 and 5. Since the voltage is doubled, the current draw of the heater is halved (its power draw is constant); it is only 150mA.

Each tube has its maximum anode voltage and maximum anode power dissipation specified by the manufacturer. Pentodes also have the maximum allowed screen grid voltage and power draw. Sometimes a maximum anode or cathode current is also specified. With triodes, the anode current is the same as the cathode current. The cathode current is always higher with tetrodes and pentodes since it is the sum of anode and screen grid currents.

The maximum allowed voltage between the heater and cathode is often overlooked by designers. There are two different figures in some cases, one when the heater is positive with respect to the cathode, the other when the heater is negative vis-a-vis the cathode.

Most small or preamp triodes such as 12AT7, 12AU7 and many others have V_{HKMAX} of only 90 or 100V. 12AX7 is unique in this respect since it has a much higher (double!) rating of $V_{HKMAX} = 180V_{DC}$!

In simple common cathode circuits used in guitar amps this rating is of no importance since cathodes are at a very low DC voltage of 2-3 Volts, but it becomes important when tubes are used in totem-pole arrangements (one on top the other) and in regulated power supplies.

Finally, the last couple of lines describe the tube's "typical operation." In this and most other cases, that is a common cathode amplification stage. We see that the anode voltage is $250V_{DC}$ (for simplicity's sake, we will write 250V instead, assuming it is DC, for AC voltage, we will write V_{AC}!) and that control grid or G_1 is at -2V with respect to the cathode. That is the negative bias voltage applied to the grid.

In preamp stages, instead of making grid negative with respect to the cathode, we usually make cathode positive with reference to the grid by a voltage drop on the cathode resistor. The grid is at ground or zero potential.

These two DC voltages determine not just the anode current through the tube, which in this case is 1.2 mA, but also the three "static" parameters. "gm" is mutual conductance or transconductance, μ (pronounced mju) is the tube's voltage amplification factor, and r_I is its internal resistance. These are not "tube constants," as some books erroneously call them but "parameters" because they change with anode current levels. Let's have a closer look at the three parameters and the three types of tube characteristics!

Transfer characteristics

Since 12AX7 is the most common preamp tube in guitar amps, let's use it as a practical example to learn about the three most important triode parameters and three types of triode characteristics.

Transfer characteristics show how anode current I_A changes with a DC bias on the control grid (V_G), with anode voltage V_A as a parameter (kept constant). There is a different curve for each value of anode-to-cathode voltage V_A, so we are really talking about an infinite number or a family of curves. Manufacturers usually publish at least two or three of them. We have six here, from 50V to 300V.

As V_A is increased, a higher negative grid bias is needed to keep the anode current I_A at the same level (draw a horizontal line through all six curves). The anode voltage has a large impact on the triode's anode current. That is not the case for tetrodes and pentodes, as we will see later. Notice the word "average" (1) printed by the manufacturer. That means that each tube's curves will be slightly different, so take them as an approximation only!

The heater voltage is also specified (2), in this case, E_F=12.6V. "F" is for filament. That indicates that the curves for the same tube (or, in this case, the "bogey" or "average" tube) will also vary with heater voltage changes!

Let's take one TX curve now as an example, V_A=200V, and position our operating, idle or quiescent point (goes by three alternative names!) at the grid bias of -1.3V (3). That "bias" will automatically determine the anode or "plate" current, which we read by drawing a horizontal line to the right until we reach the vertical or anode current axis, I_A=1.65mA (4).

❶ AVERAGE TRANSFER CHARACTERISTICS
EACH SECTION

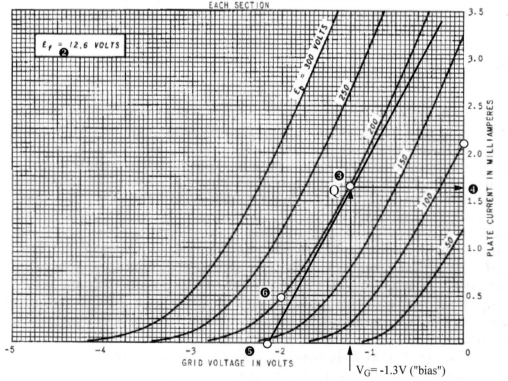

LEFT: Six representative TX curves from GE tube data sheet for 12AX7 triode.

Transconductance or mutual conductance

A derivative "d" of a function is the slope of a tangent to that function in a particular point. Most DIY constructors would rather have a root-canal job than anything to do with calculus. In the first approximation, the derivative "d" can be replaced by a small difference, for which the Greek capital letter Delta is used (symbol Δ).

Transconductance or mutual conductance (gm in English and S in many other languages, from Steilheit or steepness in German) is defined as a change in I_A (anode current) caused by a small shift in V_G (DC voltage on control grid or "bias" voltage): $gm=\Delta I_A/\Delta V_G$. That is the slope or steepness of the transfer curves in any chosen point.

You can use any two points to read gm in this point. For instance, use the "projected cutoff" point (5) and the Q point (3). So we get a change in bias from -2.2V to -1.3V, so $\Delta V_G = 0.9V$. At the same time, the anode current increased from zero in (5) to 1.65mA, so $\Delta I_A=1.65mA$! Now we can calculate $gm = \Delta I_A/\Delta V_G =1.65$ mA/0.9V = 1.83 mA/V. This means that for each Volt of control grid voltage, the anode current jumps 1.83mA. That is quite a low transconductance; some tubes have gm in the 20-30 mA/V range!

Notice that the upper part of the TX (transfer) curve is quite linear, meaning that its slope is constant, and thus gm is constant. However, the bottom part of the TX curve is very nonlinear; the slope reduces. For instance, draw a tangent and read gm in point (6) for $V_G = -2.0V$, and you will get a much lower figure.

Internal resistance and voltage amplification factor of a triode

As we have already seen with vacuum diodes, the anode curves describe the relationship between anode current and the voltage drop across the tube (between its anode and cathode). The slope of the curves is inversely proportional to the internal resistance of the tube, and we can determine such resistance graphically from the curves. The slope is low at low anode currents, meaning the internal resistance is high, and as anode current goes up, the slope rises and the internal resistance falls. Let's choose a couple of points at the same anode voltage as before (200V).

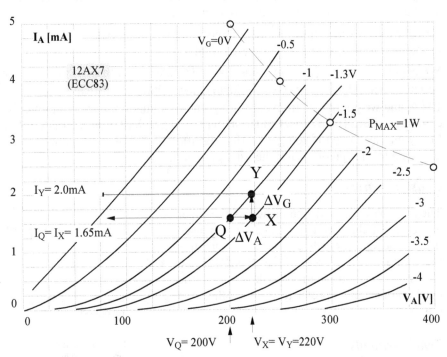

Notice that there is no curve for V_G= -1.3V, only for -1V and -1.5V. However, the curve is there, it's just that manufacturer did not publish it on this graph. We can approximately position it by copying the -1V curve and pasting it slightly closer to the -1.5V than to the -1V curve.

The Q-point is the same as before, at I_A=1.65mA! To determine the internal resistance of this tube in point Q we move to the right until we hit the V_G= -1.5V curve. We read anode voltage at that point (X), which is 225V. Then we move vertically upwards until we hit the -1.3V curve (point Y). We read the anode current at that point as I_Y= 2.0 mA.

So, our ΔV_A=220-200 = 20V, and that jump in anode voltage caused anode current to jump ΔI_A = 2.0-1.65 = 0.35 mA.

ABOVE: Estimating 12AX7 triode's internal resistance r_I in points A and B, and its amplification factor in point C, using its anode characteristics.

Now we can calculate tube's internal resistance "around" point Q as $r_I= \Delta V_A/\Delta I_A = 20/0.00035 = 57,143$ V/A or 57.14 kΩ. Again, 0.35 milliAmpere is 0.00035 A, and V/A is a unit for resistance called "Ohm" (Ω), so we have around 57 thousand ohms or 57 kiloohms (kΩ).

Tube's voltage amplification factor or "gain" is defined as the change of anode voltage divided by the change of grid voltage that caused it: $\mu=\Delta V_A/\Delta V_G$ The two identical units (volts) cancel one another, so μ is a dimensionless factor. We already have everything needed to estimate μ around point Q. Anode voltage jump ΔV_A was caused by grid voltage jump ΔV_G=1.5-1.3=0.2V. Finally, $\mu = \Delta V_A/\Delta V_G = 20V/0.2V = 100$.

The Barkhousen's equation

Now that we know how to read the tube parameters from the graphs, the good news is that we usually don't have to bother doing it. In most cases, at least for tubes intended for audio applications, manufacturers had published their μ-gm-r_I graphs.

Of course, for each anode voltage there will be a different graph. This one is for V_A=100V.

We cannot even position our point Q here since the horizontal or current axis only goes up to 1.6mA, and we had 1.65mA due to the much higher anode voltage of 200V!

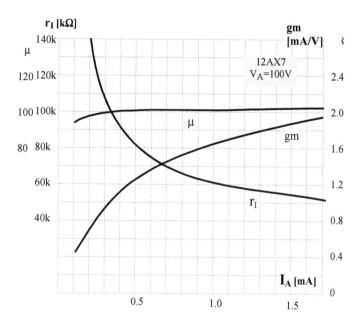

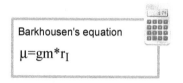

Barkhousen's equation

$$\mu = gm * r_I$$

RIGHT: How the three static tube parameters vary with anode current (for 12AX7 triode)

Notice how μ (the amplification factor) remains constant for almost the whole range of anode currents, except the very low current levels. In that low current region, the internal resistance of the tube shoots up, while gm drops rapidly off as the anode current drops. Normally, a designer would avoid using a tube in that region.

The visual relationships of the three curves indicate that gm and rI vary inversely to one another; they are a "mirror" image of each other. So, we can speculate that their product (μ) would remain more or less constant, and sure enough, another gentleman had done it all a long time ago.

The simple equation that links the three parameters, $\mu = gm * r_I$, was named Barkhousen's equation in his honor. Remember it well; we will use it very often!

We estimated the tube's gm in point Q as 1.83 mA/V, and then its internal resistance in the same point as r_I=57,143 ohms. So according to these two estimates, the voltage amplification factor μ should be $\mu = gm * r_I$ = 1.83 * 57.143 = 104.6, and we got μ=100 from our third graphic estimation. For such imprecise graphs and "eyeball" readouts from a computer screen (in my case) and a small printed graph (in your case) that is incredibly close or precise. Again, don't lose your sleep even if you get +/-10% discrepancies, this is an estimation tool only, and such accuracy is usually adequate for design or troubleshooting purposes.

Constant current curves

These aren't commonly published as the other two types of characteristics and are rarely used in any design procedures. Three parameters (N=3) mean two degrees of freedom (N-1), so choosing two automatically determines the third and designers prefer to use the other two graphs.

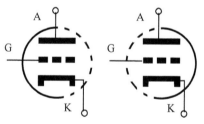

ABOVE: Two triodes inside the same tube are sometimes indicated by dotted lines.

BELOW: Symbols for a directly-heated (filamentary-type) triode (left) and indirectly-heated triode (middle). The heater of indirectly heated tubes is usually omitted from the symbols used in circuit diagrams (right) for clarity; it is assumed that it is connected to a suitable power source.

Triode symbols

Some circuit diagrams indicate the two triodes placed in the same glass bulb (one physical tube) by dotted lines around the triode symbol. For instance, as illustrated, one triode would have its left half dotted, and the other would have its right half dotted.

We will not bother with such pedantry here but will simply write next to each triode V1a and V1b, for instance, indicating that it is the same valve (3!), with two triodes, "a" and "b"!

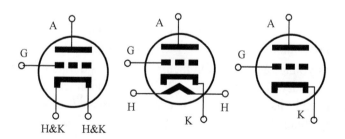

TRIODES AS VOLTAGE AMPLIFIERS

5

- COMMON CATHODE AMPLIFYING STAGE - GRAPHICAL ANALYSIS
- COMMON CATHODE GAIN STAGE - SMALL SIGNAL MODEL ANALYSIS
- THE EFFICIENCY AND DISTORTION OF THE COMMON CATHODE STAGE
- COMMON CATHODE STAGE WITH UN-BYPASSED CATHODE RESISTOR
- CATHODE FOLLOWER (COMMON ANODE STAGE)

This is one of the most important chapters in the book, so don't get scared or put off by some algebra and formulas. If you understand triode gain stages, it won't take you long to master pentode circuits as well.

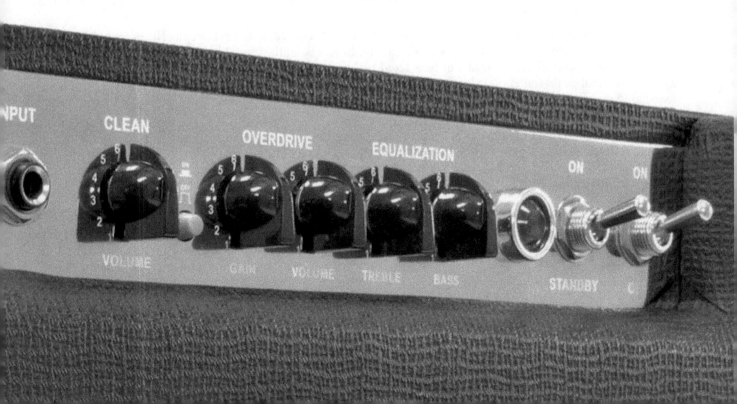

COMMON CATHODE AMPLIFYING STAGE - GRAPHICAL ANALYSIS

Three possible ways to use a triode as an amplifier

Since a triode has three active electrodes, it can be connected in three different ways. One of the electrodes is common to both the input and output circuits. Notice a pair of terminals at the input and output circuits, with a common or shared terminal. The common terminal is grounded for AC signals, but it can be at any DC voltage. This statement is profound and very important for the understanding of tube circuits!

The cathode bypass capacitor C_K in common cathode amplifier is a short circuit for AC signals, thus grounding the cathode K (zero AC potential). The DC voltage on the cathode is used as bias voltage, so cathode K is *not* at ground potential for DC currents and voltages! Correct DC voltages on all electrodes are necessary for the proper operation of a triode as an amplifier of AC signals. Thus DC conditions have to be designed, analyzed, and optimized first.

The coupling capacitor C_C (it "couples" this stage to either the following stage or external load) is also a short circuit for AC signals, as is the power supply $+V_{BB}$. It is assumed that the internal resistance of that power supply is zero (an ideal power supply), and "zero resistance" is another way of saying "short circuit for the AC signal."

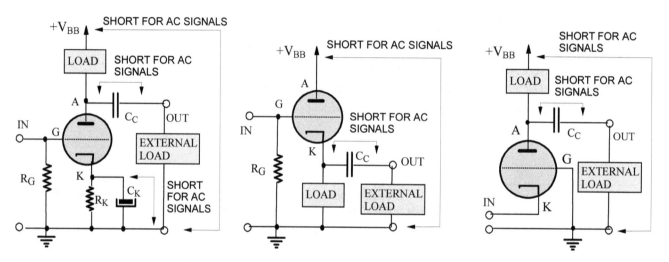

ABOVE (L-R): Common cathode (CC) stage, common anode stage, a.k.a. cathode follower (CF) and common grid stage (CG)

The meaning of input- and output-circuits and the "load" concept

There are two circuits in all amplifiers, the input, and the output circuit. The input is at the grid in the CC stage, so the input circuit is the grid-cathode circuit. The output is taken from the anode, so the output circuit is the anode-cathode circuit. The input is at the grid in the CF circuit, so the input circuit is the grid-anode circuit. The output is taken from the cathode, so the output circuit is the cathode-anode circuit.

In the CG circuit, the input is at the cathode, so the input circuit is the cathode-grid circuit. The output is taken from the anode, so the output circuit is the anode-grid circuit. In any case, the "load" is always connected to the output circuit! In guitar amplifiers, the load can be a fixed resistor or a potentiometer, a tone stack, a transformer, or a loudspeaker. Notice that anode "loads" are always present, they are part of the amplifying stage or an amplifier, but external loads are not necessary for its operation.

In CC and CG amplifiers, the load is only in the output circuit, but in the case of the cathode follower, the load is in both the input and output circuits. This leads to a specific interaction between the input and output circuits in a cathode follower, called negative feedback from the output to the input.

The biasing concept

With no DC voltage bias between its grid and cathode, a common cathode stage would behave as a rectifier - half of the input signal would be cut off, and the output signal waveform would be materially different from the input waveform. To behave as a linear amplifier, the grid potential needs to be negative with respect to the cathode. This negative voltage between the grid and cathode is called the bias voltage and can be implemented in many ways.

The amplifying principle is illustrated with the help of the dynamic transfer characteristic. The input sinewave voltage $v_G(t)$ is applied to the grid, the input of the stage, centered around the negative bias voltage (-6.0V in this case). Anode current $i_A(t)$ is the output signal, in phase with the grid voltage.

Biasing options

For proper operation, the control grid must be negatively biased in respect to the cathode. The obvious way is to connect the grid leak resistor R_G instead of ground to a source of required negative DC voltage $-V_G$. This type of bias is called a fixed or external bias.

Lithium-iron or NiMh (Nickel-MetalHydrate) batteries provide a stable grid voltage for thousands of hours since the grid current is next to zero. Both fixed and battery biasing requires the grid to be capacitively coupled to the previous stage. That way, the voltage offset in the input signal will not change the bias, and the battery cannot discharge through the output resistance of the previous stage or the guitar circuitry.

Making the grid negative is equivalent to making the cathode positive, and the following four biasing methods use that approach. The DC cathode current I_K flows through the cathode resistor RK and creates a voltage drop (V_K), making the cathode positive against the grid. That kind of bias is called self-bias, cathode-bias, or auto-bias; the tube biases itself.

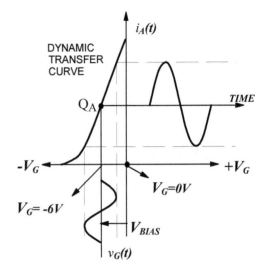

ABOVE: For linear operation, Q should be in the middle of the linear section of the transfer curve

The value of R_K depends on the required operating point Q and steady-state cathode current I_K, plus the desired value of the bias voltage. Voltage V_K is the bias voltage. If the DC current through the tube varies, the bias will change, too, which is to be avoided. The following three methods keep the cathode voltage V_K much less dependent on the cathode current I_K.

Contact bias is illustrated in d). A small grid current that always flows will develop a negative DC voltage on a very high resistance grid leak resistor (3-10M). This contact voltage is very low (under 0.5V), and so it is suitable for input stages only. It varies a lot with different tubes and with tube aging, so it is not recommended!

The internal resistance of a typical small-signal silicon diode is only around 5-6Ω, while red or infrared light-emitting diodes have a slightly lower r_I of 4-5Ω. Green and blue LEDs have a higher r_I, in the 30Ω region.

Zener diodes are ideal for keeping the cathode voltage constant. They can be connected in series to get the exact bias voltage required. Alternatively, a programmable integrated circuit can be used, where the Zener voltage is "programmed" by two resistors in a voltage divider arrangement. As illustrated below, Zener diodes are noisy and should be bypassed by a high-quality ceramic or film capacitor.

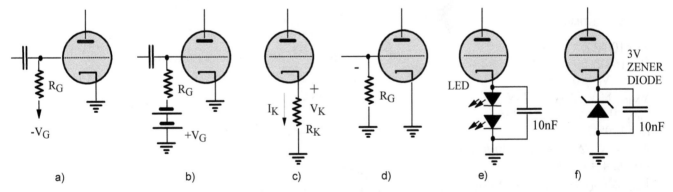

ABOVE: Six ways of providing a DC grid bias: a) fixed external negative grid bias voltage from a power supply b) negative grid bias voltage from a battery c) cathode-, self- or automatic-bias via the cathode resistor voltage drop d) contact bias e) cathode-bias via the voltage drop on silicon or LED diodes f) Zener diode or programmable voltage reference integrated circuit

Graphical analysis of the ECC83 common cathode stage

Tube circuits can be analyzed in two ways. The first is using a small signal model, where a nonlinear tube is represented by a linear model containing a voltage or current source and a resistor. This model, as its name suggests, is only valid for small signals. The other, even more important tool is graphical analysis, based on tube characteristics, either those published by tube manufacturers or our own graphs, drawn from our own measurements.

This input amplifying stage from the Fender Bassmaster amplifier (next page) is typical of most such gain stages used in tube guitar amps. We'll use it to illustrate the operational principles, characteristics, and shortcomings of triode amplifiers. The graph shows a large grid voltage signal of 2.4V$_{PP}$ (peak-to-peak), centered around the DC bias of -1.2V .

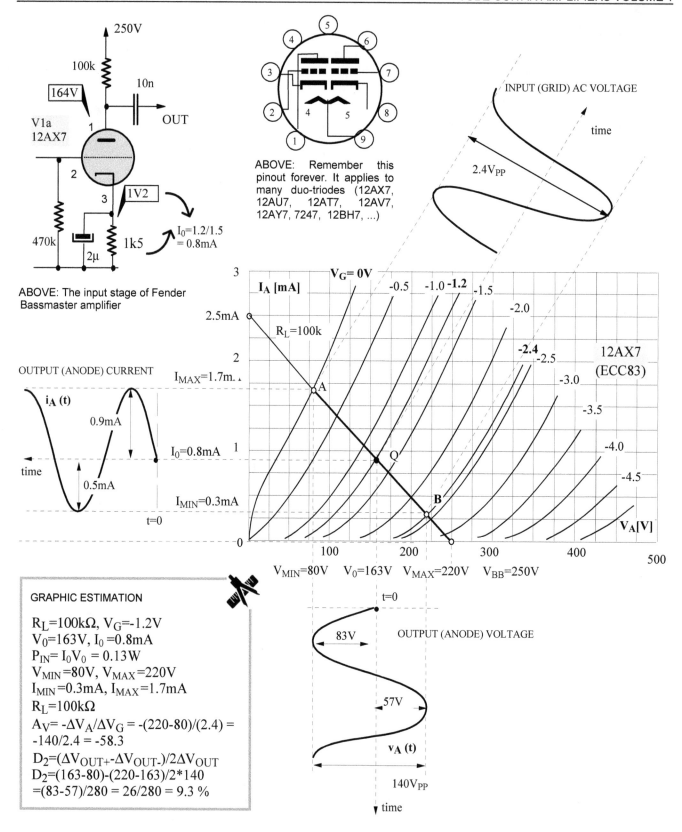

ABOVE: Remember this pinout forever. It applies to many duo-triodes (12AX7, 12AU7, 12AT7, 12AV7, 12AY7, 7247, 12BH7, ...)

ABOVE: The input stage of Fender Bassmaster amplifier

OUTPUT (ANODE) CURRENT

OUTPUT (ANODE) VOLTAGE

INPUT (GRID) AC VOLTAGE

GRAPHIC ESTIMATION

R_L=100kΩ, V_G=-1.2V
V_0=163V, I_0 =0.8mA
P_{IN}= $I_0 V_0$ = 0.13W
V_{MIN}=80V, V_{MAX}=220V
I_{MIN}=0.3mA, I_{MAX}=1.7mA
R_L=100kΩ
A_V= -$\Delta V_A/\Delta V_G$ = -(220-80)/(2.4) =
-140/2.4 = -58.3
D_2=(ΔV_{OUT+}-ΔV_{OUT-})/2ΔV_{OUT}
D_2=(163-80)-(220-163)/2*140
=(83-57)/280 = 26/280 = 9.3 %

To calculate the voltage gain, read the peak-to-peak swing of the anode voltage (80 to 220V), divide the two and you get A = $\Delta V_A/\Delta V_G$ = -(220-80)/2.4 = -140/2.4 = -58.3 times. The negative sign signifies that the two voltages are of the opposite phase (180° phase shift).

Notice that signal swing Q-A is much larger than the swing from Q to B, so the positive peaks of anode current are 0.9mA while the negative half-waves are only 0.5mA in amplitude. A similar distortion happens to the anode voltage; 83 V peaks one way and only 57 V peaks the other way. This means the harmonic distortion of this gain stage is very high, around 10 % or more!

Aguitar amp is not just an amplifier but also a musical instrument since it generates harmonics that were not present in the original signal generated by the guitar's pickups.

The AC loadline and operating conditions

The load line just drawn was a DC load line. Normally, a load would be connected after the 10n coupling capacitor - a tone stack, a gain potentiometer (illustrated), or a grid leak resistor of the next stage. Since (in the first approximation) the impedance of the coupling capacitor is minimal, it represents a short circuit for the AC signal, so the external load will be connected in parallel to the anode resistor, and the slope of the new or AC load line will change. The resistance or impedance of AC load is always smaller than the DC load, so the AC load line is always steeper than the DC load line!

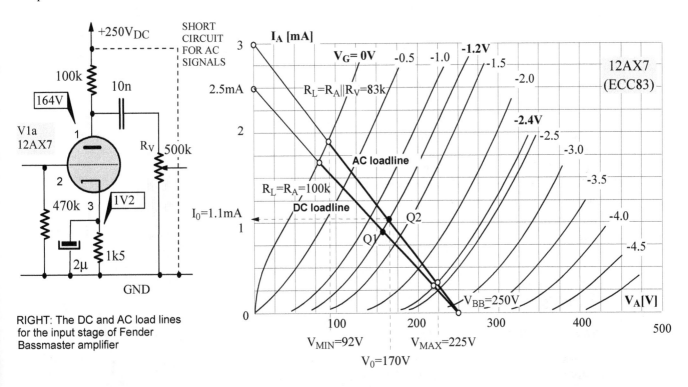

RIGHT: The DC and AC load lines for the input stage of Fender Bassmaster amplifier

Let's connect a 500k volume control pot as a load. Since (assuming a zero internal impedance power supply) the $+250V_{DC}$ point is actually at zero AC potential (dotted line), the top end of the 100k anode resistor is connected to the bottom end of the 500k pot or GND!

The total load on the tube is $100k\Omega$ in the anode in parallel with the 500k, or 83k. We will draw the 83k AC load line assuming the same bias of -1.2V. This will not be the case if we keep the same cathode resistor (the bias or cathode DC voltage would change as well), but for this discussion, let's proceed on that basis.

If you perform the gain estimation now, you will get a voltage gain lower than -58.3 (with a 100k load); the lower the load impedance, the lower the voltage gain. Also, the harmonic distortion will be higher. Another rule-of-thumb: The lower the load impedance, the higher the distortion!

Cathode- or automatic bias: choosing R_K and C_K

The capacitance of the cathode bypass capacitor is only $2\mu F$. The lower -3dB frequency of the RC filter in the cathode is the $f_L = 1/(2\pi R_K C_K) = 1/(2\pi*1,500*2*10^{-6}) = 53$ Hz, quite a high cutoff frequency, meaning this gain stage will attenuate bass frequencies more than the midrange and treble signals.

For instance, at f=1kHz the reactance (reactive impedance) of C_K is $X_C = 1/(\omega C) = 1/(2*\pi*1,000*2*10^{-6}) = 79.6\Omega$, which is much lower (almost 20 times lower) than 1k5 cathode resistance. That can be considered a short circuit for a signal of such frequency.

The cathode capacitor does not bypass signals around the cathode resistor for bass frequencies, resulting in local negative feedback (NFB) in the cathode circuit (shared between the input and output circuits). As we will see soon, NFB reduces voltage gain. Now you can see how by varying the value of the bypass capacitor, we can change the frequency response of the gain stage (and the whole amp!) and alter its voicing!

The smaller the capacitance, the higher the attenuation of the bass signals, and the brighter the amp will sound. The higher the capacitance, the more prominent the bass will be.

The modulation aspect of anode current and voltage

The total anode current is the sum of DC current and AC (signal) current; likewise, the anode voltage is the sum of the DC voltage and the AC or signal voltage.

With no AC voltage or "signal" at its control grid, the anode voltage is just a steady DC voltage of +163V!

With the signal present on the control grid, the anode current and anode voltage start changing; their waveform follows the grid voltage waveform. In this case, it is assumed that the signal is a pure sine wave of a fixed frequency.

Notice that the "zero" line of the signal sine wave on the anode is not the zero volts level, but the DC voltage in the operating point Q (in this case $V_0=163V_{DC}$)!

For emphasis, the instantaneous values of anode voltage are shaded, and those values go from the zero volts line to the AC component of the voltage. We could say that the AC signal "modulates" the DC voltage between the anode and cathode.

In an ideal case (no distortion), the dips and the peaks around the V_0 level are equal, and the average voltage is V_0 since the positive and negative dips cancel each other out. In reality, as illustrated for the ECC83 CC stage, the negative "dips" are larger than the positive "peaks," meaning the output signal is distorted.

The boundaries of the operating region

To minimize unwanted distortion, ensure reliable operation and long tube life, five regions need to be avoided:

1. Positive grid voltages, where grid current starts flowing and input resistance drops significantly. Power tubes specifically designed to operate with grid currents are an exception.
2. Above the maximum anode current level (I_{AMAX})
3. Very low anode currents, where the characteristics are very curved and distortion is very high (I_{AMIN})
4. Above the P_{AMAX} parabola, where the power dissipation of the anode is exceeded
5. Above the maximum anode voltage allowed for a particular tube (V_{AMAX}). Consult tube's datasheets.

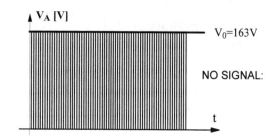

The output (anode) voltage waveform for the ECC83 common cathode stage ABOVE: with no signal voltage on its grid BELOW: with an AC signal

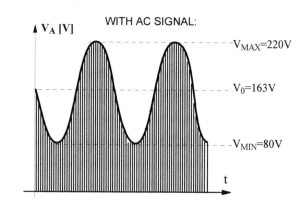

BELOW: The safe operating region of a triode is the shaded area.

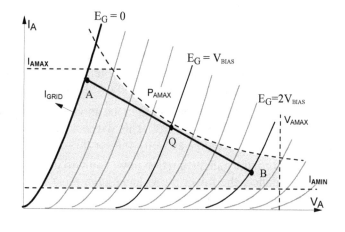

COMMON CATHODE GAIN STAGE - SMALL SIGNAL MODEL ANALYSIS

Small signal linear model (equivalent circuit) of a triode

Just as we have done with diodes, we will use the linear triode model in most deliberations from now on. Just like any model, it is an approximation. The smaller the signal amplitude around the quiescent operating point Q, the better this small-signal model fits reality.

We mainly use simplified (not entirely accurate) linear models because we know how to deal with and solve linear equations. Still, we don't know how to analyze nonlinear systems, which are described by nonlinear equations. Luckily, for most purposes in electrical engineering, such models are perfectly adequate and helpful.

Anode characteristics have anode current I_A as a dependent variable, anode voltage V_A as an independent variable, and grid bias voltage E_G (or V_G) as a parameter. With triodes, the grid can be at any bias voltage the designer chooses, so we have an infinite number of curves instead of just one. However, for practical reasons, we only draw half a dozen to a dozen such curves.

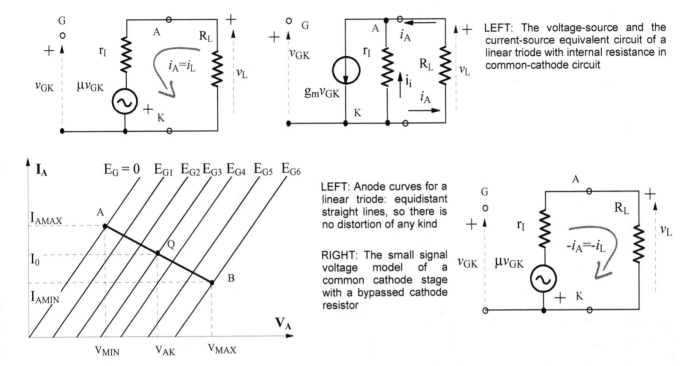

LEFT: The voltage-source and the current-source equivalent circuit of a linear triode with internal resistance in common-cathode circuit

LEFT: Anode curves for a linear triode: equidistant straight lines, so there is no distortion of any kind

RIGHT: The small signal voltage model of a common cathode stage with a bypassed cathode resistor

The voltage-source model of a linear triode had its internal resistance r_I in series with the voltage source μv_{GK}. While the voltage source model makes some types of analysis easier, the current source model is more convenient in other calculations, so that we will use both.

We've already estimated the amplification of our ECC83 stage using graphical means; let's see what results we'll get using the small-signal equivalent circuit. Normally, you don't have to use both; one estimation is sufficient.

Using a voltage model, the equation for the current loop is $i_A = -\mu v_{GK}/(r_I + R_L)$ and since $v_{OUT} = v_L = i_A R_L$ we have $v_{OUT} = -\mu v_{GK} R_L/(r_I + R_L)$. The voltage amplification is $\mathbf{A = v_{OUT}/v_{IN} = -\mu * R_L/(r_I + R_L)}$.

For ECC83 stage with R_L=100k, and 0.8mA anode current the internal resistance is r_I=67kΩ and μ=100, so A= -100*100/(67+100) = -100*0.6= -60!

The graphic gain estimation was A=-58.3, so the two results are pretty darn close! The curves are approximate, and the linear model is a simplification and, as such, assumes that I_A-V_A curves are straight lines, which they are not.

And finally, considering that parameters of real tubes vary widely from the published "bogey" or average figures, for design purposes, the accuracy under 10% is acceptable, and anything under 5% is excellent!

We have already explained how C_K bypasses the signal frequencies to the ground and how the cathode resistor R_K does not appear in the AC model. Its only purpose is to provide the negative DC bias for the grid, making the cathode positive compared to the grid, which is equivalent to making the grid negative with respect to the cathode.

You may ask yourself, how come the anode DC power supply voltage V_{BB}=+400V does not come into play either? Well, assuming the internal impedance (resistance) of this power supply is zero (another simplification), such a DC voltage source does not affrect AC signals at all.

For AC signals, point V_{BB} is at GND (COM) potential (the internal impedance of the power supply is assumed to be zero), as is the cathode K (cathode resistor fully bypassed by the cathode capacitor). This simplifies our circuit immensely.

It is interesting to see how quickly the amplification factor of the common cathode stage (A) approaches the μ of a triode. For $R_L/(R_L+r_I) = 4$, we get 80% of μ; for $R_L/(R_L+r_I) = 10$, we get A=0.9μ, without any further significant increase, which limits $R_L/(R_L+r_I)$ to the 4-10 range.

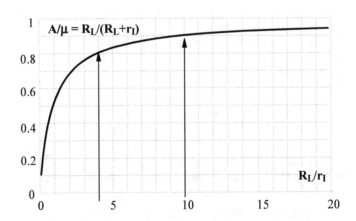

ABOVE: How the gain of common cathode stage (referenced to μ) varies with the size of the anode load R_L (referenced to r_I)

THE EFFICIENCY AND DISTORTION OF THE COMMON CATHODE STAGE

Efficiency of ideal triode stages

Let's consider an ideal triode with a resistive load R_L in the anode circuit. The quiescent point is Q, and the load line is AB. $I_{AMAX} = 2I_0 = E_{BB}/R_L$, the anode voltage in the quiescent point is half of the high voltage battery voltage V_{BB}. The efficiency is defined as η= (AC output power)/(DC input power)*100% DC input power is $P_{IN}=V_{BB}I_0$.

AC output power is $P_{OUT}= V_{RMS}I_{RMS} = (V_P/\sqrt{2})(I_P/\sqrt{2}) = V_PI_P/2$ Since the peak values are $V_P = (V_{BB}-V_{AMIN})/2$ and $I_P=(I_{AMAX} -I_{AMIN})/2$ we get

$P_{OUT}= V_PI_P/2 = (V_{BB}-V_{AMIN})(I_{AMAX} -I_{AMIN})/8$

$\eta = (V_{BB}-V_{AMIN})(I_{AMAX} -I_{AMIN})/8(V_{BB}*I_0)*100\%$

The output current's maximal swing or peak value I_P is $I_{AMAX}-I_0$ in the positive direction and $I_0-0=I_0$ in the negative direction. For an ideal triode, these are equal since I_0 is in the middle of the range ($I_0=I_{AMAX}/2$). So, the maximum efficiency is $\eta= V_{BB}2I_0/8V_{BB}I_0)*100\% = 1/4*100\% = 25\%$.

The DC anode current I_A flows through the load and dissipates power in it, which is one reason for low efficiency. In steady-state, without any AC signal, the DC power dissipated in the anode is $P_A=V_0I_0$, (the area of the light gray-shaded rectangle). The power dissipated as heat on the load is $P_L= (V_{BB}-V_0)I_0$ (the area of the dark gray-shaded rectangle). The total DC power provided by the power supply is $P=P_A+P_L= V_{BB}I_0$ (the area of both rectangles together).

In this ideal case, since Q is in the middle of V_{BB}, $P_A=P_L$, at idle (no signal) 50% of the power is wasted as heat on the tube and the other half on the load!

For real triodes, the range A-B is much smaller, and V_0 is closer to V_{BB}. Thus, due to the quiescent point Q position, the dissipation on the tube is higher than the dissipation on the load, and the overall efficiency is lower than the 25% theoretical maximum.

OUTPUT POWER END EFFICIENCY OF THE COMMON CATHODE TRIODE STAGE

$P_{OUT}= V_PI_P/2 = (V_{BB}-V_{AMIN})(I_{AMAX} -I_{AMIN})/8$

$\eta = (V_{BB}-V_{AMIN})(I_{AMAX} -I_{AMIN})/8(V_{BB}I_0)*100\%$

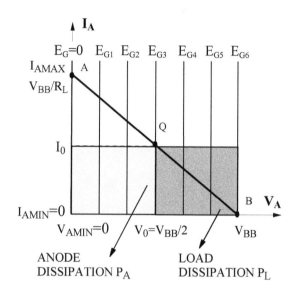

Ideal anode curves and load line for the resistive load

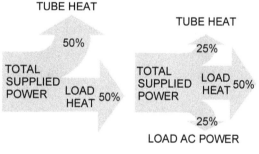

NO SIGNAL   MAX SIGNAL

With or without the input signal, half of the supplied power is wasted on the load as heat.

Common cathode stage: estimating harmonic distortion

An ideal dynamic transfer characteristic would be a straight line, and the grid signal would be "reflected" into the anode circuit without any distortion.

Compared to the ideal (linear) transfer characteristic, triodes' parabolic curve results in an asymmetric waveform of the output (anode) current. The sine wave's negative half is widened (1) and flattened, with a slight narrowing of the positive half (2).

This type of amplitude distortion results in even harmonics, with the second harmonic being dominant, resulting in that signature triode sound.

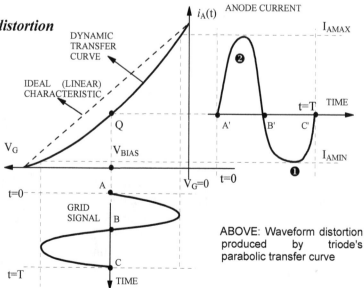

ABOVE: Waveform distortion produced by triode's parabolic transfer curve

The 2nd harmonic is dominant and can be roughly estimated by the current distortion formula (CDF):

$D_2 = [(I_{AMAX}+I_{AMIN})-2I_0]/2(I_{AMAX}-I_{AMIN})*100\%$ or $D_2 = (\Delta I_+-\Delta I_-)/2\Delta I*100\%$

Voltage estimation formula can also be used: $D_2 = (\Delta V_+ - \Delta V_-)/2\Delta V *100$ [%]

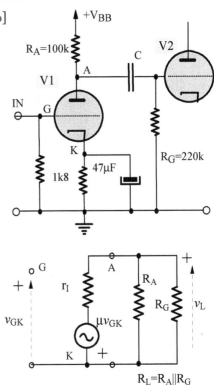

TRIODE DISTORTION ESTIMATION FORMULAS

$D_2 = (\Delta I_+-\Delta I_-)/2\Delta I*100$ [%]

$D_2 = (\Delta V_+ - \Delta V_-)/2\Delta V *100$ [%]

RC coupling between stages

For AC signals, R_A is in parallel with the grid resistor R_G of the following stage, forming a parallel load resistance of $R_L = R_A \| R_G = R_A R_G /(R_A+R_G)$.

VOLTAGE AMPLIFICATION OF RC-COUPLED COMMON CATHODE STAGES

$A = -\mu R_L/(r_I + R_L)$, where $R_L = R_A \| R_G = R_A R_G /(R_A+R_G)$

This formula is valid only if two conditions are met. First, the coupling capacitor C value is chosen to represent a short circuit at audio frequencies. Secondly, the DC power supply is assumed to have zero internal resistance; therefore, the $+V_{BB}$ point is at ground potential for AC signals.

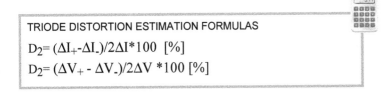

ABOVE: The small signal model of capacitive coupled stages at midrange frequencies

---

COMMON CATHODE STAGE WITH UN-BYPASSED CATHODE RESISTOR

This seemingly simple circuit is the most important of all gain stages. A dozen or so various circuits have their origins here, so we need to analyze it inside and out despite the tedious algebra. You will see why very soon. If you understand this circuit, you will also comprehend all other amplification stages and phase inverters! There are two power supplies, the positive anode supply $+E_{BB}$ and the negative cathode supply $-E_CC$. E_{BB} is always needed, but E_C is not included in many circuits, where the end of the cathode resistor R_K is grounded.

The circuit also has two possible outputs, V_A from the anode and V_K from the cathode. Again, some circuits derived from this one use V_A output only, others use V_K output only, but some circuits use both. The first step is to draw the equivalent circuit and write the loop equations. Depending on what we want to express (amplification factors, output impedances, input impedance, etc.), we will manipulate those two equations in various ways.

Amplification factor of the anode output

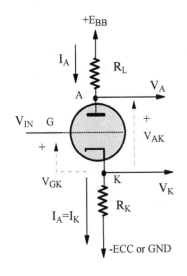

The voltage equations for the input and output loop are $V_{IN} - V_{GK} + I_A R_K = 0$ and $-\mu V_{GK} - I_A r_I - I_A R_L - I_A R_K = 0$.

Since $V_{OUT} = I_A R_L$, to eliminate V_{GK} we write $V_{GK} = V_{IN} + I_A R_K$ from the first loop and substitute it instead of V_{GK} into the second equation:

$-\mu(V_{IN} + I_A R_K)- IA(rI+RL+RK) = 0$ or $-\mu V_{IN} = I_A(r_I+R_L+R_K+\mu R_K)$

$I_A = -\mu V_{IN} /(r_I+R_L+R_K+\mu R_K)$

Now we substitute the right side of this equation into the third equation for V_{OUT} and get $V_{OUT} = I_A R_L = -\mu V_{IN} R_L/[r_I+R_L+(1+\mu)R_K]$

We finally get the anode amplification factor $A_A = V_{OUT}/V_{IN}$ or

$A_A = -\mu R_L/[r_I+R_L+(1+\mu)R_K]$

LEFT: Currents and voltages in common cathode stage with un-bypassed cathode resistor

CHECK: If R_K were zero (for AC signal) as when bypassed to ground by a capacitor, the gain would be $A_A = -\mu R_L/(r_I+R_L)$, just as we had before. So, the CC stage with a bypassed R_K is just a special case of this circuit!

Local NFB through un-bypassed cathode resistor

The voltage gain was -60 before; now, it's only -31.4. Why has almost half of the voltage gain disappeared?

Assume the signal V_{IN} is rising, making the grid more positive with respect to the cathode. This reduction in the negative bias will cause the anode current to rise. Because it also flows through the cathode resistor, the voltage drop on the cathode resistor will increase, making the cathode more positive, equivalent to making the grid more negative vis-a-vis the cathode.

The two voltages are opposing each other, resulting in a smaller output voltage, so the amplification of the stage is lower. The cathode resistor provides "degeneration" or negative feedback.

The feedback AC voltage on the cathode resistor is proportional to load current, so it's current feedback, and because it is added to the input AC signal in series, it is series-applied feedback.

The input and output circuit share R_K, so the input to the stage is no longer independent of the output (as it was when R_K was bypassed).

Gain with feedback, or "closed-loop" gain is $A_F = A/(1+A\beta)$, where A is the gain of the stage without feedback, and β is the "feedback ratio," or the fraction of the output that is fed back to the input.

The term $(1+A\beta)$ is called "feedback factor." In this case, since $A=-\mu R_L/(r_I+R_L)$ and $A_F = -\mu R_L/[r_I+R_L+(1+\mu)R_K]$, the ratio of feedback voltage to output voltage is the same as the resistance ratio, so the feedback ratio is $\beta = R_K/(R_L+R_K) \times 100$ [%].

If R_K were zero (bypassed to ground by a capacitor), β would be $0/(R_L+0) \times 100$ or 0%, there would be no feedback.

Amplification factor of the cathode output

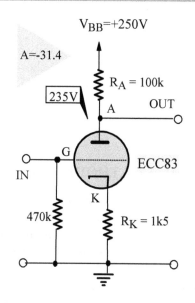

ANODE & CATHODE VOLTAGE AMPLIFICATION OF COMMON CATHODE STAGE

$A_A = -\mu R_L/[r_I+R_L+(1+\mu)R_K]$

$A_K = \mu R_K/[(1+\mu)R_K + r_I+ R_L]$

$V_{BB}=+250V$

A=-31.4

235V

$R_A = 100k$

OUT

ECC83

IN

470k

$R_K = 1k5$

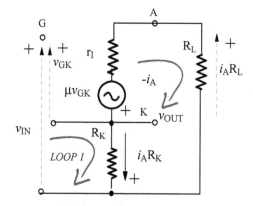

Going back to the same two loop equations: $VIN - V_{GK} + I_A R_K = 0$ and $-\mu V_{GK} - I_A r_I - I_A R_L - I_A R_K = 0$ We also know that $V_{OUT} = -I_A R_K$

From the 1st equation.: $V_{GK} = I_A R_K + V_{IN}$ and substitute it into the second: $-\mu(I_A R_K + V_{IN}) - I_A r_I - I_A R_L - I_A R_K = 0$

$-I_A(\mu R_K + r_I + R_L + R_K) - \mu V_{IN} = 0$

Since $I_A = -V_{OUT}/R_K$, we can substitute that into the eq. above and move $-\mu V_{IN}$ to the other side: $V_{OUT}(\mu R_K + r_I + R_L + R_K)/R_K = \mu V_{IN}$

The amplification factor $A_K = V_{OUT}/V_{IN}$ or

$A_K = \mu R_K/[(1+\mu)R_K + r_I + R_L]$

CATHODE FOLLOWER (COMMON ANODE STAGE)

A cathode follower (CF) is a special case of the circuit analyzed above when $R_L=0$. The anode is directly connected to a positive DC supply voltage V_{BB}; therefore, it is at the ground or common potential for AC signals. This, of course, assumes that the internal resistance of the V_{BB} power supply is zero. The input is at the grid, and the output is taken from the cathode; that is why the "official" name for a cathode follower is a common or "grounded" anode stage.

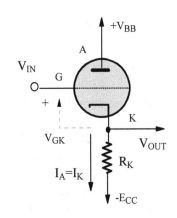

If we remove R_L from the amplification factor formula $A_K = \mu R_K/[(1+\mu)R_K + r_I + R_L]$, we get the voltage amplification for cathode follower: $\mathbf{A_K = \mu R_K/[(1+\mu)R_K + r_I]}$. To get a better feel, let's go back again to our circuit with ECC83, only this time removing R_L and taking the output from the cathode.

As before $R_K = 1k5$, $r_I = 67k$ and $\mu = 100$, so the gain is now $A_K = \mu R_K/[(1+\mu)R_K + r_I] = 100*1,500/[(1+100)1,500+67,000] = 0.69!$

We have lost all the gain! This "amplification" stage is now acting as an attenuator; the output voltage is lower than the input voltage (only 69% of it)! Why are we even mentioning such a *seemingly* useless circuit?

Just as with people, first impressions of circuits are often wrong. The output impedance of the cathode output is $Z_{OUT} = (r_I + R_L)/(1+\mu) \| R_K$. For cathode follower $R_L = 0$ so its output impedance is $\mathbf{Z_O = r_I/(1+\mu) \| R_K}$

For our ECC83 circuit $Z_O = 67,000/(1+100) \| 1,500 = 663 \| 1,500 = 460 \ \Omega$. Anything below $1k\Omega$ is a very low output impedance. For comparison, the output impedance of a common cathode stage with bypassed cathode resistor is $Z_{OUT} = r_I \| R_L = R_L r_I/(R_L + r_I)$, which for our gain stage with ECC83 would be $Z_{OUT} = 67k \| 100k = 40.1k\Omega!$

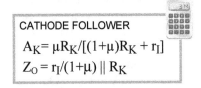

CATHODE FOLLOWER
$$A_K = \mu R_K/[(1+\mu)R_K + r_I]$$
$$Z_O = r_I/(1+\mu) \| R_K$$

Ideally, the output impedance of any amplification stage should be zero (ideal voltage source) or as low as possible. $40k\Omega$ from a common cathode stage is bad; 460Ω from a cathode follower (almost 100X lower) is much better!

Now you see that CF is not an amplifier but an impedance "transformer." Its high input impedance means it will not load the previous stage (no loading means no distortion). Furthermore, because of its very low output impedance, it will not be loaded by the following (low impedance) stage, making CF ideal for driving tone control stacks and solid state amplification stages.

The "Bootstrapping" concept

Let's consider a practical CF circuit using another excellent tube, 12AU7 duo-triode, which also goes by ECC82 and many equivalent or near-equivalent designations: E82CC, 5963, 6189, 5814, B329, CV491, E2163, E812CC, ...

12AU7 has low internal impedance and high anode dissipation (2.75 Watts), so it can operate at relatively high anode currents of 10-20mA, and high current stages generally sound better than low current ones!

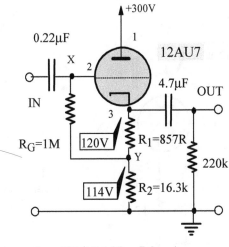

TUBE PROFILE: 12AU7 (ECC82)

- Indirectly-heated medium μ dual triode
- Noval socket, heater: 6.3V/300mA or 12.6V/150mA
- $V_{AMAX} = 300V$, $V_{HKMAX} = 100V$
- $P_{AMAX} = 2.75$ W, $I_{AMAX} = 20$ mA
- TYPICAL OPERATION:
- $V_A = 250V$, $V_G = -8.5V$, $I_0 = 10.5$ mA
- $gm = 2.2$ mA/V, $\mu = 17$, $r_I = 7.7$ kΩ

$I_K = 120/17,157 = 7.0$ mA
$V_{BIAS} = I_K * 857 = 7*0.857 = -6V$
so $V_G = 120-6 = 114V$

Bootstrapping is an American word that likens circuit action to someone pulling themselves up by their bootstraps or shoelaces. The grid resistor is bootstrapped, from point Y to the input (point X), which is equivalent to an increase of the input impedance Z_{IN} from approximately R_G to a higher value $Z_{IN} = R_G/[1-AR_2/(R_1+R_2)]$.

The higher R_2 and the smaller R_1, the higher the bootstrapping effect and the higher the input impedance. With $R_K = 17k16$, $r_I = 7k7$ and $\mu = 17$, the gain is $A_K = \mu R_K/[(1+\mu)R_K + r_I] = 17*17,160/[(1+17)17,160+7,700] = 0.90$.

The 857R and 16k3 resistors form a voltage divider, so the voltage in point Y is $16,300/(857+16,300)*V_{OUT} = 0.95V_{OUT}$, where V_{OUT} is $0.9V_{IN}$ in this case, so $V_Y = 0.95*0.9*V_{IN} = 0.855*V_{IN}$.

That means that the input signal "sees" a much-reduced voltage across the grid-leak resistor, only $0.145V_{IN}$ in this case, which is equivalent to the input impedance of $Z_{IN} = R_G/0.145 = 6.9$ MΩ, an increase of almost seven times!

The output impedance of this CF is $Z_{OUT} = r_I/(1+\mu) \| R_K = r_I/(\mu+1+r_I/R_K) = 7,700/(17+1+7,700/17,157) = 417\Omega$.

How to determine the operating point when DC voltages are not marked

The DC voltages in important points aren't marked on many circuit diagrams. Of course, should we have the physical amplifier on the test bench, the quickest and easiest option is to measure them with a multimeter set on "DC Volts." But what if we only have the circuit diagram?

Providing we know the anode supply voltage, we can figure the DC voltages out using this graphic method. We have to draw the load line on anode characteristics for the relevant tube, construct the bias line for the common cathode stage, and their intersection is the quiescent operating point Q!

As an example, to construct the DC load line for the 1st stage of Epiphone Valve Jnr. amp (below), mark the point $V_A=V_{BB}=290V$ (1). Since $290V/100k\Omega=2.9mA$, mark the maximum current of 2.9mA on the I_A axis (2). Connect the two points with a straight line, and that is your DC load line. The quiescent operating point Q will lie somewhere on that line. To determine precisely where we need to draw the bias line.

To construct the bias line, choose two current levels reasonably apart (not too close), say 0.5mA and 1.5 mA and calculate the bias voltage in each point. With 0.5mA through the 2k2 cathode resistor, the cathode voltage would be $V_K=0.5*2.2 = 1.1V$, which is equivalent to saying that the grid (at GND or)V level) is at $V_G=-1.1V$ with respect to the cathode. We don't have $V_G=-1.1V$ anode curve on the published graphs, only $V_G=-1.0V$ and $V_G=-1.5V$, so we have to approximately position point X closer to $V_G=-1.0$ (roughly 20% of the way between the two curves) along the $I_A=0.5mA$ line (3). Our estimated anode curves we draw in dotted lines on the published graph (full lines).

With 1.5mA through the 2k2 cathode resistor, the cathode voltage would be $V_K=1.5*2.2 = 3.3V$, so $V_G=-3.3V$!

Draw these two points (X and Y) and connect them with a straight "bias" line. Its intersection with the load line is the Q-point. Here it lies at the grid voltage $V_G = -1.8V$, so the cathode is at $V_K=+1.8V$. Now we draw a vertical line down to V_A axis and read $V_{A0}=210V$ (5), the voltage between pin 1 (anode) and pin 3 (cathode).

Finally, draw a horizontal line towards the I_A axis (6) and read the quiescent anode current in the Q point of $I_A=0.8mA$.

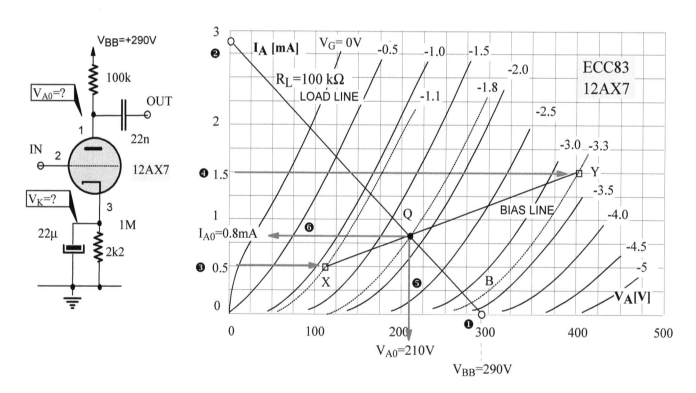

Once you have the cathode voltage (1.8V), you can calculate the anode current; you don't need to read it from the graph, so let's check if we get the same result: $I_A=V_K/R_K = 1.8/2.2= 0.82$ mA, pretty close to my eyeball estimation of 0.8 mA from the graph.

This method is not very precise; it depends on your judgment and estimation skills, so instead of $V_G= -1.8V$, different people would get results ranging from -1.7 to -1.9V, but that is fine since tube parameters vary a lot anyway.

The curves above are for an average or "bogey" tube, and real tubes can have slightly shifted curves, so if plugged into the same circuit, their bias voltages will vary a lot. So, please don't lose your sleep expecting or trying to achieve absolute accuracy; you'll never get it!

TETRODES, PENTODES AND BEAM-POWER TUBES

6

- TETRODES
- PENTODE CONSTRUCTION AND PARAMETERS
- BEAM POWER TUBES
- DESIGNING VOLTAGE AMPLIFICATION (PREAMPLIFIER) PENTODE STAGES

Most guitar amps use triodes in preamp and pentodes or beam power tubes in output stages, although many feature pentode gain stages as well. There are many similarities between triodes and pentodes, but there are also significant differences, which are explained in this chapter.

TETRODES

Tetrode = triode + screen grid

Early experiments revealed that the insertion of another grid between the control grid and the anode of a triode provides electrostatic shielding between them to such an extent that the input capacitance is reduced by the factor of 1,000 or more! Thus, tetrodes performed much better than triodes at high frequencies, so in 1928 the first commercial tetrode was introduced.

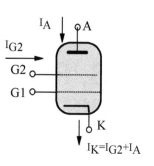

LEFT: The circuit symbol and currents in a tetrode. "Tetra" means four in Greek since there are four active electrodes: cathode, control grid, screen grid, and anode. Heater connections are not counted as active electrodes.

The addition of the second grid (G2) or screen grid means there are now three element voltages and two currents, the anode and screen current, so the behavior of a tetrode is much more complex than that of a triode. However, the screen is usually held at a constant voltage in practical applications and does not feature in AC models, so the same models as those for triodes can be used.

The screen must be positive vis-a-vis cathode; otherwise, there would be no electron flow. A positive screen takes over the electron accelerating function of the anode, and, since it is physically much closer to the control grid G1, it has much more effect on the current flow than the anode itself, almost as much control over anode current as the control grid. Depending on the spacing between the control and screen grids and the pitch of the grid wires (how closely they are wound), the degree of the screen's control can vary.

The finer the pitch and the closer the spacing between the screen grid's turns, the more control the screen grid will have over anode current. In both situations, the screen grid will intercept more electrons (allow less of them to reach the anode), and therefore in both cases, the screen current will increase.

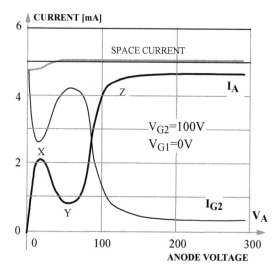

ABOVE: Screen and anode currents in a tetrode. The tube used was a 6J7 pentode connected as a tetrode (its suppressor grid was tied to the screen). This was done so we could directly compare its tetrode and pentode curves.

The reduction in input capacitance

The insertion of a screen grid seems to divide the anode-grid capacitance C_{AG} into two, which would halve it since the two capacitances are connected in series. It still doesn't explain the significant (1,000-fold) reduction in C_{AG}! A model will help, as always.

The fact that screen (S) is at cathode (K) potential is why the capacitance is drastically reduced. The dreaded Miller effect (feedback from anode to grid) and the reduction in the amplification of higher frequencies cannot occur. Thus, the capacitance is not multiplied by the amplification factor of the stage because no voltage appears across the connection of the two capacitances.

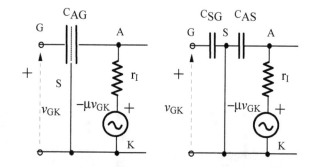

ABOVE: The AC model of a tetrode to help explain the reduction in input capacitance.

Secondary emission

The anode voltage in a tetrode may not affect the anode current, but it does determine how the total space current is distributed between the anode and the screen. Since the total space current is constant (except at very low anode voltages), the screen and anode currents are inverse or mirror images of each other.

The anode current curve has a very nonlinear shape. As the anode voltage increases, the anode current peaks and then starts dropping, only to resume its rise before reaching the plateau where it's practically constant. A tetrode exhibits a negative internal impedance in the region where the current drops with the rising voltage X-Y, which is sometimes used in its applications as an oscillator but is unwelcome in its use as an audio amplifier.

The kink is caused by a physical phenomenon called secondary emission, which begins at point X.

The primary emission is that of electrons from the cathode. The secondary emission occurs when high-energy, high-velocity primary electrons strike the anode and transfer enough energy to anode's electrons. These secondary electrons leave the anode and flow back attracted by the positive screen grid. It is also possible for the screen grid to emit secondary electrons.

To avoid this nonlinear region, the lowest anode voltage in an amplifier would have to be above point Z, which is 100V in this case. Because of that limitation, a tetrode amplifier would not achieve a wide anode voltage swing, so its voltage and power amplification would be limited. That is why tetrodes were short-lived and were replaced by pentodes only a year or two after their invention.

PENTODE CONSTRUCTION AND PARAMETERS

Tetrode + suppressor grid = pentode

To overcome tetrode's deficiencies, primarily the kink in its anode characteristics, a pentode was introduced only a year later, in 1929. An additional grid called suppressor grid or G3 eliminated the effects of secondary emission and the associated kink, further improved the shielding, and increased internal resistance to very high levels, thus increasing the amplification factors.

The suppressor is at a negative potential with regard to the screen, so it provides a retarding force that prevents secondary electrons emitted by the screen from flowing to the anode. More importantly, providing its potential is lower (more negative) than the anode, it also constrains the secondary electrons from the anode and returns them to the anode.

Since the cathode is always less positive (more negative) with respect to both the screen and anode, most pentodes have their suppressor grid internally connected to the cathode, so proper operation is assured. Some pentodes have it brought out to its own pin, most notable being the famous EL34.

Small signal models of an ideal and linear pentode

Just as we did with triodes, it makes sense to understand an ideal pentode first and then explore how, where and why real (practical) pentodes fall short of that ideal.

Ideal pentode's anode curves are equidistant horizontal lines, meaning its anode current does not depend at all on the anode voltage. It behaves as an ideal current source (infinitely high internal resistance). There are no internal losses and no distortion.

Anode characteristics are still equidistant for the linear pentode, but they aren't horizontal anymore. The internal resistance is not infinite but has some very high value r_I, modeled as a parallel resistance. There is still no distortion since the characteristics are still equidistant.

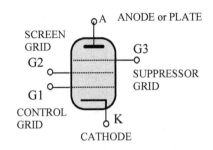

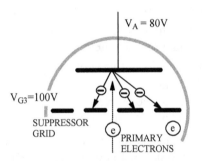

ABOVE: With suppressor grid more positive than the anode, secondary electrons are attracted by it and are not returned to the anode. This is clearly undesirable - a pentode is never used in such regime.

BELOW: As long the anode is more positive than the suppressor grid, the secondary electrons are returned to the anode. High-speed primary electrons are not affected.

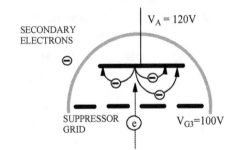

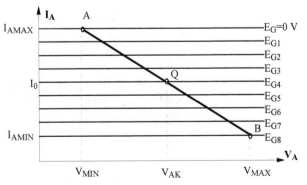

ABOVE: Anode characteristics (anode current versus anode-to-cathode voltage) of an ideal pentode

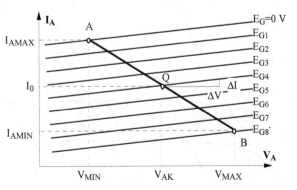

ABOVE: Anode characteristics of a linear pentode

A linear pentode behaves like a current source whose current is proportional to the grid signal v_{GK}: $i = g_m * v_{GK}$. The internal resistance of the tube is not infinite (open circuit) as with an ideal pentode but some very high resistance r_I in parallel with the current source.

Thus, the current source pushes current through both the load R_L and its internal resistance. In other words, its current gets divided, so not all current reaches the load.

Cathode current = anode current + screen grid current

The magnitude of anode voltage plays only a minor part in the operational behavior of a pentode. The current a pentode pushes through the load depends on gm, the pentode's transconductance, and v_{GK}, the AC signal between the control grid and the cathode.

However, there is one thing we cannot see from the graph (right): the importance of the screen grid (or G2) for the operation of a pentode. What anode is for a triode, the screen grid is for a pentode, so the screen grid DC voltage determines its behavior.

Real pentodes don't have an infinitely high internal impedance or resistance. The internal resistance of a pentode can be read or calculated from its anode curves: $r_I = \Delta V / \Delta I$.

The pentode's finite internal resistance means that in our model, not all of the current source's current ($g_m v_{GK}$) will reach the load so that anode current I_A will be lower than for an ideal pentode.

For instance, for EL84 power pentode $\Delta V = 80V$ and $\Delta I = 2$ mA, so $r_I = 80/0.002 = 40,000\ \Omega$. This is a much higher internal resistance than in power triode's case (typically 200 -1,000Ω)!

The slope of the curves also means that the plate voltage does have some impact on the operation of a real pentode; as V_A increases, the plate current also increases, albeit very slightly.

Real pentode's operation in three dimensions (3D)

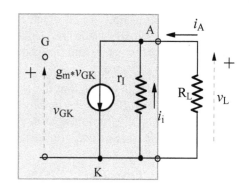

ABOVE: Linear model of a pentode. The tube behaves like a current source whose current is proportional to the grid signal v_{GK}: $i = g_m * v_{GK}$.

BELOW: Screen and anode currents in 6J7 pentode. The kink in tetrode's anode curves is gone, and the knee is at a much lower anode voltage, meaning much larger anode voltage swings are possible.

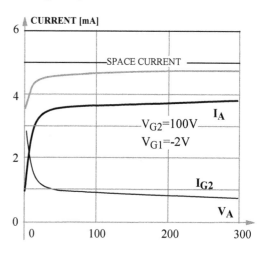

Since three electrodes impact or determine the anode current in pentode, i_A is a function of three variables, grid voltage, screen voltage, and anode voltage: $i_A = g_m * (v_{G1K} + v_{SK}/\mu_S + v_{AK}/\mu)$. The screen amplification factor μ_{G1G2} (between the control grid and screen grid) is sometimes named μ_S!

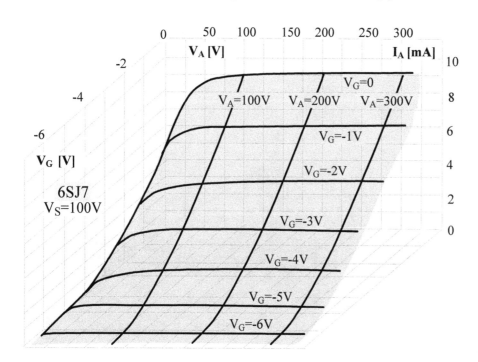

Dealing with three variables is too complicated, so parameters gm, μ, and μ_S are considered constant to simplify analysis. However, they do depend on the operating point of the pentode.

The error introduced by this simplification is not significant since other aspects of analysis and modeling introduce similar errors, as do tube datasheets. Going into tedious detail in one aspect of analysis while other factors vary widely would be insane.

LEFT: 3D characteristics of 6SJ7 small signal pentode with screen voltage of $100V_{DC}$

The operating region

The operating region of a pentode is determined in the same way as for triodes. It is limited by the $V_G=0V$ anode curve, the maximum dissipation P_{AMAX} curve and the maximum anode voltage V_{AMAX}, declared in data sheets.

The illustrated way of positioning the load line (for maximum power output) applies only to output (power) pentodes such as EL34 and EL84. The small-signal pentode stages used as voltage amplifiers (with EF86, 6AU6, 6SJ7, etc.) are designed using different criteria, as will be explained soon.

Measuring and plotting pentode's characteristics

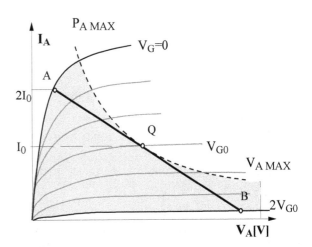

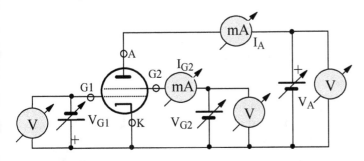

ABOVE: The boundaries of the operating region and the load line for a power pentode

ABOVE: The circuit for recording transfer, anode, and constant current curves of tetrodes and pentodes. The heater connection and its power supply are not shown.

BELOW RIGHT: How mutual conductance of a pentode varies with control grid and screen grid voltage (EF86)

Since there is an additional control electrode (the screen grid), the test circuit for tetrodes and pentodes requires an additional DC voltage source and test instruments for the screen circuit, a mA-meter and a voltmeter.

Should you wish to investigate the effect of the suppressor grid as well, another variable voltage source is needed. Although seldom used as a control electrode (it is usually kept at a constant DC potential), a suppressor grid tied to a negative voltage source (vis-a-vis the cathode) can be used to control the anode current as well.

Pentode's distortion "signature"

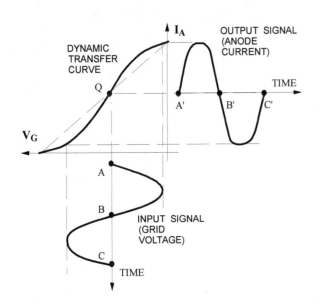

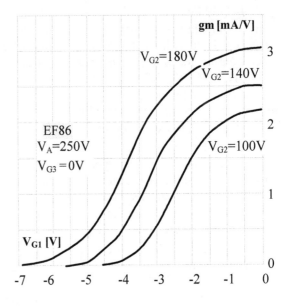

The cubic or S-shaped dynamic transfer curve of pentodes and beam power tubes produces a large 3rd and a smaller 2nd harmonic, resulting in a symmetrically distorted waveform, with flattened peaks of the sinewave. The dominant 3rd harmonic, which is discordant (unpleasant sounding) is the primary reason pentodes have never been loved by audiophiles, even in the 1960s, when they were widely used in mid-fi amplifiers of Dynaco or Eico class. Guitar players don't mind this raspy distortion; in fact, many prefer it to the soft and warm sonic signature ("even harmonic distortion") of triodes.

The second and third harmonic distortion in pentode stages can be estimated by the "5-point method".

$$D_2= [(I_{AMAX} +I_{AMIN}) - 2I_0]/[I_{AMAX} -I_{AMIN}+1.41(I_X-I_Y)]*100\%$$
$$D_3= [I_{AMAX} -I_{AMIN} - 1.41(I_X-I_Y)]/(I_{AMAX} -I_{AMIN}+1.41(I_X-I_Y))*100\%$$

The amplification factor of a pentode

The overall amplification factor of a pentode can be expressed as a product of the amplification factors between the control grid and screen grid and between the screen grid and anode: $\mu=\mu_{G1G2}\mu_{G2A}$ The alternative symbol for screen amplification factor μ_{G1G2} is μ_S.

For example, for EF86 the figures are $\mu_{G1G2}=38$ and $\mu_{G2A}=122$, so $\mu = 38*122 = 4,625$

For power pentodes, the figures are lower. EL34 has $\mu_{G1G2}=15$ and $\mu_{G2A}=11$, so $\mu = 15*11 = 165$.

When a pentode is strapped as a triode, the amplification factor is approximately equal to that of the screen, so $\mu_{TRIODE}\approx\mu_S$! Screen transconductance gm_S and anode transconductance gm are proportional to the ratio of screen and anode currents: $gm_S=gm_{(pentode)}*I_S/I_A$.

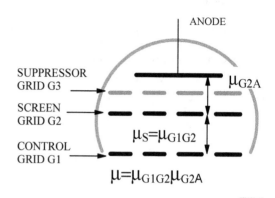

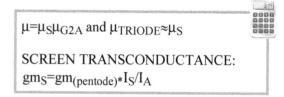

$\mu=\mu_S\mu_{G2A}$ and $\mu_{TRIODE}\approx\mu_S$

SCREEN TRANSCONDUCTANCE:
$gm_S=gm_{(pentode)}*I_S/I_A$

Beam power tubes

The advances in tube design did not stop in 1930 with pentodes. Just as adding a third grid to a tetrode has done in pentodes, the addition of electron beam-forming plates also eliminated the dreadful kink. It also brought a few additional benefits that made beam power tetrodes superior to power pentodes.

In beam power tubes or "kinkless tetrodes," the beam-forming plates restrict the operation of the tube to the regions where the space-charge is effective. The cloud of primary electrons acts as a virtual suppressor and prevents the flow of secondary electrons from the anode to the screen. This phenomenon requires the space charge cloud to be at a much lower potential than the anode, requiring a large spacing between the screen and anode.

To limit the excessive increase in screen current, tube designers placed the screen grid windings in the shadow of the control grid. This deflects electrons away from the screen and allows more of them to pass through the screen.

The region of negative internal resistance and the "kink" in the anode curves is thus eliminated, prompting the whole series of European output tubes to be named kinkless tetrodes: KT66, KT77, KT88, KT90, KT100, KT120, KT150. Of course, Americans were not to be outdone with their masterpieces such as 6L6 (plus all its variants, including the industrial version 5881), 6550, 7027, 6AR6, and others.

The rounding of the pentodes' characteristics limits the lowest anode voltage achievable under dynamic conditions and causes distortion. This rounding results from the inefficiency of suppressor grids at lower anode voltages, when some secondary electrons evade the pickup by the suppressor grid.

Beam power tubes have steeper (faster rising) curves in the low anode voltage region. V_{MINA} is closer to zero than V_{MINB} for pentodes, which results in a wider anode voltage swing for beam power tubes, higher power outputs, and improved efficiency. Their curves are also closer to horizontal in the higher V_A range, leading to KT tubes' better linearity and lower distortion.

For the sake of simplicity and linguistic brevity in the further text, we will use the term pentode for both true pentodes and for beam power tubes (which, strictly speaking, are tetrodes). This is justified because, for practical purposes, their in-circuit behavior is all but identical.

Let's dissect one 6L6 beam power tube. If you have a faulty tube of any kind, open it up and study its construction, you will appreciate its inner beauty!

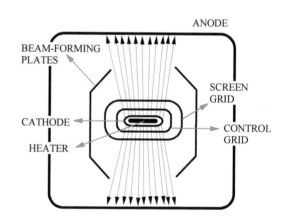

ABOVE: Horizontal cross-section of a beam power tube. The electron flow is indicated by arrows.

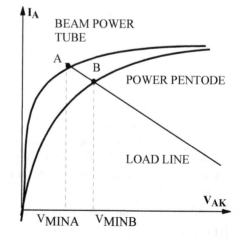

ABOVE: Beam power tubes versus pentodes

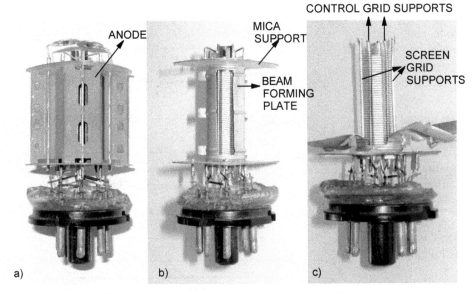

ANODE

MICA SUPPORT

CONTROL GRID SUPPORTS

BEAM FORMING PLATE

SCREEN GRID SUPPORTS

a) b) c)

LEFT: Striptease, 6L6 style

a) The internals of Russian-made 5881 beam power tube with the glass removed

b) Beam-forming plates became visible (the cylinder with the window in the middle) once the anode was removed

c) Two grids (screen and control grid), with the white-colored cathode (from the oxide coating) visible at the center.

Providing DC voltage for screen grid biasing

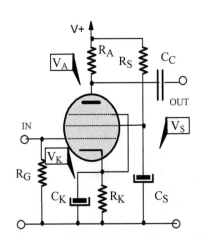

The standard way of providing screen voltage (R_S and C_S)

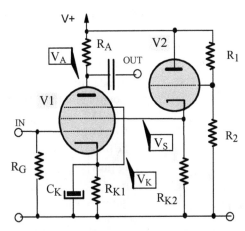

Triode cathode follower providing a stable screen DC voltage, so R_S and C_S are not needed.

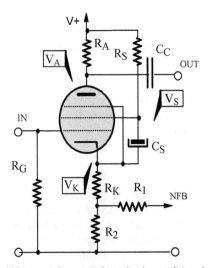

When un-bypassed cathode resistor is used, the screen capacitor C_S should be returned to the cathode!

Since the maximum allowed DC screen voltage in small-signal pentodes is much lower than the allowable anode DC voltage, it is usually supplied from the anode supply through a voltage dropping screen resistor R_S. This is akin to biasing a triode by a cathode resistor through the voltage drop the cathode current makes on that resistor.

However, that means the screen DC voltage will vary as the screen current changes, which is clearly undesirable due to its high effect on the pentode's operating conditions. For best results, the screen should be supplied from a fully regulated DC voltage source, or at least stabilized to some extent by the use of a cathode follower, a Zener diode, or a voltage regulator tube.

The screen capacitor C_S needs to bypass any signal voltage from the grid to the ground, so there is no AC signal across it; otherwise, local negative feedback would develop and significantly reduce the stage's voltage gain!

In output stages, audio power pentodes were specifically designed to have the same maximum allowed voltage on the anode and the screen, so the two are connected together to a high voltage source V_{BB}. Many tubes not designed for audio use have much lower V_{G2MAX}. They require a dedicated lower screen supply voltage source, which is better filtered and regulated than the anode supply.

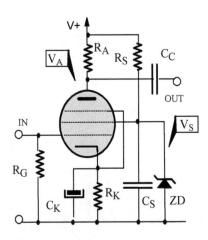

ABOVE: A Zener diode ZD or VR tube can keep the screen voltage V_S steady, but must be bypassed to ground by a film capacitor C_S.

The real pentode example: 6SJ7

The anode characteristics for pentodes include I_A-V_A curves and I_{G2}-V_A curves, the screen grid current curves. Two operating points are given in the table (from data sheets) for two different anode voltages (100V and 250V).

Notice how the anode voltage does not impact the operation of a pentode, providing the screen voltage is the same (as it is here), the anode current change (2.9 to 3.0mA), and the screen current change (0.9 to 0.8mA) are minimal and can be neglected.

The anode and screen currents ratio is relatively constant (providing the operating point is above the "knee," that the anode voltage is higher than 50V for 6SJ7).

In this case, 2.9/0.9 =3.22 or 3/0.8 = 3.75, so for anode voltages between 100 and 250V, we can use a ratio of 3.5 to estimate screen currents. However, notice a discrepancy between the table and the graphs, redrawn from the same GE datasheet!

The graphs show that the screen current at both operating points is around 1.6mA, not 0.8-0.9mA. In our analysis, we will assume that the graphs are correct.

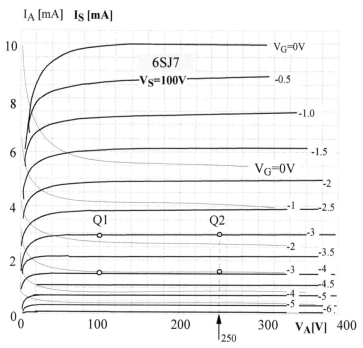

ABOVE: Operating parameters for 6SJ7 pentode at two anode voltages.

6SJ7	Q1	Q2
Anode volts [V]	100	250
Control grid bias [V]	-3	-3
V_{G2} [V]	100	100
V_{G3} [V]	0	0
I_A [mA]	2.9	3
I_S [mA]	0.9	0.8
r_I [kΩ]	700	1,000
gm [mA/V]	1.58	1.65

TUBE PROFILE: 6SJ7

- Indirectly-heated sharp cutoff pentode
- Noval socket, 6.3V/825 mA heater
- Maximum heater-cathode voltage: 90V_{DC}
- Anode/Screen dissipation: 2.5W/0.7W
- Max. anode/screen voltage: 300V/125V
- R_{GMAX}=1MΩ

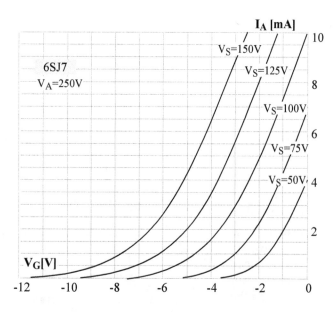

Anode transfer characteristics of a pentode (6SJ7)

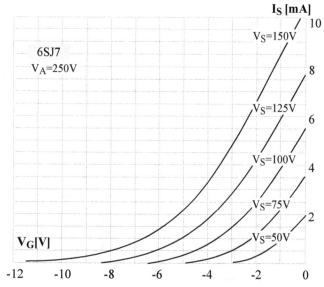

Screen grid transfer characteristics of a pentode (6SJ7)

DESIGNING VOLTAGE AMPLIFICATION (PREAMPLIFIER) PENTODE STAGES

Constructing approximate anode characteristics for any screen voltage

Pentode datasheets only publish anode characteristics for a few screen voltages, often only one, better ones up to three or four. However, you can construct approximate anode curves for any screen voltage, providing you have a static transfer curve for that voltage. There are five transfer curves published for 6SJ7 pentode for screen voltages of 50, 75, 100, 125, and 150V. Even if there is no published curve for your desired screen voltage, say 60V, you can first estimate the transfer curve and then construct anode curves from it. The method is illustrated below.

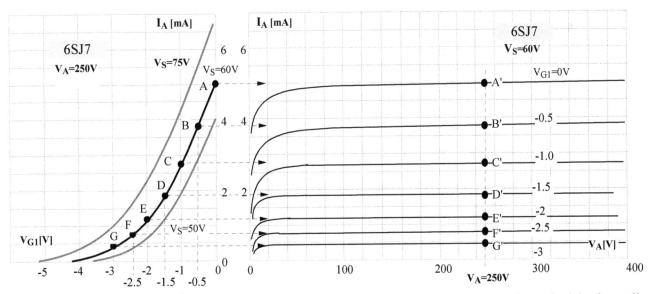

Say we need anode curves for 60V screen voltage. Starting with V_G=0V estimate point A and transfer it horizontally onto the I_A-V_A chart. Then, draw a vertical line through V_G=-0.5V from point B. Transfer it horizontally onto the I_A-V_A chart and draw the second curve. Repeat the process for control grid voltages V_G=-1V, -1.5V, -2V, etc. The points A', B,' C,' etc., must lie on the anode voltage specified on the transfer graph, in this case, V_A=250V!

Choosing the load resistance

Let's look at three possible choices for an anode resistor. With R_L=100kΩ, Q needs to sit in the middle of the grid swing at V_{G1}=-1.5V. With the grid voltage swing ΔV_G=1V, the anode voltage swing is ΔV_A= 40-180= -140V, so the amplification of the stage is A=ΔV_A/ΔV_G= -140/1= - 140. The negative sign indicates that the stage inverts signal's phase.

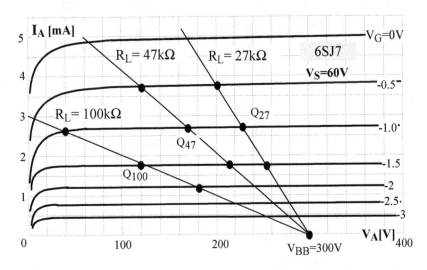

Reducing R_L to 47kΩ, Q sits at -1V. With the grid voltage swing ΔV_G=1V, the anode voltage swing is ΔV_A= 120-215= -95V.

The amplification of the stage is lowered to A= -95! With an even smaller R_L of 27kΩ, Q sits again at -1, and with the same grid voltage swing ΔV_G=1V, the anode voltage swing is ΔV_A= 200-255= -55V, so the amplification of the stage is A= -55!

The gain is highest with 100kΩ load, but so will also be the stage's output impedance. The compressed curves near the top will result in clipping and distortion. This compression will be less pronounced with a lower load of 47kΩ. The gain has dropped from -140 to -95, but the distortion and the output impedance will be lower.

The load line of the 27kΩ load is entirely above the knee in the curve, so the characteristics are not bunched at the top end anymore. Their spacing is similar to triode's, and the sound will move towards the "triode tone." The third and other odd harmonics are reduced, while the 2nd and even harmonics dominate. There is no rapid increase in screen current anymore on positive signal peaks.

The stage gain of 55 may seem low for a pentode, but that is still higher than we usually get from a 12AX7 common cathode stage, so still higher than with any triode!

The load line and the quiescent point

Designing a pentode stage is not much different from that of the triode amplifier. Let's use datasheet curves for the screen voltage of 100V and the anode resistance value of 47kΩ. Assuming you can choose V_{BB} freely, draw the load line, so that point A does not sit in the curved part of the characteristics.

To get a decent input swing (from -2V to -4V), we need to have V_{BB} of around 300V. Thus, Q needs to sit in the middle of the grid swing or at $V_{G1}=$ -3V. This has determined the quiescent anode voltage as $V_0=160V$.

The maximum grid voltage swing (AB) is $\Delta V_G=-2-(-4) = 2V$, the anode voltage swing is $\Delta V_A= 70-230= -160V$, so the amplification of the stage is $A = \Delta V_A/\Delta V_G = -160/2 = -80$.

Even higher amplification factors are possible with larger anode loads and higher V_{BB}.

To determine the values of screen and cathode resistors we need to know the screen current in the quiescent point Q.

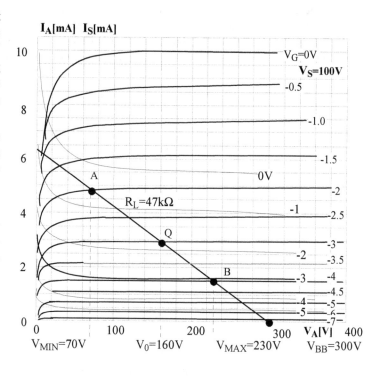

Designing the screen circuit

The graphs show that screen current at the quiescent point Q is around 1.6mA. If an RC circuit is used for the screen voltage, we know that $V_{BB}=300V$, that $I_S=1.6mA$ and that V_S needs to be 100V. Thus the value of the resistor R_S must be $R_S=(V_{BB}-V_S)/I_S = (300-100)/1.6 = 200/1.6 = 125kΩ$.

The first rule-of-thumb for sizing C_S (the screen bypass capacitor) is that its reactance at the lowest desired operating frequency (-3dB) should be ten times lower than R_S. Let's say that is $f_L=5Hz$, then $\omega_L=2\pi f_L=10\pi=31.42$ So $X_C=1/\omega_L C_S= R_S/10 = 125,000/10 = 12,500$, or $C_S=1/X_C\omega_L=1/(12,500*31.42) = 2.55\mu F$.

Another rule-of-thumb some designers use is the time constant of the screen circuit $\tau=R_S C_S$. From here (again, for $f_L=5Hz$) we have $C_S=1/(2\pi R_S f_L) = 0.26\mu F$. This is a ten times lower value than obtained by the first method and is the absolute minimum we should use.

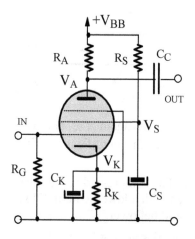

ABOVE: RC screen grid bias of a pentode stage

If a voltage divider is used for the screen voltage, we know that $V_{BB}=300V$, $I_S=1.6mA$, and that V_S must be 100V. We can choose the value of R_2 so the total current drawn by the screen biasing circuit does not exceed a certain value.

This will depend on how much spare current capacity we have in our power supply for the preamplifier circuit. Let's say we want a total of $I_T=10mA$. Thus, 10mA will flow through R_1 and $I_T-I_S= 10-1.6 = 8.4mA$ through R_2. Thus $R_1 = (V_{BB}-V_S)/I_T = (300-100)/10 = 200/10 = 20kΩ$ and $R_2 = V_S/(I_T-I_S) = 100/8.4 = 11.9kΩ$.

These are not exact formulas, and there is a degree of error. The total resistance is $R_1+R_2 =31.9kΩ$, so the total current will be $300V/31.9kΩ = 9.4mA$, while we assumed 10mA. This is close enough. Since 20kΩ and 11.9kΩ aren't standard resistor values anyway, we will choose 22k and 12k, so that the total current will be $300/34k = 8.8mA$.

Now 22k*8.8mA=193V, so our screen voltage may end up being 300-193=107V instead of 100V, which is acceptable. Alternatively, increase R_1 to 25k and you will get down to around 100V on the screen.

Cathode circuit

The cathode current is the sum of the screen and anode currents $I_K=I_S+I_A=$ 1.6+3 = 4.6mA. The control grid must be at -3V with respect to the cathode, so the cathode must be at +3V! Since $V_K=I_K*R_K$, the required cathode resistor is $R_K=V_K/I_K= 3/4.6 = 0.652k\Omega$. The standard value is 680Ω.

The cathode bypass capacitor is sized in the same way as the screen capacitor. For f_L=5Hz we have $C_K=1/(2\pi R_K f_L)$ = 46.8μF Use standard value of 47μF, or, for lower f_L, use 100 or 220μF.

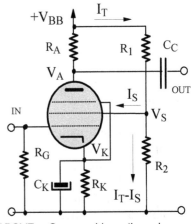

ABOVE: Screen bias through a resistive voltage divider

The effect of screen impedance on pentode's amplification

An AC signal on pentode's control grid causes changes in both anode and screen currents. These current changes will result in a voltage drop across any impedance in the screen circuit and thus also between screen and cathode. Assuming that the usual way of biasing the screen is used, a voltage dropping resistor R_S and a capacitor C_S, the impedance in the screen circuit is the internal dynamic screen resistance r_S in series with the parallel combination of the screen resistor and capacitor, as depicted in the equivalent circuit.

This voltage drop across the screen impedance reduces the gain of the pentode stage, acting as a kind of negative feedback. In an ideal case with Z_S=0 (the reactance X_C of the capacitor is so low that it practically represents a short across R_S) only the internal resistance r_S affects gain, so the ratio of this maximum amplification A_{MAX} and the actual amplification factor A is $A/A_{MAX}=r_S/(r_S+R_S)$.

Typically R_S is much larger than r_S, so their parallel combination is only slightly lower than r_S, so for complete bypassing, the capacitor C_S must have a reactance at least ten times lower than r_S!

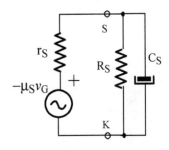

ABOVE: The equivalent circuit of the screen grid

Side-by-side: Triodes versus pentodes

To get a better feel for the difference between triodes, imperfect voltage sources, and pentodes, imperfect current sources, let's compare the 12AU7 triode and EF86 pentode! Typically, triodes have a higher anode power rating and work with higher anode currents (10mA vs. 3mA here).

The transconductance gm values are similar, but since the pentode's internal resistance is 10-20 times higher, its amplification factor μ is also 10-20 times higher, making pentodes much better voltage amplifiers than triodes.

However, pentodes require minimal control grid bias, meaning they cannot take as large input signals as triodes can.

	12AU7	EF86
P_{MAX} [W]	2.75	1
V_A[V]	250	250
Bias [V]	-8.5	-2.0
V_{G2} [V]	N/A	140
V_{G3} [V]	N/A	0
I_A [mA]	10.5	3
r_I [kΩ]	7.7	100
μ	17	190
gm [mA/V]	2.2	2.0

ABOVE: Side-by-side comparison of typical operating conditions of preamp triode (12AU7) and a pentode (EF86)

Shielding pentode tubes

As any police detective would tell you, clues are all around us; you just need to look in the right places. The extremely high m (amplification factor) makes pentodes susceptible to hum and noise. Anything present on the control grid will get amplified hundreds of times, including low frequency induced voltage from tube heaters (50 or 60 Hz), high-frequency radio (RF) signals, and other interference.

Notice how the EF86 electrodes are wired. G1 (pin 9) is as far as possible from heater pins (4 and 5) and the anode (pin 6). There is also the internal shield taken out to pin seven between them. Anode and cathode pins are also away from one another, and another shield pin connection (pin 2) is between the cathode and high voltage-connected G2 (screen grid) on pin 1.

Some amplifier manufacturers even install metal shields at tube sockets to minimize hum, usually to shield heater pins and wires from the rest of the pentode's pins.

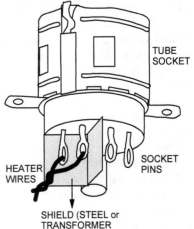

TUBE SOCKET

HEATER WIRES

SOCKET PINS

SHIELD (STEEL or TRANSFORMER LAMINATION MATERIAL)

The most common preamp pentode in guitar amps: EF86

Developed in the early 1950s by Mullard in the UK, EF86 was described as a low noise pentode for use in input stages of high gain amplifiers, microphone preamplifiers, and magnetic tape recorders. Although Mullard touted EF86 as low noise and anti-microphonic, they still recommended mounting it on a vibration-resistant base, such as a rubber mount. Their marketing hype aside, the tube is noisy and microphonic, especially in combo amps, whose whole cabinet and chassis vibrate like crazy.

The UF86 version is electrically identical except for the 12.6 Volt 100 mA heater.

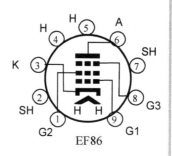

TUBE PROFILE: EF86 (6F22)

- Indirectly-heated pentode
- Equivalents: 6F22, Z729, 6267, 6BK8, CV2901, 6CF8, CV8068, CV10098, 6J8 (Chinese copy)
- Noval socket, Heater: 6.3V, 200mA
- V_{AMAX}=300V, P_{AMAX}=1W
- V_{SMAX}=200V, P_{SMAX}=0.2W
- I_{KMAX}=6mA
- Typical operation: V_A=250V, V_S=150V, V_{G1}= -2V, I_A=3mA, I_S=0.6mA
- gm=1.85 mA/V, μ=4,070, r_I=2M2

L-R: Chinese 6J8 (from military stocks, notice the star of the People's Liberation Army), Russian (no brand marked), Philips Miniwatt (Holland) , Valvo (German Philips), Mullard (UK) and Telefunken (Germany).

Do they sound different in the same amp? You bet!

PENTODE INPUT STAGE: BIG WHITE MONKEY FG15 (INSPIRED BY MATCHLESS)

The boys marketing BWM amps, based in Queensland, Australia (one cannot say "making BWM amps" since they were made in China by others) describe this head as "... our take on the classic Matchless remake of the Vox, it's basically the EF86 side of a DC30." Apart from only two EL84s in the output stage (instead of four), the circuit is practically identical to DC30's EF86 channel.

Although that monkey logo isn't everybody's taste, the top and rear chrome plated grilles looked very classy. As if one wasn't enough, the same symbol is also printed on the control panel, between the pilot light and the gain knob.

1) Gain (pull switch)

2) Cut

3) Tone

4) Volume (pull switch)

5) Bright switch (Br1-PASS-Br2)

6) Two speaker output jacks

7) Speaker impedance selector (4-8-16Ω)

8) Half power switch

9) No serial number (prototype unit?) and 230V version sent instead of 240V unit!

The steel chassis is attached by four bolts to the bottom of the cabinet and slides out once the rear cabinet cover is removed.

The topology is correct, the large 5AR4 rectifier next to the power transformer and a pair of output tubes, all three held by tube retainers, despite being in an upright position.

Internal wiring

The ground bus (1), a piece of solid copper wire, runs parallel with the terminal board. Although BWM says that "there is no PCB in sight," this board is actually a PCB from which almost the whole copper layer has been stripped, except two wide copper tracks. The ground bus is connected to the chassis in the same point (2) as the CT of the heater secondary, the speaker output, and the power transformer's electrostatic shield.

The standby switch is connected to the + of the first elco (3), in fact, two 33m elcos in parallel, together with the CT of the output transformer (anode supply +V1) and one end of the filtering choke.

The tone control switch has six positions, with six caps and five resistors wired in a messy fashion (4).

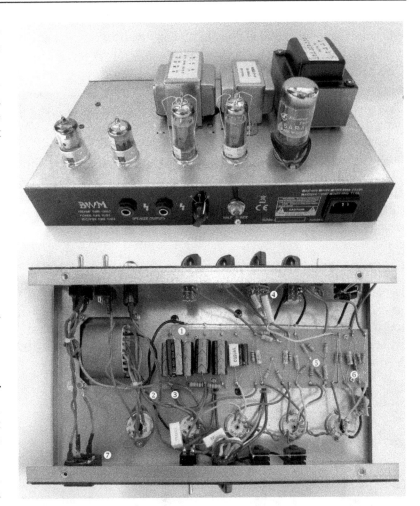

Apart from that, the circuit is fairly standard and easy to identify visually. You can immediately recognize 3+1 resistors in the long-tail inverter pair (5) and cathode elements of the 1st stage (6).

The cutout for the mains IEC inlet (7) is too large, so someone used silicon to glue the IEC socket to the chassis (without success); the socket can be pulled out by hand.

The input stage

The topology of the whole amp is basically identical to Matchless DC30. However, the values of some components are different: 22k grid stopper (DC30 uses 68k), 4k7 cathode resistor and bypass cap (2k2 & 2m5 on DC30), and the values of six coupling resistors in the switch-selectable *Tone* circuit.

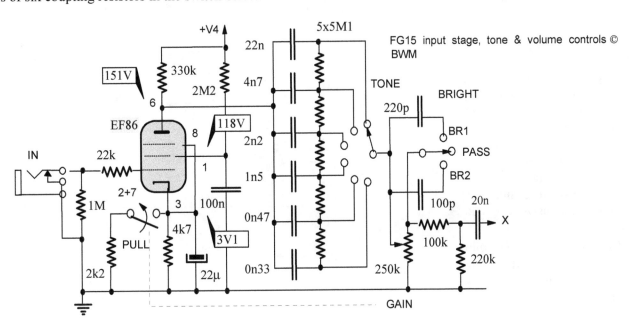

When the *Gain* pot is pulled, the 2k2 and 4k7 resistors are in parallel (1k5 total), increasing the stage's gain.

While DC30 has a fixed bright bypass capacitor, FG15 has a 3-position switch that selects from no bypass to two different bypass caps, 100p and 220p, for two different levels of brightness, a good idea.

The 6-position Tone switch selects one of the six coupling capacitors, whose capacitance, together with the 250k *Gain* potentiometer's resistance, determines the turnover or lower -3dB frequency of the high pass CR filter thus formed. The highest capacitance, 22nF, results in the lowest limit (more bass frequencies), while the lowest capacitance, 330pF) results in the brightest sound (most bass cut).

However, calculating these frequencies isn't simple; notice how the signal at capacitors' output isn't brought to one fixed end of the 250k pot but its wiper instead. This makes the resistance between the capacitor's end and GND variable; as the wiper moves towards lower volume (towards GND), the resistance is progressively lower, and the -3dB frequency gets higher and higher. As a result, the bass frequencies are much more attenuated at low Gain settings than at higher *Gain* levels!

The phase inverter and output stage

The cathode coupled phase inverter uses the same R&C values as DC30's, but the output coupling caps are 22n instead of 47n on DC30. That increases the lower -3dB frequency of FG15 from DC30's 15Hz to 33Hz. The Half power feature is the standard pentode-triode switching arrangement. The Gain control is normally inactive; pulling the knob closes the switch and connects the 1M potentiometer between two output grids. In its minimum position (CCW), the wiper shorts the two signals out for zero output.

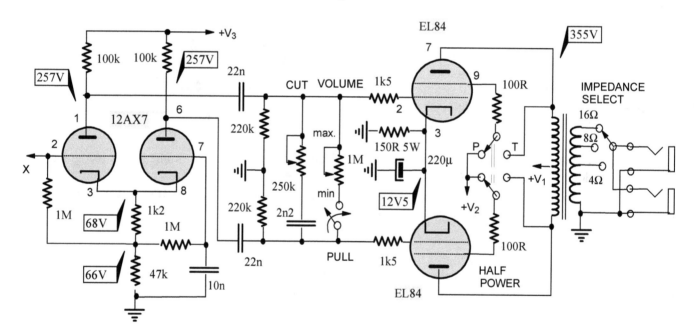

The power supply

GZ34 rectifier is too powerful for a 15 Watt amp and can be replaced by 5V4 for a slightly larger power sag or a 5Y3 rectifier for a much larger voltage drop. Such substitutions are the easiest ways to change this amp's responsiveness and even voicing.

A decent power supply, with Standby feature and filtering choke, nothing to modify here.

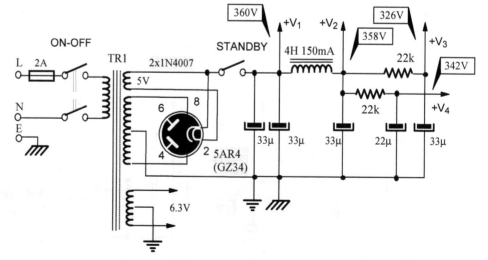

FG15 power supply, phase inverter and output section, © BWM

INPUT CIRCUITS AND STAGES

7

- 1/4" PHONE JACKS (PLUGS AND SOCKETS) - A CRASH COURSE
- INPUT CIRCUITS, FILTERS AND ATTENUATORS
- COOL AND HOT INPUTS: FENDER RAMPARTE (PAWNSHOP SERIES)

Input circuits are much more important than they'd seem at first. Apart from enabling a guitar lead (cable) to be plugged into a guitar amp, they perform other functions.

For instance, they act as resistive voltage dividers to match the amplifier's sensitivity to wide-ranging voltage signals produced by various types of guitar pickups, whose voltage can vary from 20 or so mV to over a Volt!

Many also incorporate series and shunt capacitors, thus acting as input filters, shaping amplifier's frequency response to achieve various designer's aims, such as a "Bright" input, a "Bass" input, and many others.

1/4" PHONE JACKS (PLUGS AND SOCKETS) - A CRASH COURSE

The official or proper name for guitar connectors or "jacks" is 1/4" phone jacks. "Phone" comes from the fact that they were designed in the 19th century (yes, they are an ancient technology, much older than tubes themselves!) for use in manually operated telephone exchanges and switchboards. Thus, they had to be reasonably large (to fit snuggly into a palm of the operator's hand) and robust to withstand thousands of plugging & unplugging operations.

1/4" indicates their diameter, around 6.3mm. The symbols for various 1/4" sockets are based on the basic symbol (above). The rectangle represents the "female" sleeve contact, basically a tube that fits snuggly around the "male" sleeve. Likewise, the arrow-like male tip clicks into the appropriately shaped and flexible tip contact.

Of course, hookup cables inside an amplifier are not soldered directly onto the sleeve or the tip contact, but to the lugs located on the other ends of their metal parts, shaped as "lugs", meaning either as "eyelets" as in the photo, or as thin tips in the other type of sockets, which are smaller, fully enclosed and which usually feature auxiliary switching contacts (pictures on the next page).

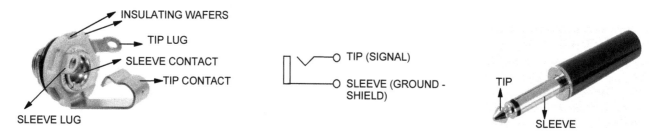

ABOVE: The simplest 1/4' socket (left) and plug (right), also called "mono" or TS jack. T denotes a TIP which is "hot" or + of the signal, S denotes a SLEEVE, which is the negative of the signal or GND (ground). A simple single-core shielded cable is used, with the inner core connected to the tip and the shield connected to the sleeve.

Open type sockets

There are many other open 1/4" sockets, such as the mono switched socket, unswitched stereo socket, stereo switched socket, etc. All switched sockets operate on the same principles, so let's look at the simplest one, illustrated above. When the male plug isn't plugged in, the tip of the shunt contact is touching the tip contact, which can be called an NC or "Normally Closed" contact. This is often used at the input of guitar amps, with a tip shunt lug wired to GND. That way, when a guitar cable isn't plugged in, the tip is grounded, meaning the grid of the input tube is also grounded, so it cannot pick up any stray signals or interference, keeping the amp quiet.

When a guitar cable is plugged in, the tip pushes the tip contact downwards and breaks the mechanical connection between the tip shunt contact and tip lever.

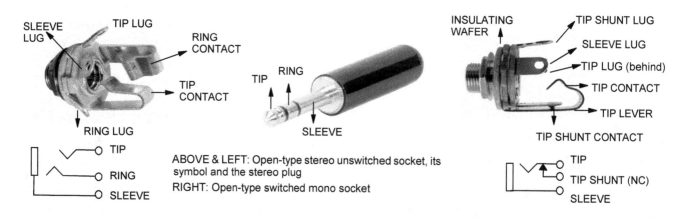

ABOVE & LEFT: Open-type stereo unswitched socket, its symbol and the stereo plug
RIGHT: Open-type switched mono socket

Closed type sockets

The stereo socket illustrated (next page) has two C/O (Change Over) contacts. Apart from lug 9 (sleeve or GND), all lugs are at the back. Lugs 4 and 8 are the tip and the ring, two "hot" or signal + lugs. The three lugs at the top (1, 2 & 3) and the three at the bottom (5, 6 & 7) are the two auxiliary switches.Contacts 6 & 7 are normally closed and are broken apart (opened) by the insertion of the stereo plug, which then pushes contacts 5 and 6 together, so they close. Ditto for contacts 2&1 and 2&3 on the other switch.

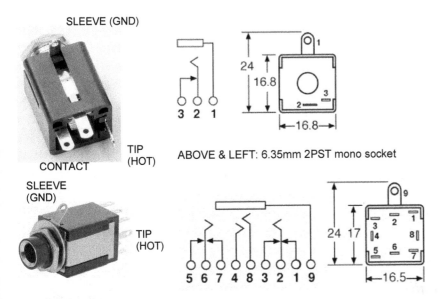

ABOVE & LEFT: 6.35mm 2PST mono socket

Remember that none of the six lugs or contacts ever come into contact with the actual tip or ring; they are only mechanically actuated by them.

The simplest closed type mono-socket has only two lugs, the tip and the sleeve (GND). The next complexity level is the mono-socket with one NC contact, as illustrated on the left. Lug 1 is the sleeve; lug 2 is the tip, and lug 3 is the normally closed aux. contact.

6.35mm 3PST stereo socket (far left), its wiring diagram (middle) and rear view with dimensions and lug numbers (left)

Both open and closed type sockets can be insulating or non-insulating. That refers to the relationship between the mounting chassis (almost always metal) and the socket's sleeve. In non-insulating sockets, the metal sleeve ring touches the chassis, so there is a galvanic connection between the two. This may be undesirable if the amp builder does not want to use the metal chassis as the ground bus and uses a star grounding system instead. An insulating socket would then ensure that there is no double grounding, a situation that can cause a ground loop and result in an annoying hum at the amp's output.

Finally, there is a third socket type, illustrated below, easily recognizable. Each active element has its own normally closed (NC) contact or shunt, which opens when the plug is inserted. A mono-socket has two switchable contacts; one is normally closed to the tip, the other to the sleeve, designated by DPST (double pole single throw), DP meaning the two poles or contacts are separate (no galvanic or mechanical connection between them). A stereo socket has three pairs of lugs or legs and three "ST" or "single throw" contacts, thus 3PST in its catalog name.

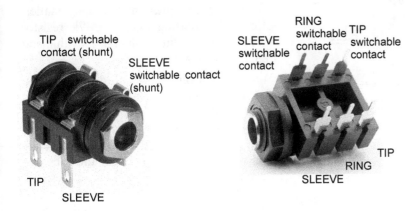

FAR LEFT: 6.35mm DPST non-insulated mono socket
LEFT: 3PST non-insulated stereo socket
BELOW: 3PST stereo socket symbol and contacts

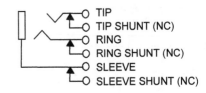

INPUT CIRCUITS, FILTERS AND ATTENUATORS

Input circuits are much more important than they seem at first. Apart from enabling a guitar lead (cable) to be plugged into a guitar amp, they perform other functions. For instance, they act as resistive voltage dividers to match the amplifier's sensitivity to wide-ranging voltage signals produced by various types of guitar pickups, whose voltage can vary from 20 or so mV to over a Volt!

Many also incorporate series and shunt capacitors, thus acting as input filters, shaping amplifier's frequency response to achieve various designer's aims, such as a "bright: input, a "Bass" input, etc. There are many variations on this theme, but two examples will show you how to analyze them and estimate their behavior quickly.

CASE STUDY: The input network of Fender Twin Reverb

This legendary vintage amp had two pairs of inputs, one marked "Normal" and the other "Vibrato." Both are identical, and both have two input jacks each, labeled "In 1" and "In 2". When there is no guitar plugged in, the normally closed (NC) contact on the "In 2" jack connects the two 68k resistors in parallel, while the (NC) contact on "In 1" jack short-circuits that point (X) to GND.

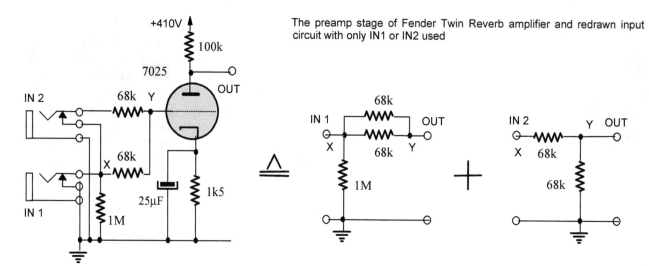

The preamp stage of Fender Twin Reverb amplifier and redrawn input circuit with only IN1 or IN2 used

CASE STUDY: The input network of Univox 1236

This input circuit is much more complex, so its configuration and behavior depend on which input a bass or lead guitar is plugged in. So let's redraw the inputs one by one, assuming that only one guitar is plugged in at a time!

BASS: At very low frequencies, the reactance of the bypass capacitor to GND is high, so it can be considered an open circuit. The attenuation is around 0.55, meaning the output amplitude is only 45% of the input signal's amplitude.

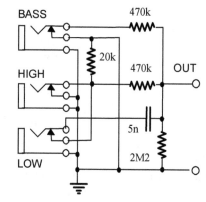

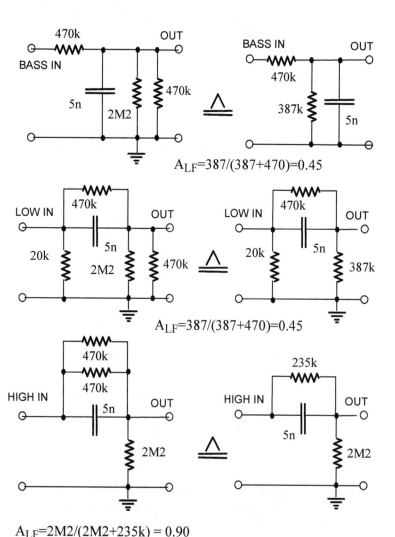

$$A_{LF}=387/(387+470)=0.45$$

$$A_{LF}=387/(387+470)=0.45$$

$$A_{LF}=2M2/(2M2+235k) = 0.90$$

As the frequency increases, the reactance of the 5n cap becomes smaller and smaller, effectively shorting out the 387k resistor more and more, thus reducing its value and reducing gain at HF!

With a guitar plugged into the "LOW" jack, the 5nF cap is now in parallel with the series 470k resistor. At low frequencies, the signal voltage is divided the same way as with BASS input. Still, as the frequency increases, the reactance of the 5n cap becomes smaller and smaller, effectively shorting out the series resistor. The attenuation is reduced as more and more of the input voltage reaches the output, thus increasing gain at HF!

With a guitar plugged into the "HIGH" jack, the 5nF cap is now in parallel with the series 470k resistor. However, the other 470k resistor is also switched in parallel, for an effective halving of the double 470k resistance to 235kΩ.

The voltage divider works the same way as for the "LOW" input; the only difference is that the attenuation is only around 10% now (90% of the signal passes through at LF), compared to only 45% at the "LOW" input. Thus, the HIGH and LOW designations refer to the signal amplitude and, ultimately, loudness.

COOL AND HOT INPUTS: FENDER RAMPARTE (PAWNSHOP SERIES)

Outside looks and features

Priced at AU$639, Ramparte looked and sounded good, but the competition was stiff; for that money, one could almost have Vox AC15. We got one in pristine condition from a USA eBay seller for US$250 plus US$75 shipping. Upholstered in a two-tone brown & gold fabric, Ramparte looks like a classy cinema speaker from a 1930s movie theater. Alas, beauty and practicality usually don't go together (a rare combination, just like a beautiful and classy-looking woman who can cook a feast).

Corners aren't protected, so if you want to preserve this beauty for next generations, forget about lugging it around.

The top-facing control panel is the most Spartan of all modern combo amps. Two inputs, hot and cool, but a guitar has to be unplugged from one and plugged into the other, the most idiotic arrangement ever. A single input and a simple (foot)switch would have been infinitely better.

Also, notice Fender's insistence on the minimalist approach, no tone control, no presence, no gain boost, no reverb, no tremolo, nothing.

One possible modification is to remove the HOT input jack (1) and install a tone control pot in its place, then convert the HOT volume pot (2) into the master volume control.

Alternatively, reposition the large pilot light (3) to the other side of the on-off switch and install the tone control pot in its place.

Small toggle switches can be installed at (4), (5), or (6), giving you boost or bright controls. Pentode/triode (or "half power) switch, standby switch, and NFB on-off switch can go somewhere close to the on-off switch. Alternatively, one or more of those can be mounted at the back panel (7) so as not to make the smallish top panel too crowded.

SPECS

FENDER RAMPARTE
- Two channels, "Cool" and "Hot"
- Individual inputs and volume controls for each channel
- 2x 12AX7 preamp tubes and a 6L6 power tube
- Solid state rectification
- PCB construction
- Output power: 9 Watts
- 12" (30cm) Speaker
- External 8Ω speaker output
- Dimensions: 9.8 x 15.8 x 17.3 inches
- Weight: 25.5 pounds (16.8 kg)

ABOVE: The retro-looking Ramparte is one of the most beautiful combos (if you like that sort of soft "haberdashery" look).

BELOW: The minimalist control panel is Ramparte's downfall. No tone control, no gain boost, no master volume, nothing!

The insides

Ramparte's Taiwanese-made output transformer has a relatively high quality factor, indicating a wide frequency range. However, it is only rated at around 14 Watts, a borderline case for an amp that can produce 9 Watts of audio power. It will saturate and distort heavily at high power levels, especially in the low frequency (bass) region.

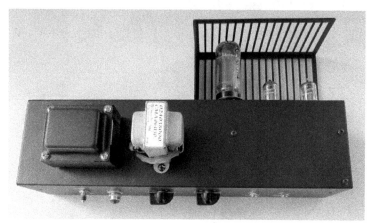

OUTPUT TRANSFORMER:

- EI57 laminations
- a=19 mm, stack thickness S=20mm
- Center leg cross section $A=aS = 3.8cm^2$
- Power rating: $P=A^2= 3.8^2= 14W$
- Primary DCR: $R_P=255\Omega$
- Primary inductance @1kHz $L_P= 13.5H$
- Leakage inductance @1kHz $L_L=9.5mH$
- Quality factor: $QF = 13.5/0.095 = 1,421$

LEFT: The top view of Ramparte's chassis. A tiny output transformer and no filtering choke.
BELOW: The inside view of the chassis.

The main PCB (1) carries input jacks and two volume control pots. It is reasonably well laid out and not too crowded, so modifications should not be difficult. The smaller PCB (2) carries tube sockets and a few resistors and capacitors. Two relatively long pairs of wires carry the mains voltage to that PCB (3) only because it holds a single mains capacitor (under the four connectors). That is a bad design idea from the hum-reduction point of view. Shorten and run these wires directly from the on-off switch to the mains transformer, away from the PCBs.

The volume control pots are far from the preamp tubes and the PCB-mounted components, so a shielded cable was used. However, notice how the two high voltage wires from the output transformer's primary winding (4) are bundled together with the heater wires and high voltage supply pair (5) from the mains transformer, a big no-no!

There is room for a small HV filtering choke under the chassis (6) next to the on-off switch (7) and IEC power cable inlet (8). Alternatively, a small step-down transformer or autotransformer can fit in there.

The circuit

While the signal from Hot input (next page) passes through three 12AX7 gain stages (with a volume control after the first stage,) the "Cool" signal has its own input stage and volume control and then shares the 2nd and 3rd stage with the "Hot" channel. There is no master volume before the output stage, another major blunder by the amp's designers, even more unforgivable than the lack of any tone controls.

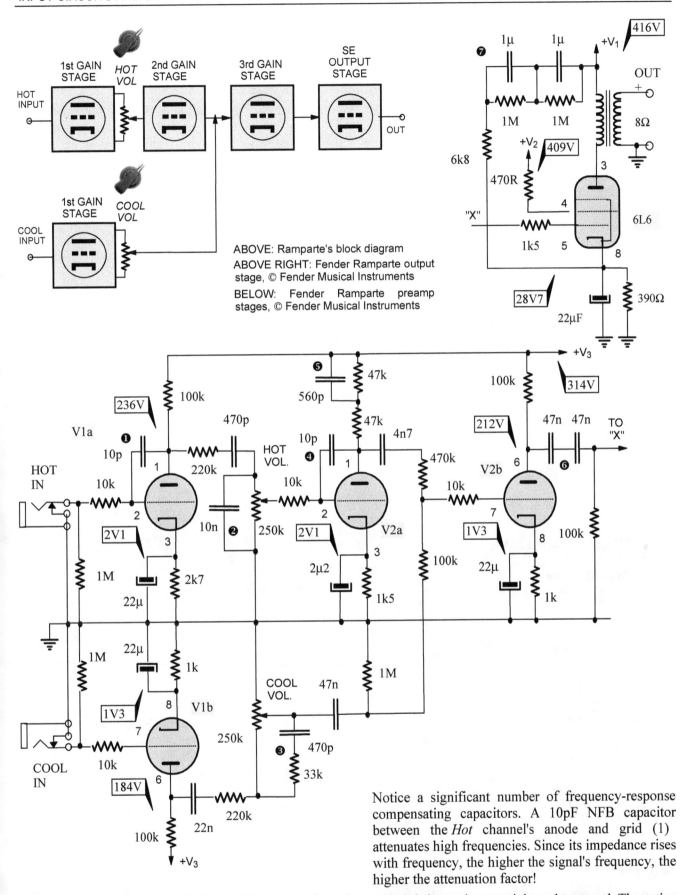

ABOVE: Ramparte's block diagram
ABOVE RIGHT: Fender Ramparte output stage, © Fender Musical Instruments
BELOW: Fender Ramparte preamp stages, © Fender Musical Instruments

Notice a significant number of frequency-response compensating capacitors. A 10pF NFB capacitor between the *Hot* channel's anode and grid (1) attenuates high frequencies. Since its impedance rises with frequency, the higher the signal's frequency, the higher the attenuation factor!

The 10n cap (2) bypasses high frequencies around the volume pot and shunts them straight to the ground. The active part of the Cool channel's volume pot (between the signal input and the wiper) is bypassed by an RC combination, 33k resistor, and 470p capacitor (3), aiming to achieve a specific voicing the amp's designer wanted.

The anode resistance of the 2nd stage was split into two 47k resistors, and only one is bypassed for HF by a 560k capacitor (5), possibly another attempt by the designer to tame the treble response. Usually, this is done to reduce the phase shift at HF and avoid potential instability and oscillations when strong negative feedback is used.

However, this amp uses no global negative feedback, so the only possible reason for this measure would be to tame the unpleasant HF shrills and peaks.

The two 47n caps in series at the output of the 3rd stage (6) have a total equivalent capacitance of 23.5nF (half); why would Fender use two caps when one would do is a question only they can answer.

A 220k-250k voltage divider follows both input's 1st stages, almost a 2:1 attenuation of their output signals. In contrast, the 2nd stage's output goes through a 470k-100k-1M signal attenuator for another significant reduction in signal level. Since there are three amplification stages in the Hot channel and too much gain, the designer had to reduce it via those voltage dividers.

Notice that the lower -3dB limit of the 470p/250k RC filter (2) is very high, 1,355 Hz, assuming the Hot Volume pot is at its maximum. The lower the volume setting, the higher that -3dB turnover frequency and the weaker the bass. In any case, this amp's bass response is seriously suppressed! A similar situation exists in RC coupling between the 2nd and 3rd stages.

The analysis of Cool and Hot input stages

While the hot anode current is 2.1/2.7=0.78mA, the cool anode current is 1.3/1=1.3mA, almost double. The external load on both stages is a 220k fixed resistor in series with a 250k volume control pot, a total of 470k. There is also a 22nF film cap in series in the cool channel and a 470pF cap in the hot channel. At 1 kHz, their respective reactances are 7k2 and 339k, so in both cases, the external load has a much higher impedance than the anode load of 100k (identical in both stages). Thus, we will simplify our analysis and work with 100k load-lines.

GRAPHIC ESTIMATION -
COOL CHANNEL

$R_L=100k\Omega$, $V_G=-1.3V$
$V_0=183V$, $I_0=1.3mA$
$V_{MIN}=100V$, $V_{MAX}=260V$
$\Delta V_A = 160V$
$A_V= -\Delta V_A/\Delta V_G = -160/2.6 = -61.5$
$D_2=(\Delta V_{OUT+}-\Delta V_{OUT-})/2\Delta V_A$
$D_2=(183-100)-(260-183)/2*160$
$=(83-77)/320 = 1.88 \%$

GRAPHIC ESTIMATION - HOT CHANNEL

$R_L=100k\Omega$, $V_G=-2.1V$
$V_0=234V$, $I_0=0.78mA$
$V_{MIN}=100V$, $V_{MAX}=315V$
$\Delta V_A = 215V$
$A_V= -\Delta V_A/\Delta V_G = -215/4.2 = -51.2$
$D_2=(\Delta V_{OUT+}-\Delta V_{OUT-})/2\Delta V_A$
$D_2=(234-100)-(315-234)/2*215$
$=(134-81)/430 = 12.3 \%$

Positioned at -1.3V grid bias, the cool quiescent point Q_C is almost in the middle of the maximum voltage swing (point A to point B), from 0V to -2.6V on the grid, and as a result, even with the maximum grid signal swing the distortion is fairly low, under 2%!

In contrast, notice how the hot operating point Q_H was positioned way down, much closer to point C than to point A, so the Q_H-A swing is much larger than the Q_H-C swing, resulting in the 2nd harmonic distortion above 12%!

Easy mods

Installing a *Hot/Cool* input selector switch isn't difficult (1). It is even possible to install a 3-position switch (*Hot-Cool-Both*), which would feed the input signal to both V1a and V1b grids, so both *Hot* and Cool signals could be mixed in various proportions.

Equally easy would be adding a *Master Volume* control (2). A *Triode-Pentode* and *NFB ON-OFF* switches could also be implemented without much fuss (not shown). Around 4 Watts of output power is achievable in the triode mode, which could thus be called a "half-power mode."

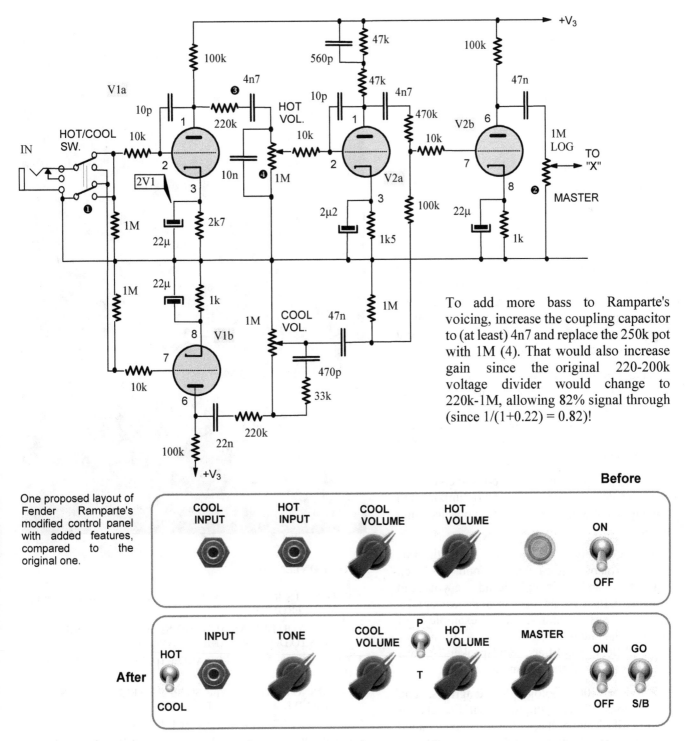

To add more bass to Ramparte's voicing, increase the coupling capacitor to (at least) 4n7 and replace the 250k pot with 1M (4). That would also increase gain since the original 220-200k voltage divider would change to 220k-1M, allowing 82% signal through (since $1/(1+0.22) = 0.82$)!

One proposed layout of Fender Ramparte's modified control panel with added features, compared to the original one.

Before

COOL INPUT　HOT INPUT　COOL VOLUME　HOT VOLUME　ON　OFF

After

HOT COOL　INPUT　TONE　COOL VOLUME　P T　HOT VOLUME　MASTER　ON OFF　GO S/B

The hum issue

Despite a few easy fixes to Ramparte's wiring and wire dressing, the annoying hum persisted. Suspecting that it could cause hum and buzz (perhaps there was no internal electrostatic shield between its primary and secondary windings?), we even replaced its 115V mains power transformer, all to no avail.

We could not see the purpose of the intricate feedback network in the output stage, marked (7) on the diagram two pages back. Two paralleled 1μF/1MΩ RC networks, connected in series with a 6k8 resistor, all between the HV supply point +V1 and the 6L6's cathode. Perhaps it was an attempt by the designer to feed some of the ripple (hum) on the power supply line (+V1) back to the cathode, hoping that such local negative feedback would reduce or eliminate hum.

Perhaps a SPICE simulation would produce some insights into this network's purpose, but we disconnected it and observed (or rather listened to) the results to keep things simple. The cathode DC voltage or bias changed slightly since some DC current flowed through the high resistance network and into the 390R cathode resistor. However, the hum was still there and the output power and amp's tone were unchanged.

BY THE SAME AUTHOR:

Sound Improvement Secrets For Audiophiles: Get Better Sound Without Spending Big

Publisher: Career Professionals
Year published: 2021
Language: English
Paperback: 328 pages
ISBN: 978-0648298205

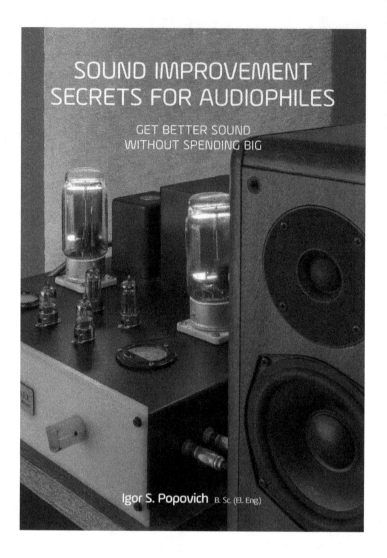

Avoid the hit-and-miss approach and stop wasting money on overpriced high-end products in the blind hope of sonic improvement. Achieve the ultimate audio synergy and get more enjoyment from your audio system by making it as good sounding as possible.

"Sound Improvement Secrets for Audiophiles" will teach you how things work, why some circuits, designs, and technologies sound the way they do, and how to make them sound even better through simple modifications and improvements.

It is like having an audio and acoustic consultant by your side to guide you through optimizing and voicing your audio system and your listening room.

While relatively technical and in-depth, this practical manual goes way beyond "a dozen quick tips" and the simplistic advice you read elsewhere. Instead, the focus is on dozens of DIY projects, case studies, and examples of commercial audio components – turntables, preamplifiers, amplifiers, loudspeakers, power supplies, and acoustic treatments.

With over 400 photographs, diagrams, and illustrations, "Sound Improvement Secrets for Audiophiles" makes it easy for you to understand and comprehend complex technical concepts and issues.

The author does not shy away from many controversial and hotly debated topics. Tubes vs transistors, objectivists vs subjectivists, measurements vs listening, and digital vs analog: all of these are discussed in detail.

The money invested in this book would not even buy you a budget-priced pair of cables: it will prove to be one of the best financial investments you ever make. Even if you implement only a few improvements from the hundreds described within its pages – you will never look back!

BOOK CONTENTS:

1. WHY YOU SHOULD READ THIS BOOK AND HOW YOU WILL BENEFIT FROM IT
2. BEFORE YOU BUY AN AUDIO SYSTEM OR COMPONENT - THINGS TO DO & MISTAKES TO AVOID
3. WHAT DO WE LISTEN FOR AND WHAT DO WE ACTUALLY HEAR?
4. CLEANING UP THE POWER SUPPLY TO REDUCE NOISE, HUM, AND INTERFERENCE
5. CABLES, FUSES, CONTACTS, AND CONNECTIONS
6. UPGRADING & FINE-TUNING THE SOURCES: OPEN REEL RECORDERS, TURNTABLES, PHONO STAGES AND CD PLAYERS
7. AUDIO AMPLIFIERS - HOW THEY WORK AND HOW TO IMPROVE THEIR SOUND
8. HEADPHONES AND HEADPHONE AMPLIFIERS
9. LOUDSPEAKER TYPES, TESTS, AND IMPROVEMENTS
10. COMPONENT MATCHING AND AUDIO SYSTEM INTEGRATION ISSUES
11. LOUDSPEAKER POSITIONING
12. OPTIMIZING THE ACOUSTIC PERFORMANCE OF YOUR LISTENING ROOM
13. ACOUSTIC TREATMENTS
14. MINIMIZING UNWANTED VIBRATIONS & OSCILLATIONS
15. TROUBLESHOOTING YOUR AUDIO SYSTEM

TONE CONTROLS

- SIMPLE FILTERS - BUILDING BLOCKS OF COMPLEX TONE CONTROLS
- THE CLASSIC TONE CONTROL CIRCUIT
- ONE KNOB TONE CONTROLS
- TONE CONTROL STACK
- INDEPENDENT AND INTERACTIVE TONE CONTROLS: Legacy Blues Twin (Epiphone Blues Custom 30)
- TONE CONTROLS IN THE NEGATIVE FEEDBACK LOOP

Although the guitar amplifier's main task is power amplification, a guitar amp is also a tone control center since its tonal balance and voicing determine the ultimate guitar sound even more or at least as much as the guitar itself.

Tone controls are such a crucial aspect of amplifier design that it is hard to understand why so many commercial amplifier models have no tone control capability at all!

SIMPLE FILTERS - BUILDING BLOCKS OF COMPLEX TONE CONTROLS

To fully understand tube amplifiers' inner workings and outward behavior, especially tone control circuits, we must first review the basics of linear systems theory. Unfortunately, we cannot study this topic without some pretty heavy mathematics. If you get lost, skim through this section and review its main conclusions and their practical consequences. Remember the rules of thumb, and you should be able to get by without fully understanding the "whys" behind them.

Zeroes and poles of a transfer function

A linear system or network (a filter or an amplifier, for example) does something to the input signal. The "transfer function" of the system H(s) determines how the system changes the input signal and can be expressed as a ratio of two polynomials: $H(s)= Z(s)/P(s)$

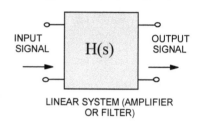

LINEAR SYSTEM (AMPLIFIER OR FILTER)

Polynomials with real coefficients can be factorized so we have

$$H(s)= H[(s-z_1)(s-z_2) ... (s-z_M)]/[(s-p_1)(s-p_2) ... (s-p_N)]$$

The operator "s" is an independent variable, a complex number of the form $s = \sigma + j\omega$, which can be represented graphically in a complex plane with s on the X-axis and w on the vertical or imaginary axis. Since we are primarily interested in system response to sine wave signals, we will replace the complex operator "s" with "jω".

The numerator is a polynomial of the M-th degree; the denominator is a polynomial of the N-th degree. The zeroes of the numerator, z_1 to z_M, have a dimension of frequency, just as do zeroes of the denominator, p_1-p_N. To distinguish between the two sets of frequencies, we call the zeroes of the upper polynomial zeroes of the transfer function (z_1 to z_M) and zeroes of the denominator we call "poles of the transfer function" (p_1 to p_N). These can be located on the complex s plane; small circles are used for zeroes and crosses for the poles.

A transfer function is a complex number; it has a modulus (amplitude) and phase, often studied separately. Since there are two possible input and two possible output signals, four possible transfer functions are outlined in the table. Of most interest in tube amplifies is the voltage amplification factor $A_V(s) = v_{OUT}/ v_{IN}$, but everything we say about $A_V(s)$ applies to the other three types of transfer functions. We will write A instead of A_V to simplify writing, so keep in mind that this is the voltage amplification.

	INPUT SIGNAL	OUTPUT SIGNAL	H(s)	NAME
The four types of transfer functions	v_{IN}	v_{OUT}	$A_V(s)=v_{OUT}/v_{IN}$	VOLTAGE AMPLIFICATION FACTOR
	i_{IN}	i_{OUT}	$A_I(s)=i_{OUT}/i_{IN}$	CURRENT AMPLIFICATION FACTOR
	v_{IN}	i_{OUT}	$G(s)=i_{OUT}/v_{IN}$	TRANSFER ADMITTANCE
	i_{IN}	v_{OUT}	$Z(s)=v_{OUT}/i_{IN}$	TRANSFER IMPEDANCE

Bode plots

Henrik Bode was one of the pioneers of linear system analysis. The amplitude-frequency and phase-frequency plots using asymptotes are named in his honor. They will enable you to quickly determine the behavior of any system, providing you know its transfer function. The rules are simple.

A constant (frequency-independent term) is represented by a horizontal line. A pole introduces a downward slope at -20dB/decade, starting at its angular frequency ωP. A zero introduces an upward slope at +20dB/decade, starting at its angular frequency ωZ. And finally, a double pole or double zero has a double slope, +/-40 dB/decade.

Where the pole and zero asymptotes overlap, the resultant slope is their difference. In this example (right), with increasing frequency, the asymptote of the pole (1) is active first, and the resultant characteristic is equal to its slope. At ωZ the zero introduces the upward slope (2), the two slopes cancel each other, and the resultant (3) is a horizontal line from that frequency onward!

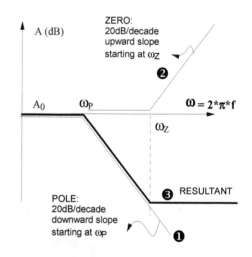

ABOVE: How the overall Bode plot is constructed from individual zero and pole characteristics

The best way to ease into this relatively complex mathematical treatise is to analyze basic filters, the building blocks of all circuits.

1st-order low pass RC filter

This filter is a resistive-capacitive voltage divider. For low frequencies, the reactance of the capacitor shunting the output (load) is high, and all input signal passes through the series resistor R and reaches the load. As the signal frequency increases, the reactance $X_C = 1/\omega C$ gets smaller and smaller, meaning more and more of the input signal is shunted (diverted) to the ground (COM).

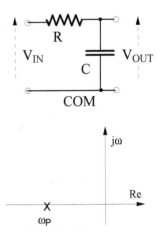

The voltage transfer function is $A_V = V_{OUT}/V_{IN} = A(s) = X_C(s)/[R + X_C(s)] = (1/sC)/(R + 1/sC) = 1/(1 + RCs)$ We divide both the numerator and denominator by RC and get $A(s) = (1/RC)/(s + 1/RC)$

For sinusoidal signals $s = j\omega$ so we have $A(j\omega) = 1/(1 + j\omega RC) = 1/(1 + j\omega/\omega_P)$, where the pole frequency is $\omega_P = 1/RC$, this is also the upper -3dB frequency f_U where the output drops -3dB! This frequency is also called a "corner", "break", "turnover" or "half-power" frequency.

The pole-zero diagram for 1st order RC network

To draw the Bode diagram and its asymptotes, first, we assume that $\omega \ll \omega_P$. In that case A=1, or in dB A=20log1= 20*0 = 0 dB this represents a horizontal line ❶.

If $\omega \gg \omega_P$, $A = \omega_P/\omega$ or in dB $A = 20\log(\omega_P/\omega) = 20\log\omega_P - 20\log\omega$ For $\omega = \omega_P$ A= 0 dB This represents a downward sloping asymptote with a slope of -6 dB per octave or -20dB per decade ❷

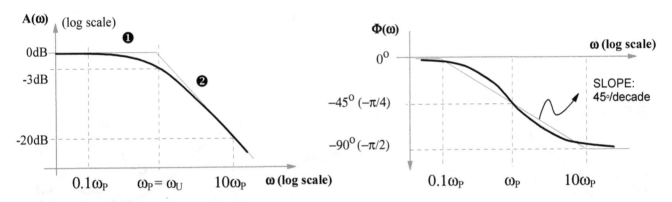

ABOVE: Bode diagrams for the modulus $A(\omega)$ and phase $\Phi(\omega)$ of the A_V transfer function
BELOW: The RC network's response to square wave signals of various frequencies with reference to the upper -3dB frequency f_U

Since the output voltage lags behind the input voltage, this circuit is also called a phase lag circuit.

With a square wave signal at its input, the output of the low pass filter will depend on the frequency of the signal. For $f_{SIG} \ll f_U$, the output voltage will have a slight rounding of the corners, as illustrated. As the frequency of the input signal is increased, the closer it gets to the upper -3dB frequency f_U, the more rounded the output signal will get until it resembles a sawtooth shape. Because the output signal is an integral of the input signal, this circuit is also called the *integrating circuit*.

For $f_{SIG} \gg f_U$ (where f_{SIG} is more than ten times f_U), the output voltage degenerates into a practically linear sawtooth shape, as illustrated.

This filter and its behavior are extremely important in audio since it models the behavior of a guitar amplifier at high frequencies.

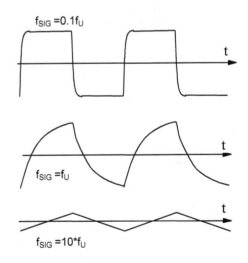

The meaning of the time constant

The time constant τ (Greek letter "tau") of a circuit is the reciprocal value of the angular frequency ω of its pole (Greek letter "omega"). If a DC voltage or a square wave signal is applied to a fully discharged capacitor, the time constant is the time required for the voltage on the capacitor to reach 63% of its final value. At the same point, the voltage drop across the resistor is 100-63 = 37%.

As a "charging" and "discharging" parameter, it depends on the capacitance and resistance value. For the CR filter just discussed, $\omega_P = 1/RC$, so $\tau = 1/\omega_P = RC$!

Say a DC voltage V_{IN} ("step-function") is applied to this circuit at t=0. The capacitor is fully discharged. At t=0, the full V_{IN} is across the resistor, and the output voltage is zero. The capacitor starts charging, and the charging rate is determined by τ - the larger the time constant, the slower the charging rate!

After t=t, the output voltage (across the capacitor) is 63% of the DC voltage at the input. The voltage across the resistor also follows an exponential curve and falls at the same rate, so after t=τ, it is 1-0.63 or 37% of V_{IN}!

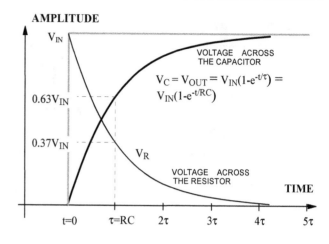

1st-order high pass RC filter

$A_V(s)=R/[R+ZC(s)] = R/(R+1/sC) = RCs/(1+RCs)$

Divide both the numerator and denominator by RC and get $A(s)=s/(s+1/RC)$ The transfer function has a form $A(s)=s/(s+\omega_P)$ There is one zero at $\omega_Z=0$ and one pole at $\omega_P=1/RC$, this is also the lower -3dB frequency f_L where the output drops -3dB!

For sinusoidal signals s=jω so we have $A(j\omega)=j\omega RC/(1+j\omega RC) =j\omega/\omega_P/(1+j\omega/\omega_P)$, where the pole frequency is $\omega_P=1/RC$

To draw the Bode diagram and its asymptotes, first we assume that $\omega\gg\omega_P$. In that case A=1, or in dB A=20log1= 20*0 = 0 dB This represents a horizontal line ❶.

For $\omega\ll\omega_P$, A =ω_P/ω or in dB A = 20log(ω_P/ω) = 20logω_P - 20logω

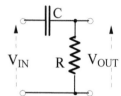

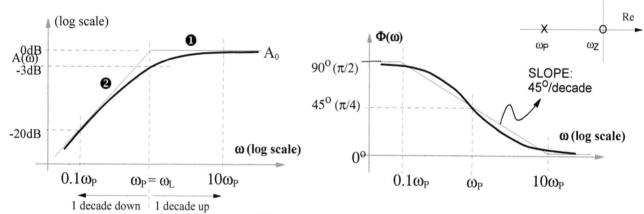

ABOVE: Bode diagrams for the 1st-order high pass RC filter
BELOW: The response of the CR network or high pass filter to square wave signals of various frequencies

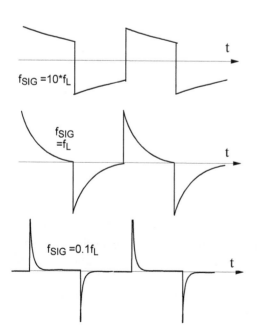

For $\omega=\omega_P$ A= 0 This represents a upward sloping asymptote with a slope of +6 dB per octave or +20dB per decade ❷

Since the output voltage leads the input voltage, this circuit is also called a phase lead circuit.

With a square wave signal at its input, the output of the low pass filter will depend on the frequency of the signal. As illustrated, for high frequencies, where $f_{SIG}\gg f_U$, the output voltage will have a small, almost linear droop of the upper edge.

As the frequency of the input signal is lowered, the closer it gets to the lower -3dB frequency f_L (the pole frequency), the more pronounced the droop. Because the output signal is a derivative of the input signal, this circuit is also called the differentiating circuit.

For $f_{SIG}\ll f_U$ (where f_{SIG} is more than ten times f_U), the output voltage degenerates into sharp pulses, as illustrated.

This filter and its behavior is very important in audio since it models the behavior of an audio amplifier at low frequencies.

Limited phase lag network (bass lift circuit)

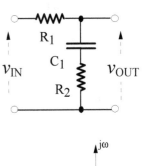

This important network is sometimes called a basic bass lift circuit. The language suggests that this circuit somehow boosts bass frequencies, which is not the case; it actually cuts or attenuates treble frequencies. The transfer function has a pole and a zero at frequencies ω_P and ω_Z: $H(j\omega) = V_{OUT}/V_{IN} = (1+j\omega C_1 R_2)/[1+j\omega C_1(R_1+R_2)]$ $= (1+j\omega\tau_Z)/[1+j\omega\tau_P)]$.

At the frequency of the zero, the capacitive impedance Zc or Xc is equal to the resistance R_2 or $1/(\omega_Z C_1) = R_2$. The time constant of the zero is $\tau_Z = 1/\omega_Z$, so $\tau_Z = R_2 C_1$! At the frequency of the pole, the capacitive impedance Zc or Xc is equal to the total series resistance R_1+R_2, or $1/(\omega_P C_1) = R_1+R_2$, so $\omega_P = 1/[(R_1+R_2)C_1]$ and $\tau_P = (R_1+R_2)C_1$.

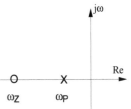

The final output level A_X is equal to the ratio $\omega_P/\omega_Z = R_2/(R_1+R_2)$ or, if the logarithmic scale is used, $A_X = 20\log[R_2/(R_1+R_2)]$.

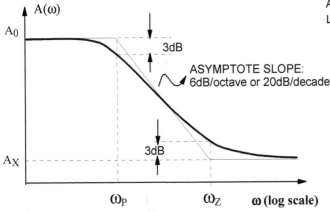

ABOVE: The pole-zero diagram for the limited phase lag network
LEFT: Bode plot for the limited phase lag network

LIMITED PHASE LAG NETWORK
$\omega_Z = 1/(R_2 C_1)$
$\omega_P = 1/[(R_1+R_2)C_1]$
$A_X = 20\log[R_2/(R_1+R_2)]$

Limited phase lead network (bass cut circuit)

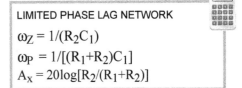

This network, also known as the bass cut circuit, also has one pole and one zero at frequencies ω_P and ω_Z: $H(j\omega) = V_{OUT}/V_{IN} = R_2(1+j\omega\tau_Z)/[(R_1+R_2)(1+j\omega\tau_P)] =$ $R_2(1+j\omega C_1 R_1)/[(R_1+R_2)(1+j\omega C_1(R_1\|R_2))]$.

At the frequency of the zero, the capacitive impedance Z_C or X_C is equal to the resistance R_1 or $1/(\omega_Z C_1) = R_1$. The time constant of the transfer function's zero is $\tau_Z = 1/\omega_Z$ and $\omega_Z = 1/(R_1 C_1)$, so $\tau_Z = R_1 C_1$.

At the frequency of the pole, the capacitive reactance X_C is equal to $R_P = R_1\|R_2$, the parallel combination of $R_1\|R_2 = R_1 R_2/(R_1+R_2)$, so $\omega_P = 1/[(R_P C_1)]$ and $\tau_P = R_P C_1$.

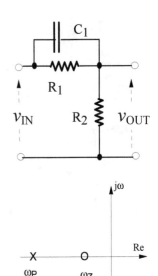

The final output level A_X is equal to the ratio $\omega_Z/\omega_P = (R_1+R_2)/R_2$, or, if logarithmic scale is used, $A_X = 20\log[(R_1+R_2)/R_2]$.

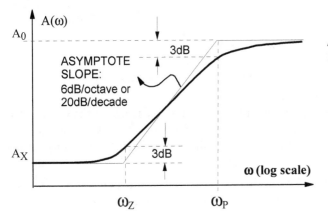

ABOVE: The pole-zero diagram for the limited phase lead network
LEFT: Bode plot for the limited phase lead network

LIMITED PHASE LEAD NETWORK
$\omega_Z = 1/(R_1 C_1)$
$\omega_P = 1/[(R_P C_1]$
$R_P = R_1 R2/(R_1+R_2)$
$A_X = 20\log[R_2/(R_1+R_2)]$

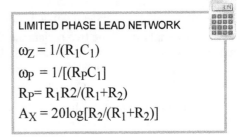

THE CLASSIC TONE CONTROL CIRCUIT

Variable bass lift and cut circuits combined

Now that we understand the basic passive filters, we can use them as building blocks for various tone control circuits. Let's start with the "classic" tone control, very common in vintage hi-fi but also used in many guitar amps. By adding a variable resistor in parallel with C_1 in the bass lift circuit, we can control the degree of bass lift. We can arrange the RC in series instead of a parallel RC branch in the bass cut circuit. We get the variable bass-cut circuit if we add the parallel (shunt) resistor as before and make it a variable one (a potentiometer).

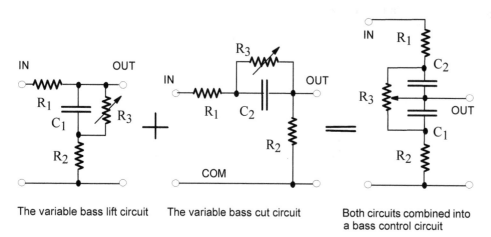

The variable bass lift circuit   The variable bass cut circuit   Both circuits combined into a bass control circuit

Notice the commonalities of the two filters. R_2 is the shut resistor to COM or common terminal in both cases, and R_1 is the series resistor. C_1 is the shunt cap, C_2 is the series cap in both, and R_3 is the potentiometer.

Thus, logically, the two circuits can be elegantly combined into one, the bass control that can both cut and lift bass frequencies!

Variable top (treble) lift and cut circuits combined

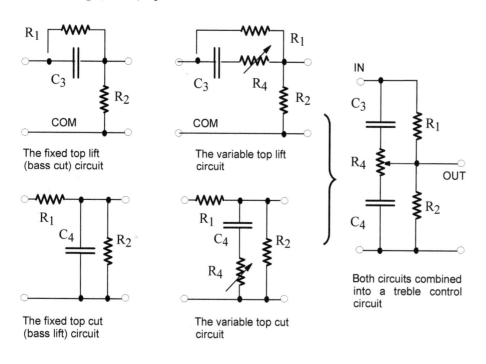

The fixed top lift (bass cut) circuit

The variable top lift circuit

The fixed top cut (bass lift) circuit

The variable top cut circuit

Both circuits combined into a treble control circuit

We approach the design of treble cut and boost controls in the same way. Starting with the bass-cut circuit (only now we call it "treble lift"), we add the control pot R_4 in series with the capacitor C_3.

Likewise, starting with a fixed top-cut circuit, we add a variable resistor in series with the shunt capacitor and get a variable top-cut circuit.

As with bass controls, R_2 is the shunt resistor to COM in both cases, and R_1 is the series resistor. C_4 is the shunt cap, C_3 is the series cap in both, and R_3 is the potentiometer. So, we combine them into one circuit that can provide both treble cut and lift!

The classic tone controls synthesis - bass and treble controls combined

Finally, we combine the two filters, R_1 and R_2 are common to both, and now all fall into place, and you can see why we numbered the resistors and capacitors the way we did (next page).

The audio signal "sees" the two resistances R_1+R_2 as being in parallel with the internal resistance of the tube. Thus, their total resistance should be reasonably high, so the tube does not work into a low load. The rule-of-thumb says that its value should be at least twice the internal resistance of the tube. At 0.8mA, the internal resistance of the 12AX7 triode is around 68kΩ, so R_1+R_2 should be higher than, say, 150kΩ!

On the other hand, the total value R_1+R_2 shouldn't be too high either; otherwise, the high-frequency response will be affected due to their interaction with the tube's output capacitance and various stray parasitic capacitances.

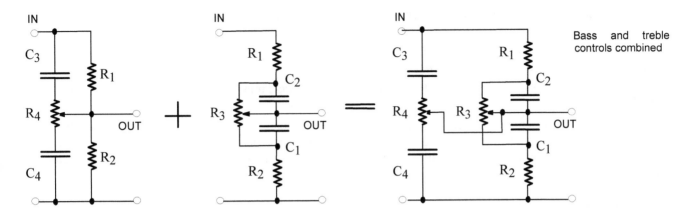

A quick analysis of the classic tone controls

Now that we understand the genesis of these tone controls let's say we've never seen them before and must figure out how this circuit works. The trick is to place the wipers of the four pots in their end positions, study those extreme cases, and then interpolate the circuit behavior in-between. There are two pots and four possible end combinations.

When R_3 is in maximum boost position (the upper end), C_1 is short-circuited and C_2 is in parallel with R_3. At low frequencies the reactance of C_2 is much larger than R_3 and can be considered an open circuit. The output impedance is R_2+R_3 is much greater than R_1, so the output impedance is a high percentage of the total input impedance.

At high frequencies, the reactance of C_2 is much smaller than R_3, and the output impedance is approximately $\sqrt{(R_2^2+X_{C2}^2)}$.

R1=100k, R2=10k, R3=R4=1M, C1=C4=2nF, C2=20nF, C3=0.2nF

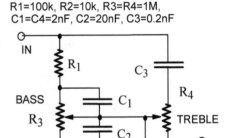

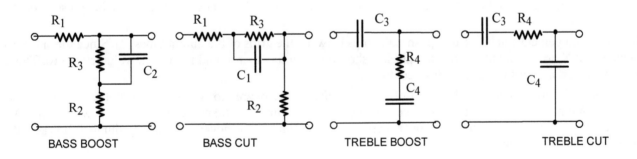

BASS BOOST   BASS CUT   TREBLE BOOST   TREBLE CUT

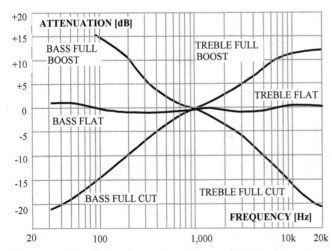

The attenuation curves for the classic tone control circuit

Thus, the attenuation is $A=\sqrt{(R_O^2+X_S^2)}/\sqrt{(R_I^2+X_S^2)}$, where $R_O=R_2+R_S$, $R_I=R_1+R_2+R_S$, $R_S=R_3X_{C2}^2/(R_3^2+X_{C2})$ and $X_S=R_3^2X_{C2}/(R_3^2+X_{C2})$.

When R_3 is the max. cut position, C_2 is short-circuited and C_1 is across R_3. At low frequencies the output resistance is approx. equal to $\sqrt{[(R_1+R_2)^2+X_{C1}^2]}$. The attenuation is $A=R_2/\sqrt{(R_I^2+X_S^2)}$.

When treble control R_4 is in max. boost position the output impedance is R_4 and C_4 in series. At high frequencies the reactances of C_3 and C_4 are low and Z_{OUT} is approx. equal to R_4.

At low frequencies the output impedance is approximately $\sqrt{(R_4^2+X_{C4}^2)}$, since the reactances of C_3 and C_4 are high.

Attenuation: $A=\sqrt{(R_4^2+X_{C4}^2)}/\sqrt{(R_4^2+(X_{C3}+X_{C4})^2)}$

When R_4 is the max. cut position, the output impedance is only that of C_4. The attenuation is $A=X_{C4}/\sqrt{(R_4^2+(X_{C3}+X_{C4})^2)}$.

The filter cannot really "boost" anything, so even when both controls are in their maximum or "boost" positions, there is still an overall loss of at least 15dB and sometimes much more!

Remember, attenuation or amplification in dB is $A_{dB}=20*logA$, so if $A_{dB}=-15$ (attenuation is indicated by negative and amplification by positive dB figures!), $A=0.18$!

This means that to compensate for the lossy tone controls, the next stage must amplify $1/0.18= 5.6$ times (or 15dB!) just to bring the signal back to the level it had before the tone controls!

Again, "boost" means lower attenuation in this context, while "cut" means higher attenuation. Notice that 0dB in the graphs usually published, such as the one below, is for both controls' "flat" position, where potentiometers R3 and R4 are in the middle of their range. That position is usually indented, so there is a positive feel when the center is reached.

ONE KNOB TONE CONTROLS

There are dozens of tone control circuits devised over the decades the tube technology has been around. The classic 2-knob tone control circuit is relatively complex (eight components in total), and such precise control isn't always needed in a guitar amp. Thus, the circuit was more prevalent in vintage hi-fi amplifiers.

A simple RC circuit or "treble cut" tone control

The simplest type of tone control, usually found in smaller amps, is a single knob treble cut (1). As the tone control pot is turned CCW (to the left, "0" or "MIN" position), less and less of the 1M resistance is in series with the 150p capacitor, and more and more treble frequencies are shunted to GND. So, in the MAX position (fully CW), the treble is most prominent (bright), and in the MIN position (fully CCW), the bass is most prominent (deep).

Such a treble cut tone control can also be implemented in the anode circuit of the power tube (2) across the output transformer's primary.

In fully CCW or "MIN" position, the 50k pot is short-circuited, and 47n capacitor is across the primary, thus fully attenuating treble. As the pot is turned clockwise towards the MAX position, progressively higher resistance is added in series with the 47n cap, and treble frequencies are attenuated less and less.

One disadvantage of this location for tone control is that a high DC anode voltage will be present on the potentiometer.

The same arrangement can be used in push-pull amps between the grids of the output tubes (3). Since the signals at points X and Y are equal but of opposite phases, shorting them together would reduce the volume to zero. This is sometimes used as a primitive "Standby" control.

If a simple RC filter is connected instead of the short circuit, the degree to which high or treble frequencies are shorted or attenuated can be controlled. Since the tone control circuit in (1) takes its signal from the wiper (slider) of the preceding Volume control pot, the upper section of the Volume pot is never capacitively shunted. It is added in series with the 22k resistor that precedes it.

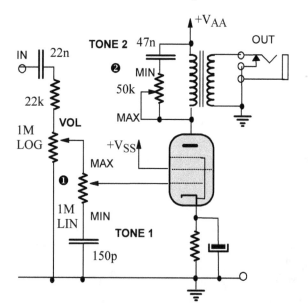

ABOVE: Two simple treble cut controls, one in the grid circuit, the other in the anode circuit of the power tube

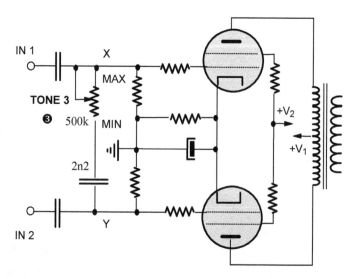

ABOVE: An equivalent treble cut control in a push-pull output stage. "MAX" means fully CW position of the pot, "MIN" means "0" or fully CCW

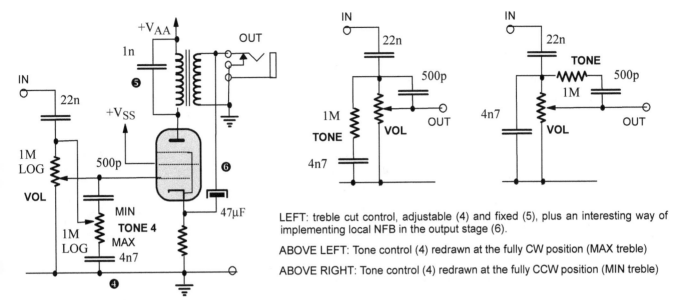

LEFT: treble cut control, adjustable (4) and fixed (5), plus an interesting way of implementing local NFB in the output stage (6).

ABOVE LEFT: Tone control (4) redrawn at the fully CW position (MAX treble)

ABOVE RIGHT: Tone control (4) redrawn at the fully CCW position (MIN treble)

The arrangement in (4) is a bit more complex. If you cannot intuitively understand its operation, use the trick that never fails - redraw the circuit in extreme situations (next page). In this case, the extremes are the two positions of the Tone control pot, the fully CCW or MIN position, and the fully CW or MAX position.

In the MAX position, the slider is at the junction of the 500pF capacitor and the upper end of the Tone potentiometer. That point is connected to the upper end of the Volume pot. The 500pF capacitor thus shunts the upper portion of the Volume control and boosts treble by diverting it across that section. The lower frequencies are not diverted and are attenuated by the voltage divider formed by the upper and lower sections of the pot.

In the fully CCW or MIN position, the whole 1M of the Tone control pot is in series with the 500pF capacitor, and that RC branch is shunting the upper portion of the Volume pot. However, due to such a high resistance in series with it and its own low capacitance, the 500pF cap does not change the frequency balance here. The 4n7 capacitor is now fully shunting the volume control pot and diverting some HF to GND, so a treble cut results.

In (5), a capacitor is hooked up in parallel with the output transformer's primary, resulting in a high-frequency shunting effect, so they are not amplified as much as the lower frequencies. The circuit seems simple, but its theoretical analysis would be very complex due to audio transformers' frequency-dependent parameters and behavior. Since it also depends on the amplifier and speaker characteristics, the only way is to determine the value of such a capacitor experimentally, which means by ear!

Instead of the usual way of connecting the cathode bypass capacitor of the output tube directly to GND, the arrangement (6) shows a possibility of strapping it in series with the loudspeaker's voice coil. Again, this seems a simple enough circuit, but its analysis would be an even more complex and time-consuming undertaking.

I have better things to do with my life, and I'm sure you do too, so suffice to say that this circuit provides a degree of local negative feedback (since part of the output voltage is brought back to the power tube's cathode) and thus reduces distortion and maximum achievable output power.

However, since a complex, frequency-dependent voice coil impedance is now in series with the cathode bypass elco, it also has a noticeable effect on the tone balance and voicing of the amp. Since it is so easy to try it in any amp and so easy to reverse if you don't like its effects, try it and make your own judgment

A treble cut circuit can also be installed on the load side of the output transformer (7), between the speaker + and GND (in parallel with the speaker). In the MIN or fully CCW Tone pot's position, the pot is out of the circuit, and the 47nF capacitor is shunting high frequencies (HF) away from the speaker.

At the CW or MAX end, the whole of 50k is in series with the shunting capacitor, and not much HF shunting takes place, so the full treble (bright voicing) is achieved.

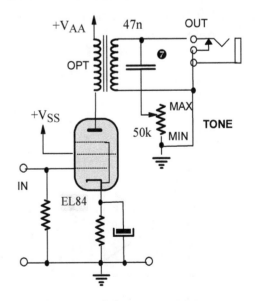

ABOVE: A simple treble cut tone control on the load side of the output transformer

Bass cut tone controls

The RC-coupling between amplifier stages is a simple 1st order filter whose -3dB or corner frequency can be varied by selecting the values for R (the grid resistor of the next stage) or C, the coupling capacitor.

This arrangement has gained popularity among some modern "boutique" amp makers. A switch with four or more positions switches different values of coupling capacitance and changes the -3dB of the high pass filter. We have seen that the angular frequency of the pole is $\omega_P=1/RC$, meaning the lower -3dB or "corner" frequency is $f_L=\omega_P/2\pi = 1/(2\pi RC)$.

With 22n & 1M the corner frequency is 7.23 Hz, while with 1n & 1M it raises to 159 Hz.

So there is no loud "pop" during switching, the switch should be of the "make before break" type, meaning the next contact should be made before the previous contact is broken.

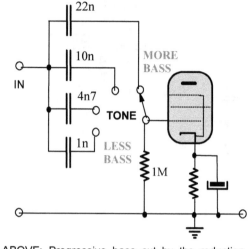

ABOVE: Progressive bass cut by the reduction in coupling capacitance

BELOW: The low-loss tone controls as described in US patent #2,680,231

TWO-KNOB TONE CONTROLS

Low-loss tone controls

Tone controls attenuate signals significantly and require at least one additional amplification stage to bring the signal back to the original level. The mid-band attenuation of this simple circuit is only around 9dB, compared to 15 or more dB for conventional tone controls. A detailed explanation of the circuit's operation would take too much space here; you can easily download the full document. Search online for US patent #2,680,231!

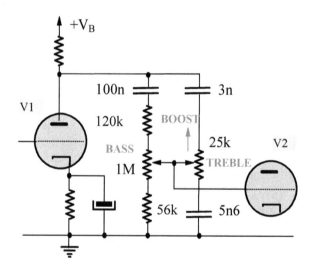

Tone control in the negative feedback loop (Baxandall circuit)

First published in 1952 in his Wireless World (now Electronics World) article titled "Negative Feedback Tone Control - Independent Variation of Bass and Treble Without Switches", this tone control method has been named after Peter James Baxandall. It offers quite a few advantages over the passive circuit.

It is fully symmetrical, meaning the boost and cut ranges are equal, and, with the controls in the mid-position, the frequency response is flat.

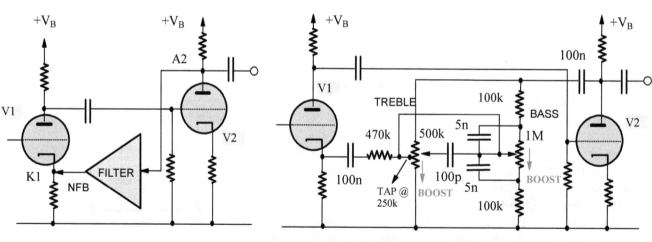

The operational principle behind the Baxandall circuit: the filter is in the NFB branch from A2 back to K1

The Baxandall tone control circuit

Unlike the tone control stack, there is no interaction between the bass and treble controls. A high degree of boost or cut is available at the frequency extremes, but the middle frequencies are unaffected.

Finally, it is an active circuit, comprising not just the passive filter itself but the following tube (V2), from whose anode the negative feedback signal is taken and filtered before it is brought back to the cathode of the previous stage (K1). Thus, the voltage gain of the V2 stage can be chosen so that there is no overall loss or attenuation.

However, what seems desirable on paper may not be so advantageous in practice. While the circuit had some success in vintage hi-fi amplifiers, guitar players did not like some of its "advantages," namely the heavy bass emphasis and lack of interactivity. Very few guitar amps used this type of tone control, and those that did were not successful or popular models. As always, what ultimately matters, is the overall feel and tone guitar players experience.

TONE CONTROL STACK

A tone control stack or just "tone stack" is a passive filter RC network with two (Treble & Bass) or three controls (Treble, Bass, and Middle). The topology is the same; in the 2-knob tone stack, the Middle potentiometer is replaced by a fixed value resistor.

So, in the further analysis, we will assume a more general case, that of the 3-knob tone stack.

There are various ways of implementing a tone stack. Still, by far the most common has been the Fender version, which was also adopted by other major amp manufacturers such as Marshall and Vox, and is sometimes referred to as the FMV tone stack.

In most Fender amps, the tone stack is driven directly from the anode of the preceding stage's tube, inevitably a 12AX7 triode (or 7025, its lower noise variant). Unfortunately, while 12AX7 has the highest gain of all triodes ($\mu=100$) it also has the highest internal impedance, ranging from 50-100kΩ, depending on its operating point.

66kΩ is usually used as an estimate in calculations. This is the same order of magnitude as the input impedance of the tone stack, so this way of driving the stack causes significant attenuation (signal loss) due to the voltage divider action between the internal impedance of the tube and the tone stack.

Marshall and some Fender amps, such Bassman 5F6A, used the output of a cathode follower to drive the tone stack. Since the output impedance of a cathode follower is under 1kΩ, the loss of signal on the tube's internal impedance is negligible.

When anode driven, tone stack "slope" resistor is usually at least 100kΩ, but when cathode follower driven, values of 33-56kΩ are used, as in the three examples illustrated above.

We will not perform a detailed analysis of this complex circuit. That would make us none wiser. Since such heavy linear algebra involving multiple loop equations and matrices would annoy most readers, we will only focus on the fundamental aspects of tone stack's behavior.

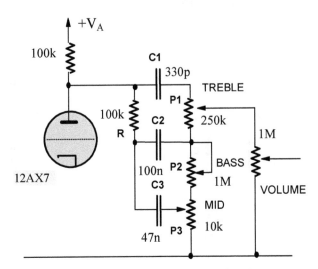

ABOVE: Fender Super Twin tone control stack

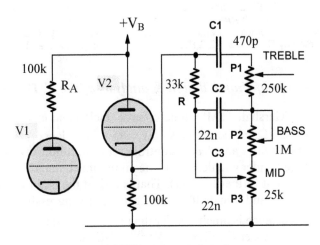

ABOVE: Marshall JTM50 tone control stack
BELOW: Fender Bassman 5F6A tone control stack

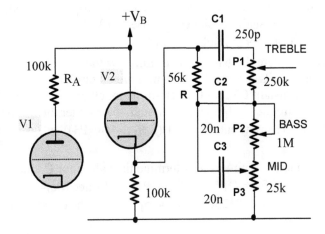

Breaking down the FMV stack into three basic control circuits

The in-principle schematics below illustrate the basic building blocks of each of the three control circuits. The Mid potentiometer varies the attenuation of the bandpass filter formed by R, C3, and P3, but it also acts as a variable output attenuator of the whole tone stack. In other words, it affects the Bass and Treble settings as well.

The Bass control works in the same fashion, only this time the bandpass filter it controls is more complex since it includes not only R-C2-P2, but P3 as well. The functioning of the Treble control is even more complex - it features in two filters. One is the high pass filter formed by C1 and the sum of the series-connected control pots (P1+P2+P3), the other being the complex bandpass filter formed by R, C2, C3, P2 & P3. The Treble pot acts as a balance or blend control between the two filters.

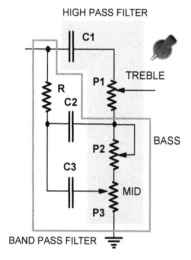

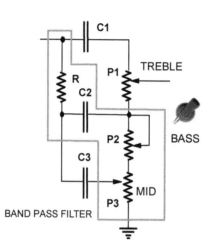

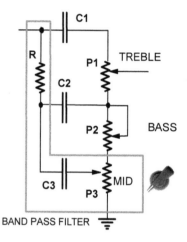

ABOVE: TREBLE pot acts as a balance control between the HP filter and the complex bandpass filter formed by R, C2, C3, P2 & P3.

ABOVE: BASS pot controls f_{LBP} (lower -3dB frequency) of the bandpass filter formed by R, C2, P2 & P3, but it also impacts the TREBLE circuit's -3dB frequency.

ABOVE: MID pot varies the attenuation of the bandpass filter formed by R, C3 and P3, but it also acts as a variable output attenuator of the whole tone stack.

The rule-of-thumb design & analysis of the FMV tone stack

Using Marshall JTM50 tone control stack as an example, a designer would choose P3's resistance first. Marshall chose 25kΩ. Since the very low output impedance Z_{OUT} of the driver stage (cathode follower) can be neglected, assuming *Bass* and *Treble* controls set at minimum (zero) and the *Mid* control at maximum (min-max-min), this voltage divider will determine the minimum attenuation of the *Mid* control. This means that midrange attenuation can only be cut further (deeper). That is one of the main factors behind Fender's low gain, clean tone with scooped up midrange, and Marshall's higher gain, more aggressive tone.

C3's reactance is much smaller than R or P3, so it can be considered a short circuit. The attenuation A of the resistive voltage divider (in dB) is $10^{(A/20)} = P3/(Z_{OUT}+R+P3) = P3/(R+P3)$. Since A=7.3 for Marshall's stack, we get R= $P3*(10^{A/20}-1) = 25*(10^{7.3/20}-1) = 25*(10^{0.365}-1) = 25*(2.317-1) = 25*1.317 = 32.9$ or 33kΩ, exactly the value marked on the diagram.

Now, C3 can be determined by choosing f_{UBP}, the bandpass filter's upper -3dB frequency. $C3=1/(2\pi f_{UBP}*R)$ and since Marshall designers chose 219 Hz for this corner frequency, the required value of C3 was $C3=1/(2\pi*219*33,000) = 22nF$!

To complete the bandpass design, now the value for C2 needs to be chosen. It controls the degree of attenuation when the *Bass* control is at its minimum (zero), meaning the resistance of P2 will not feature in the calculations. The time constant C2 and C3 determine together with R and P3 must be high enough that its corresponding corner or cutoff frequency f_{LBP} is low enough to be under the lowest frequency produced by a guitar or under 80 Hz. Marshall used 62Hz as its cutoff: $C2=1/(2\pi f_{LBP}*(R+P3)) - C3 = 1/(2\pi*62*(33,000+25,000)) - 22nF = 22nF$, the same value as for C3.

All that is left now is to determine the values for C1 and P1, and since 250kΩ was chosen for P1, C1's value can be calculated from $C1=1/(2\pi f_{LHP}*P1)$, and since Marshall used 1.4kHz as the turnover frequency for this filter, we get $C1=1/(2\pi*1,400*250,000) = 455pF$. Since the closest standard value is 470pF, that is the capacitor Marshall used in the amplifier.

The attenuation curves for a typical FMV tone stack

To get a feel for the frequency-dependent behavior of a typical tone stack, the eight attenuation-Vs-frequency curves below illustrate all combinations of extreme (min and max.) settings of its three control knobs.

With *Bass & Treble* at minimum and *Mid* at maximum, the attenuation curve is almost flat, except the bass attenuation that starts between 100 and 200 Hz (1). If *Mid* control is also wound down to a minimum, we get a camel's hump curve (2), the stack is a narrow bandpass filter with the lowest attenuation of around -15dB falling between frequencies of 100 and 300Hz.

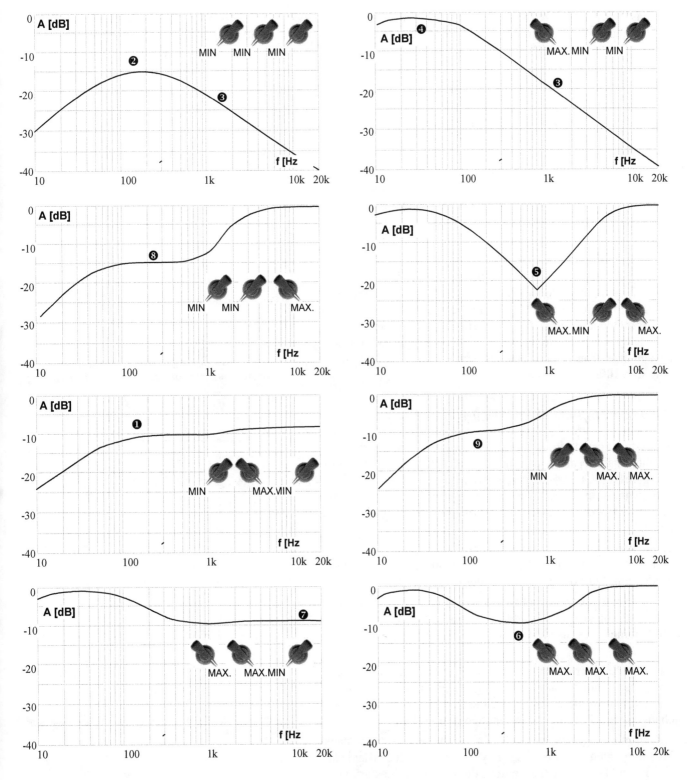

ABOVE: Qualitative frequency response graphs for eight extreme setting of the three tone control potentiometers, BASS-MIDDLE-TREBLE (from left-to-right)

Increasing the *Bass* control to its maximum, the falling high-frequency slope remains (3), but the bass is boosted (4) in the 10-100Hz range.If we keep the *Bass* at maximum, *Midrange* control at the minimum, and boost the *Treble* setting to its maximum, the low-frequency LF attenuation curve remains the same, but at a turnover frequency (5), the HF boost results in a typical notch curve.

If the *Mid* control is also increased to its maximum (max-max-max), the cardioid notch curve smoothes out into a bathtub curve (6) with LF and HF "humps" and a mild dip (up to 10dB) at upper bass-lower mid frequencies, 200Hz-2kHz.

Winding back the *Treble* control now to its minimum would flatten the HF hump of the previous case and result in relatively flat downward shelf characteristics as in (7).

The last two cases have the same basic shape, an upward raising shelf-curve. The "min-min-max" case (8) shows a typical high pass filter curve in the bass region, a flat "shelf" between 100Hz and 600Hz, and then a similar high pass curve resulting in another shelf above 4-5kHz.

The "min-max-max" case (9) shows less attenuation in the bass region. The first shelf is at -10dB (compared to -15dB before), but it is relatively narrow (only between 100-300Hz), when the 2nd shelf starts its (slower) rise.

Fine tuning the filter's performance

We could look at each of the components in the tone stack and investigate how changing its capacitance or resistance would impact the amplitude-vs-frequency characteristic of the tone stack, but that would take too much time and space here. Let's just do it for C1, arguably the most crucial capacitor of all three.

Since its values typically range from 250 to 500pF, the curves show four curves for C1 ranging from 100 to 1,000pF (1nF0).

First, it is immediately apparent that as its capacitance is reduced, the midrange dip becomes deeper and deeper, from -20dB attenuation with 1nF to -25dB with 100nF. Second, the frequency of the minima shifts towards higher with lower capacitance, from approx. 200 Hz with C1=1nF to 800-900 Hz with C1=100pF.

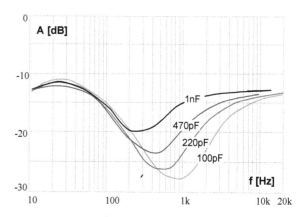

ABOVE: The effect of changing C1 capacitance on the dip in midrange attenuation

Blackstar's Infinite Shape Feature

Blackstar's promotional literature describes their "ISF" or Infinite Shape Feature in terms of tone: "When turned towards 0, it gives tight, focussed, percussive response, characteristic of American amps. When turned towards 10, the response becomes woody, crunchy and warm, like traditional British amps."

While that cryptic summary could mean any number of things, their UK Patent Application GB 2446188, available online, sheds more light on the technicalities of this circuit.

In addition to the three traditional tone stack controls, ISF adds the fourth. Instead of a fixed "slope" resistor R, a dual-ganged potentiometer is used as per the diagram (right), named "sweep."

VR1a shifts the middle or center dip frequency, while VR1b compensates for the accompanying change in the depth of the mid-cut that would occur if only the slope resistor's value was changed, thus keeping a constant depth of mid-cut.

All circuit diagrams available online show this tone stack driven by integrated circuits (in hybrid amps such as Blackstar HT-1 and HT-5). For that reason, the resistor values are very low (10k instead of 100k); we could not find an example of this circuit being used with tube drivers.

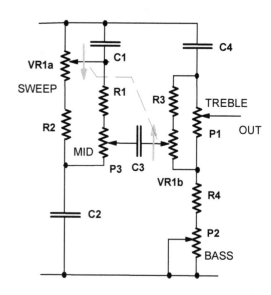

ABOVE: The ISF tone control stack by Blackstar Amplification

INDEPENDENT AND INTERACTIVE TONE CONTROLS: Legacy Blues Twin a.k.a. Epiphone Blues Custom 30

Just as with its smaller brother, Legacy Valve Junior, this amp seems to be an Aussie version of Epiphone Blues Custom, a 30 Watt combo first released in July 2006 by Gibson.

The Epiphone had an MSRP (Manufacturer Suggested Retail Price) of AU$1,199.- in Australia, which was quite expensive for an amp that retailed for around US$580 in America. Legacy Blues Twin sold for AU$300 cheaper, a significant saving.

The only possible difference could be in the speakers used. While the Epiphone version had two 12" Eminence "Lady Luck" speakers, the speakers in our Legacy amp had no markings whatsoever and may not be Eminence drivers at all.

The amp was very heavy; 70 lb. was a figure mentioned online. Once the rear panel was removed (a breeze), the reason became apparent. Two huge transformers and a choke probably weigh more than 15 lb. by themselves. The large steel chassis, two 12" speakers, and a thick cabinet all contribute to its backbreaking weight.

Once the four thick bolts that hold the chassis to the top panel are removed, the chassis sits securely on the side rails and cannot accidentally drop, as can happen with many other amps. It is a simple and effective safeguard that other manufacturers either don't seem to comprehend or don't care.

I like Legacy/Epiphone amps. They are well designed, decently made, and a great value for money. We got our used amp in pristine condition for only AU$280 (US$200 at the time). The transformers/chokes and tubes only would cost you more than that, not to mention the chassis, the cabinet, and the speakers.

The whole circuit is on a single large PCB. Compared to amps with three of four smaller boards and a messy spaghetti of interconnecting wires, that is a welcome change; it certainly makes the amp's internals neat. The components have plenty of space around them, so multimeter and soldering iron access is easy. Although a double-sided PCB was used, most modifications can be carried out without taking the board out.

The power section of the back panel features the IEC power inlet with an integral primary fuse, a H.T. ("High Tension" or high voltage) fuse, and a triode-pentode switch.

The triode position is marked "15W Class A", and the pentode is "30W Class AB". Although all references to Epiphone had been manually scraped off the PCB (#4 in the photo), the screen-printed "Epiphone model" was left in place at the back.

Legacy Blues Twin

- All tube combo, PCB construction
- 30 Watts Class AB pentode or 15 Watts Class A triode
- EQ: Two modes- independent and interactive
- Tubes: 5x12AX7, 2x5881, 5AR4 rectifier
- Speakers: 2x12" nondescript
- Tube driven spring reverb
- Channels: 2
- Outputs: 2x4 Ω, 2x8 Ω and 16 Ω

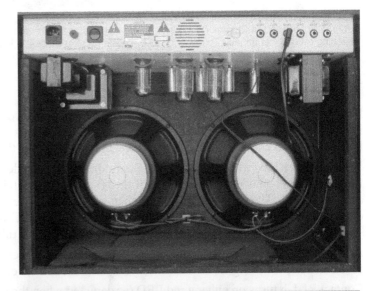

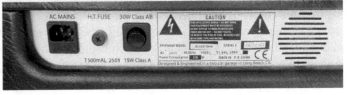

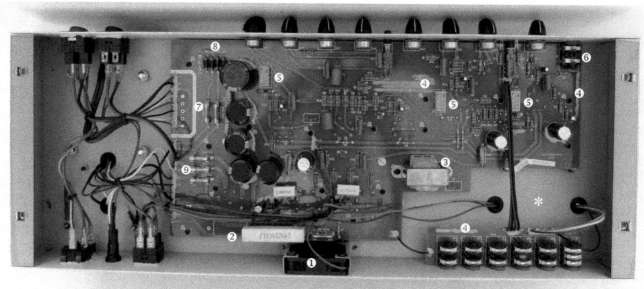

ABOVE: 1) cooling fan 2) cathode resistor of the output stage 3) reverb driver transformer 4) scraped out references to Epiphone 5) switching relays 6) input jack 7) heater fuses 8) heater rectifier diodes 9) HV rectifier fuses (AC and DC side)

The preamp and reverb stages

It took almost a whole page to depict only the preamp stages of this amp (next page), not even including the phase inverter, let alone the output stage and the power supply. This is by no means a simple amp! It uses relay switching between its two channels, Ch. 1 is the clean channel, Ch. 2 is the overdrive or dirty channel.

When relay RL3 (whose contact RL3-a is drawn in the de-energized position) is activated (energized) by a footswitch, its contact shorts the grids of the reverb recovery tube V4 to the ground (pins 2 and 7 since both triodes are paralleled). The relay is energized when the footswitch connects the top end of its coil to GND. The bottom contact of its coil is permanently connected to the positive DC voltage V6 (as is the case with the other two relays). The same voltage V6 ($6.3V_{DC}$) is used for heaters of preamp tubes and the fan.

As for relays RL1 and RL2, their coils are in parallel, so both relays operate simultaneously. Instead of one relay with four contacts, the designer used two relays with two contacts each.

Contact RL1-b turns on either a red LED (as depicted when the relay is de-energized) or a green LED when the relay is energized. In its de-energized state, contact RL1-a does nothing; it simply shorts itself. With RL1 energized, that contact connects point C to GND, shorting tube V1b's grid to GND. RL2 is also energized at that time, so its contact RL2-a switches point A to point B, which is the grid of gain stage V2b.

Thus, the clean channel is active when relays RL1 and RL2 are energized, and the green LED is on. Gain stages V1b and V2a are bypassed. Contact RL2-b shorts their output (point D) to GND.

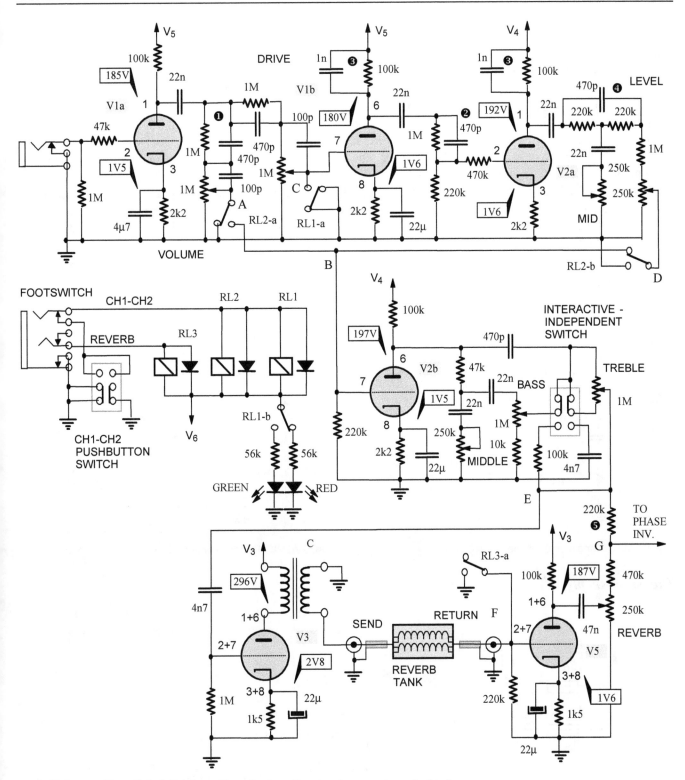

ABOVE: Legacy Blues Twin & Epiphone Blues Custom: Preamp stages and reverb circuit

As drawn, the diagram shows the overdrive channel in operation, with the red LED on and the signal passing through all stages.

There are no less than four frequency shaping or compensating film capacitors after the 1st stage (1), two 100p and two 470p caps. Then notice a 470p cap after the 2nd stage (2) and that both anode resistors in stages V1b and V2a are bypassed by 1n caps (3). The output of the 3rd stage (V2a) is followed by a bridged T filter (4), in whose vertical arm is the *Mid* control pot, and the *Level* pot.

The *Middle* and *Bass* controls are positioned after the 4th gain stage (V2b), the circuit we will discuss next. The input signal for the reverb section is taken from one end of the 220k resistor, point E, and the delayed or "wet" signal is returned to its other side, point G, so the 220k resistor is the mixing resistor (5)!

The independent and interactive tone controls

While most amps have only one tone control circuit, Legacy Blues Twin's tone control circuit can work in two different ways, with independent or interactive *Bass-Mid-Treble* controls. Since tracing the circuit through the switch contacts can be tedious and confusing, let's redraw it in each position of the changeover switch to see how the Legacy/Epiphone designer(s) implemented this duality.

In the *Independent* mode, starting from point Z, the anode of the 12AX7 tube, we have the first branch, the *Middle* control, to ground, RCR, 47k-22n-250k pot. The *Bass* branch is a high-pass CR filter 22n-1M-10k, whose output goes to the "mixing" point X through a 100k resistor. The output (wiper) of the *Treble* control pot is connected to the same point X. The CRC *Treble* branch also starts in point Z, 470p-1M pot - 4n7 to ground.

In the *Interactive* mode, the 4n7 cap from the *Treble* circuit and the 100k resistor from the *Bass* circuit are switched out. At the same time, the wiper of the *Bass* pot is connected to its top end (signal input) from the 22n cap, and to the bottom end of the *Treble* pot. Now the *Bass* and *Treble* controls will affect each other, so the effectiveness and overall behavior of one will depend on the setting of the other. A very clever design indeed!

Being a hi-fi designer first and foremost, I cannot for the life of me understand why a guitar player would prefer such interactivity (unless it's a matter of being used to it).

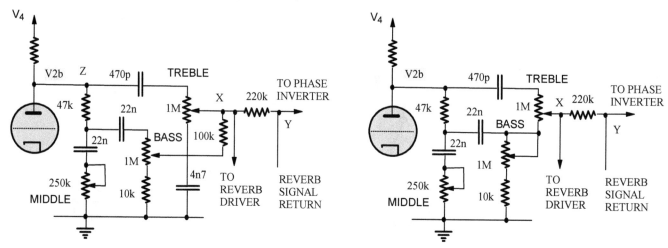

ABOVE LEFT: Tone controls in the "Independent" mode   ABOVE RIGHT: Tone controls in the "Interactive" mode

The power supply

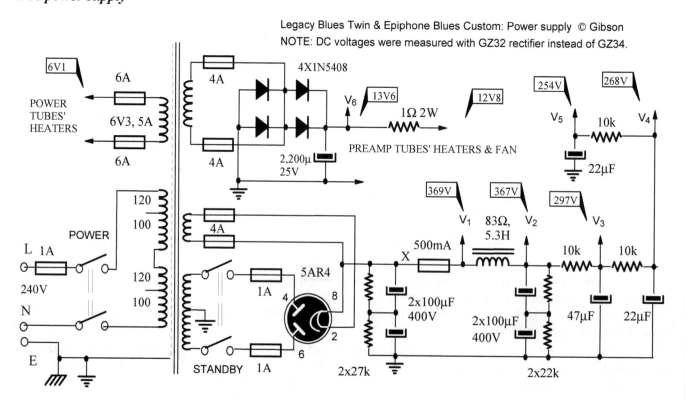

Legacy Blues Twin & Epiphone Blues Custom: Power supply © Gibson
NOTE: DC voltages were measured with GZ32 rectifier instead of GZ34.

In this book, we have seen and analyzed vintage amps without even a mains or power fuse. This amp's designer went to the other extreme; there are almost a dozen fuses on its massive PCB. Apart from the mains and HV fuses, all other circuits a double-fused, with fuses in both secondary legs. This is unnecessary; one in each circuit would do.

The power tubes are AC heated. The voltage is a bit low, only 6.1V, but that is within acceptable tolerances. However, the preamp tubes' heaters and the fan are powered by a full-wave rectifier bridge and a single capacitor for filtration, and their voltage is a whopping 13.6V! The preamp tubes on this amp will not last very long. So, we removed one of the 0Ω resistor links on the PCB and soldered in a 1Ω 2Watt resistor, which brought the voltage down to 12.8V.

The *Standby* switch is on the AC side of the GZ34 indirectly-heated rectifier, not the best way of implementing the standby feature. Plus, a double pole switch had to be used. Usually, a single-pole switch is located in point X, after the first filtering capacitor. That way, the first elco (or two in series here) stays fully charged. This blooper can easily be rectified (no pun intended) on this amp.

The phase inverter and output stage

With 28.4V on the common cathode resistor, the combined cathode current of the output tubes is 114 mA, or 57 mA per tube, of which around 55mA is anode current. The anode-to-cathode voltage is 366-28.4 = 337.6V.

Now we can calculate the anode dissipation as $P_A = I_A * V_{AK} = 55mA*337.6 = 18.6$ W, which is way under the maximum limit of 30 Watts for 6L6-GC. Even the original 6L6 and 5881 with the dissipation of 23 Watts would be cruising in this amp.

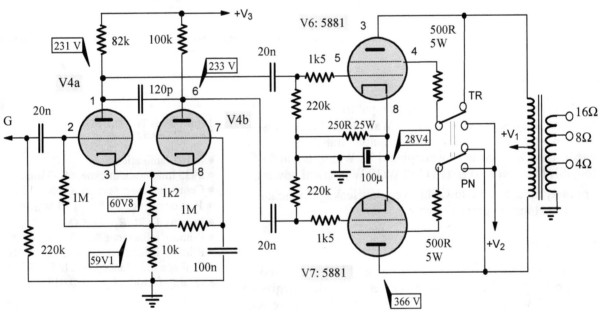

Legacy Blues Twin & Epiphone Blues Custom: Phase inverter and output stage, © Gibson

Adding Master Volume and Line/Headphone Output

Since there are so many speaker outputs and the internal speaker is plugged into one of them (as would be an external speaker), a relatively primitive manual arrangement, our solution for implementing the line and headphone output will also remain relatively simple.

Choose the output you use the most, say an 8Ω speaker jack, and add three resistors as per the diagram. Once a speaker (internal or external) is unplugged, an 8Ω dummy load resistor will replace it as a proper load for the amp. That will give you a "silent run" with only headphones and line out active. If you want to retain the speaker while headphones and line out are being used, omit the 8Ω dummy load from the circuit.

The dummy load will get very hot, so make sure you bolt it onto the chassis. There is some real estate next to the speaker-out board, marked with (*) on the photo of the amp's innards.

With 30 Watts on the 8Ω tap, the maximum signal voltage is 15.5V, so when the total resistance of 130Ω is added in parallel, the current through it is approx. 15.5/130=120mA. The voltage will be divided, so only 10/130 = 0.077 or 7.7% is developed across the 10R output resistor, which is a 1.2V line out signal.

The voltage across the 120R resistor is then 15.5-1.2= 14.3Volts. The power dissipated on the 120Ω resistor is 14.3*0.12= 1.7 Watts, so a 3Watt resistor is needed! The power dissipated on the 10Ω resistor is only 1.2*0.12= 0.15 Watts, so even a 1W resistor would stay cool as a cucumber!

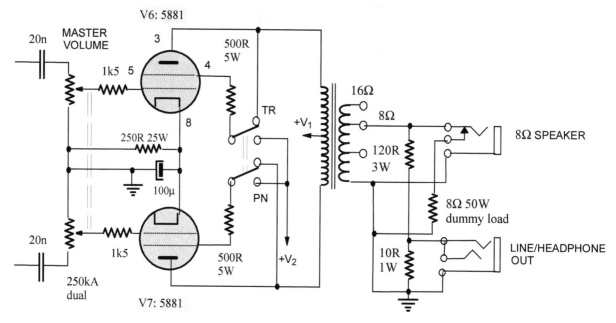

ABOVE: Legacy Blues Twin & Epiphone Blues Custom: the output stage with added master volume and Line-Headphone output

The output transformer

Blues Twin features one of the largest output transformers of all guitar amps, not just those in the 30-50 Watt range. Most amps in the 100 Watt range have far smaller output transformers. Its magnetic core is rated at a whopping 150 Watts, or, in hi-fi jargon, it is five times oversized (150 W/30 W nominal power).

The primary inductance is low, but a very low leakage inductance of only 13mH improves the Q-factor.

How relays work

An electromagnetic relay has at least one coil and one or more separate contacts. When the electromagnet coil is de-energized, no current flows through it, and no magnetic field is created inside it. Once the nominal voltage is applied across the relay's coil ($12V_{DC}$ relay shown here), the coil is energized; its magnetic field exerts a mechanical force on the movable plunger or lever, which in turn actuates (moves) the contact.

In a de-energized state, the changeover contact shown here connects the common point X with terminal Y, while in the energized state, point X is connected to terminal Z. This remains the case for as long as the coil stays energized. The relay's contact will revert back to its de-energized state as soon as SW is opened again.

With one switch (mechanical or electronic, such as a transistor), relays enable control of many contacts and loads. The switch can be very small (low power rating) and cheap, while relay contacts can have a high current and voltage rating.

Finally, the switch SW and the relay coil and contacts (loads) can be far apart, so the loads can be switched from a distance (not in guitar amps, of course), even using radio telemetry, as in industrial control systems.

Legacy Blues Twin OUTPUT TRANSFORMER:

- EI105 laminations
- a=35 mm, stack thickness S=35mm
- Center leg cross section A=aS = 12.25cm^2
- Power rating: P=A^2= 150 Watts
- Primary DCR: R_P=18.5 Ω + 24.9 Ω
- Primary inductance @120Hz L_P= 8.5H
- Primary inductance @1kHz L_P= 7.6H
- Leakage inductance @1kHz L_L=39mH
- Quality factor: QF = 7.6/0.013 = 576

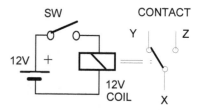

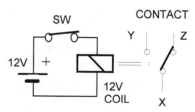

ABOVE: Relay coil de-energized
BELOW: Relay coil energized

ANALOG EFFECTS (TREMOLO, VIBRATO, REVERB) AND EFFECTS LOOPS

- TREMOLO
- TWO STAGE CATHODE-COUPLED TREMOLO CIRCUIT: SILVERTONE 10XL (model 1421)
- ADDING TREMOLO DEPTH (STRENGTH) CONTROL: GIBSON SKYLARK GA-5T
- MAGNATONE'S VIBRATO CIRCUIT
- REVERB
- EFFECT LOOPS
- CATHODE FOLLOWER-DRIVEN EFFECTS LOOP: VHT SPECIAL 6 ULTRA

Compared to a wide variety of digital effects available, the analog effects with tube circuits are limited to just a few. While tremolo (amplitude modulation of the signal) and reverb (delay) have been relatively common, very few vintage amplifiers used a vibrato effect (frequency modulation), which required much more complex circuitry.

Effect loops enable external effects to be used, so it is crucial to understand how they work and how to design and modify them.

TREMOLO

Amplitude modulation

Tremolo is the amplitude modulation of an audio signal. The modulating low frequency (LF) signal is produced by the LF oscillator, whose output is usually fed to a *Depth* or *Strength* potentiometer, so the depth of modulation is made adjustable by changing the amplitude of the modulating signal.

The frequency of the modulating voltage is also usually adjustable and is marked "Rate," "Speed," or "Frequency".

Many commercial brands, most notably Fender, erroneously call this effect vibrato, which is wrong. As we will soon see, Vibrato is the frequency-modulation of the signal (FM), a very different effect from tremolo. The two can be likened to the AM and FM analog radio bands!

Phase-shift oscillator

The simplest low-frequency oscillators for tremolo circuits are based on the phase-shift principle. An active element, tube, FET, or bipolar transistor works as an amplifier whose anode output feeds a passive CR or phase-shift network back to its own grid.

Each CR combination provides around 60 degrees of phase shift, so when part of that anode signal gets back to the tube's grid, it is 180 deg out-of-phase. Since the original anode signal was already 180 deg out-of-phase, the returned signal is actually in phase with the original grid signal that started the whole process. A positive feedback loop is closed.

Providing there is enough voltage gain (that is why a high gain tube such as 12AX7 is usually used), the stage becomes a free-running LF oscillator.

Cathode-coupled tremolo circuit

There are a few ways the modulating LF voltage can be brought into the audio circuit. Instead of changing the DC bias on the grid of a preamp tube, its cathode voltage could be varied instead, as in the circuit depicted on the right. The audio tube V1 and the oscillator tube V2 share the cathode RC circuit. As V2's cathode voltage pulsates in the rhythm of the LF voltage, so will V2's cathode, changing the gain of the whole stage and thus modulating the amplitude of its anode output!

In this case, the "depth" of the modulating signal is varied by changing the anode resistance of V2 between 100k and 350k. Its oscillating frequency ("Rate") isn't affected, but the amplitude of the cathode voltage variations is.

RIGHT: Another tremolo topology, this time modulating V1's cathode voltage through the shared cathode resistor and bypass capacitor

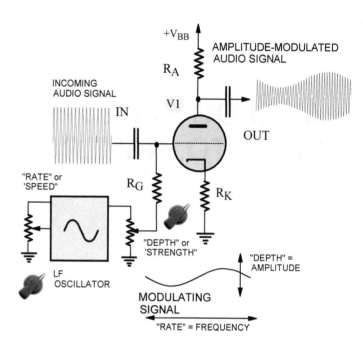

ABOVE: One possible tremolo topology, the output of the low frequency oscillator modulating the grid voltage of the preamp tube V1 (grid-coupling)

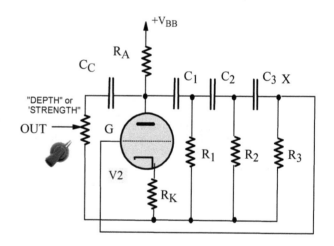

ABOVE: The most common phase shift oscillator with a single tube. Both triodes and pentodes can be used.

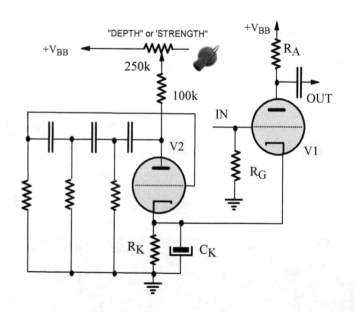

Add a tremolo module using a transistor phase shift oscillator

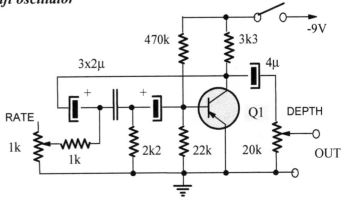

Adding a tremolo effect to a tube guitar amp does not necessarily require a tube for a phase-shit oscillator. The same circuit can be built using a transistor, either a FET or a bipolar type.

While solid-state components in the signal path affect the overall sound or voicing of an amp, there is no such impact when they are used in the tremolo circuit. This simple oscillator whose output voltage signal modulates the signal's amplitude from one of the audio tube stages.

The RATE or frequency of oscillation with component values indicated can be varied between 3 and 8 Hertz (cycles per second). The output potentiometer varies the DEPTH or the amplitude of the modulating signal. The lack of transistor designation is not an omission; any PNP transistor with a current gain (β) above 50 will work.

Notice, however, the polarity of the three phase-shift capacitors. The two end units are electrolytic type, but the middle capacitor must be non-polarized, either a bipolar elco or a film type (polyester, polypropylene, paper).

Of course, if you are using one film cap here, you may as well use all three of that same type and dispense with the elcos altogether. The same goes for the 4mF output coupling cap; a film type can be used instead.

Since a PNP transistor is used, a -9V power supply is required. The first option is to use the amp's bias supply, assuming the host amplifier uses a fixed negative bias for the output tubes. The current draw of this whole circuit is in the order of only 1 mA, so even bias supply circuits, which are invariably of a very low current capability (since grids of tubes in class A1 or AB1 draw minimal current), will be able to provide such current.

Alternatively, use a 9V battery. This way, any possibility of unwanted interaction of this module and other parts of the amp through the common power supply will be eliminated. Plus, assuming you don't forget to switch the tremolo circuit off once you've finished playing, that battery will last a long, long time.

TWO STAGE CATHODE-COUPLED TREMOLO CIRCUIT: SILVERTONE 10XL (1421)

Silvertone is the brand name Sears, Roebuck and Company were using for their range of guitars and guitar amps in the 1950s and 60s. Sears outsourced manufacturing to various amp makers such as Danelectro, National, Harmony, Valco, Thomas and Kay, and later Japanese Teisco.

Model 1421 was sold from 1969 to 1972, and since Danelectro ceased its operations by 1969, it is uncertain who manufactured this and other models in the XL series of amps for Sears.

While its design was sound, the cabinet was lightweight, with a flimsy back panel made of pressboard.

The Oxford speaker sounds good at low power levels, but breaks up very early when driven by 7189, a higher voltage industrial version of EL84 (6BQ5) pentode.

The output transformer was tiny, EI41 laminations (13mm center leg width) with 13mm stack with, so A=1.3x1.3= 1.7 cm^2, meaning its power rating was only P=A^2 = 1.7^2= 2.9 Watts! And that is what this baby is, a 3 Watter. To call it a 10 Watt amplifier (even at peak power levels) was a huge stretch of the imagination!

SPECS

SILVERTONE 10XL (1421)

- 10W peak power, SE combo amplifier
- Construction: point-to-point, PCB (later models)
- Rectifier: 6X4 tube
- Reverb: No, Tremolo: Yes
- Volume and tone control, tremolo speed and strength
- 2 inputs, 1 channel
- Tubes: 2 x 12AX7, 1 x 7189
- Speaker: Oxford 8" alnico, 4Ω

The tremolo circuit

V2a works as an LF oscillator. Any AC voltage present on the output (anode) of the tube at pin 6 passes through three CR (high pass filters), starting with 22n (1) and a series connection of a 68k fixed resistor and a 1M "Speed" pot, followed by 10n and 560k (2) and the third filter comprised of 10n cap and 4M7 grid leak resistor (3).

In this case, a buffer stage is used (V2b). That triode's cathode is coupled (4) with the cathode of the triode in the 2nd stage (V1b). The Tremolo effect is free-running by default unless a footswitch turns it off by short-circuiting the buffer tube's grid (pin 2 of V2b) to GND.

RIGHT: The front and rear (cover removed) view of Silvertone 10XL

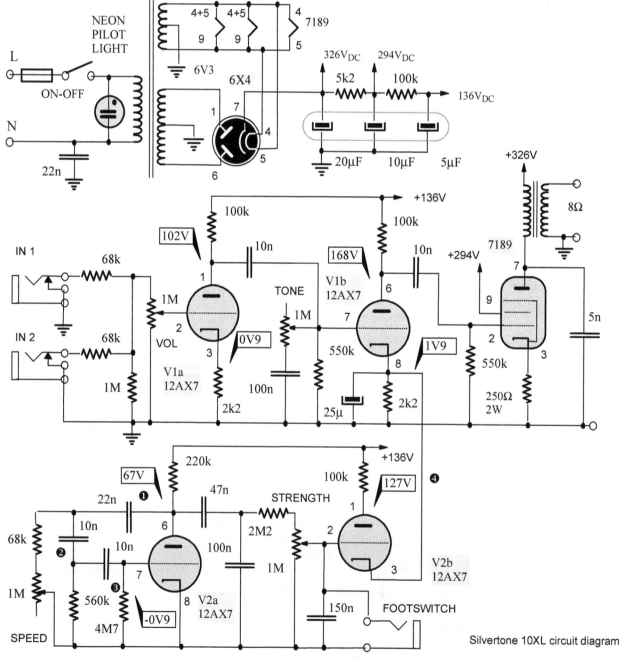

Silvertone 10XL circuit diagram

ADDING TREMOLO DEPTH (STRENGTH) CONTROL: GIBSON SKYLARK GA-5T

Gibson Skylark is a small vintage push-pull combo guitar amp. Two 6AQ5 beam power tubes produce up to 10 Watts in the class AB regime.

6EU7 duo-triodes were used instead of the more common 12AX7. The only controls are *Loudness* or volume control and tremolo speed control marked *Frequency*. There is no external speaker output, no tone control, or tremolo depth control.

The original circuit diagram

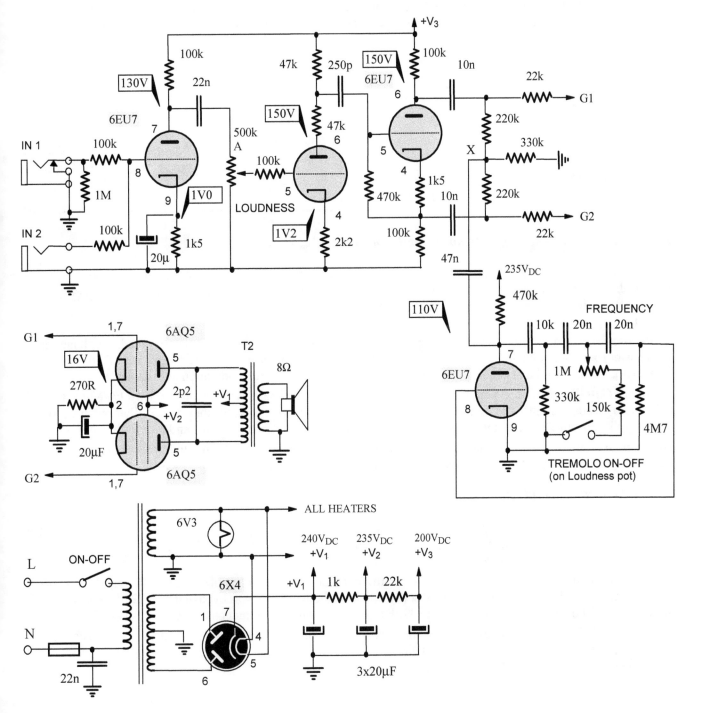

The first two stages are fairly standard. Since there is no tone control, the designers thought that the overall gain was too high, so they reduced it by lowering the gain of the 2nd stage. Notice that its cathode resistor was not bypassed, for approx. -3dB (29%) reduction in voltage gain. Also, its anode resistor was split into two 47k resistors, forming a voltage divider; the output of the 2nd stage is taken from that point. Less than half of the 2nd stage output signal is fed into the input of the split-load phase inverter (3rd stage). That type of inverter provides no voltage gain.

Notice the very low value of the coupling cap between the 2nd and 3nd stage only 250pF, which, together with the 570k grid resistor to ground (470+100) forms a high-pass filter whose -3dB frequency is $f_L = 1/((2\pi RC) = 1,117$ Hz, which is incredibly high. This amp has no bass whatsoever! Replacing that coupling cap with a higher capacitance unit is one of the first mods needed here.

The output idle current is approx. $16V/270\Omega = 60mA$, or 30 mA per each 6AQ5 tube.

Specifications, topology and output transformer

The output transformer was tiny, EI48 laminations (16mm center leg width) with 16mm stack width, so A=1.6x1.6= 2.5 cm^2, meaning its power rating was only $P = A^2 = 2.5^2 = 6.2$ Watts! Clearly, this is one of the weakest points of this amp. The primary's DCR was 320Ω (plate-to-plate), the secondary measured at 0.6Ω.

The primary inductance was 18H at 120Hz, dropping to 12.5H at 1kHz. The leakage inductance at 1kHz was relatively high for such a small physical size stack, 60mH, resulting in the low quality factor of Q=208.

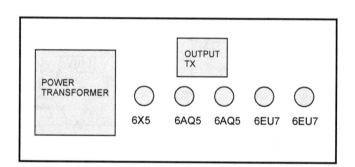

The tremolo circuit - adding tremolo strength control

The tremolo oscillator is fairly standard as well. The 4M7 resistor between the grid and ground provides grid leak bias for the 6EU7 triode, just as with Silvertone 10XL amp. However, the way tremolo signal modules the audio signal differs. The oscillator's output is capacitively coupled to point "X," which would normally be grounded (GND).

Notice also the common 330k resistor between point "X" and GND. The tremolo signal varies the DC voltage across that resistor (between point "X" and GND) and modulates the audio signal's amplitude. Replacing this fixed resistor with a variable one is all that is needed to make the strength or depth of the modulation adjustable.

The rotary *On-Off* switch (next page) was replaced by a 250k potentiometer with an integral *On-Off* switch, to work as *Tremolo Strength* control (1). Since tremolo circuits are not sensitive to hum (they are oscillators anyway) and don't' amplify the audio signal, they can be placed close to power supply components and on-off switches.

The tremolo strength was controllable over a wide range, from a faint hint of amplitude modulation to the deep staccato signal-chopping effect with psychedelic overtones.

Adding tone control and other improvements

Although this isn't a highly desirable vintage amp, to preserve the integrity of the front panel and make the mods reversible not just electrically, but cosmetically as well, we decided to use only the existing holes for two new controls, the *Tone* and *Tremolo Depth*.

Input 2 isn't usually used, so we installed a *Tone* control potentiometer (8) in its hole. The two 47k anode resistors in the 2nd stage were changed to one 100k resistor (2), and its cathode resistor was bypassed by a 22mF capacitor (3) to increase gain. The coupling cap was raised to 47nF (4), and the final coupling caps were increased to 22n (5).

The 2-pronged plug and the old power cord were ditched, a new three-core cable and earthed plug were installed, and the chassis earthed (6). The capacitance of all three filtering caps in the power supply was increased to 47µF (7). No other improvements were needed in this simple power supply.

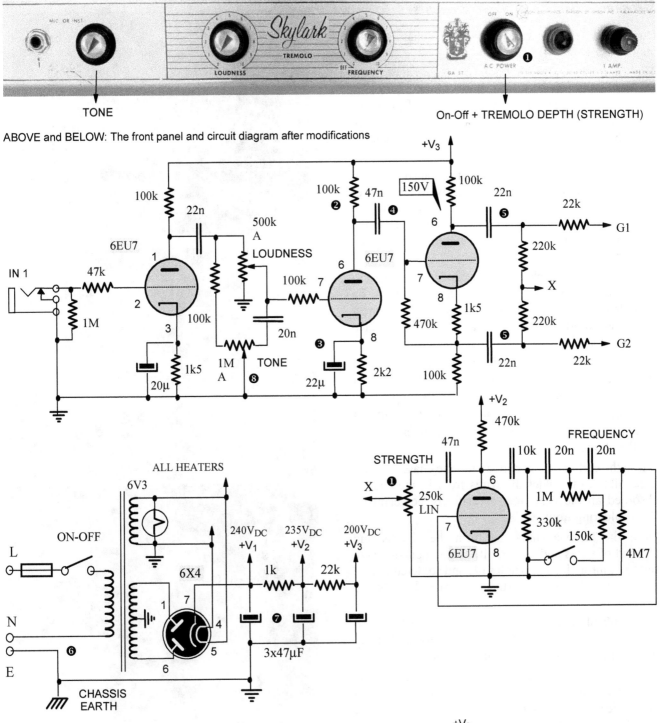

TONE

On-Off + TREMOLO DEPTH (STRENGTH)

ABOVE and BELOW: The front panel and circuit diagram after modifications

The fixed voicing filter between the 1st and 2nd stages

The original circuit diagram illustrated a couple of pages back was based on the original Gibson drawing downloaded from the Internet. Our amp was wired slightly differently, but it is not clear when that was done; most likely, that was not the original factory wiring but a subsequent mod by one of the amp's previous owners.

In essence, the two 220k resistors and the added 500p and 3n3 film capacitors act as a "fixed" tone control before the Loudness potentiometer. Also, its wiper was not hooked up directly to pin 7 (the grid of the 2nd stage) but through a 100k grid stopper resistor. It seems that somebody voiced the amp specifically to their taste.

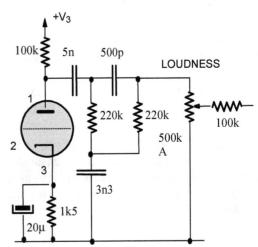

Mechanical issues and previous repairs

Due to its inferior design, removing the chassis was a major hassle.

The four back screws that hold the back panel to the cabinet had to be removed, followed by the two long bolts (1) which attach the top rear panel to the chassis. Some dimwit installed a large electrolytic capacitor (2), whose height prevented the chassis from sliding backward.

However, a timber edge at the front also made it impossible to remove the chassis backward. The power transformer hits the front baffle, so the only way was to remove the speaker baffle (3) by unbolting it from the rest of the cabinet and then sliding the chassis forward.

But, the internal speaker was permanently soldered to a terminal strip, so those two wires had to be unsoldered first. We added the quick-disconnect terminals, so now the speaker can easily be detached from the chassis.

The newly-added multisectional can-type elco was removed and reused on another project. Only three of its four 20µF sections were used, and the 60µF total filtering capacitance was a bit low for this type of design, so we replaced it with three 47µF axial elcos as per the photos.

While the earth lug was bolted onto the chassis, it was impossible to unsolder the wires from it; the chassis absorbed all the heat from a relatively powerful 65 Watt soldering iron.

Once the lug was unbolted from the chassis and cleaned of wires and solder, all the components were fed back through its two holes (4) and soldered before attaching the lug back onto the transformer bolt chassis (5).

The 220k bleeder resistor (6) between the first elco (rectifier output) and the ground was added by whoever messed around with the amp before.

Once the amp is turned off, the filtering capacitors discharge through that resistor in a matter of minutes, a good idea that facilitates repair and minimizes the possibility of a nasty shock.

The transformer CT (center tap) and three negative leads of the three elcos connect to that star point, five leads in total.

The long red wire coming from the anode circuits of the input stages on the left (7) had to be attached to one of the elcos, so we added a single lug terminal strip.

The leads of the original series resistor in the filtering circuit (8) were not insulated, so we ensured that all the positive leads of elcos were insulated (9) to completely remove any possibility of a short no matter how unlikely.

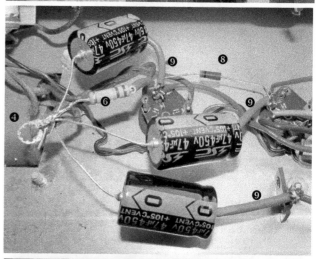

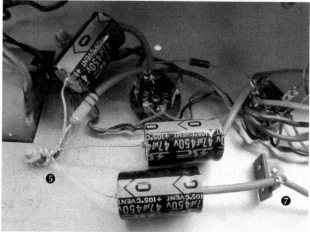

MAGNATONE'S VIBRATO CIRCUIT

RC bridges were common test instruments in the halcyon days of tube technology. In 1953 Robert C. Moses devised a phase measurement method for oscilloscopes based on a resistance-capacitance bridge as a phase-shifting circuit. Don L. Bonham used this circuit as a basis for his "Vibrato circuit comprising a bridge having nonlinear impedance elements," U.S. patent 2,988,706, available for download online (filed in 1956, approved in 1961). The nonlinear components Bonham used were silicon carbide varistors.

Varistors

Although varistors are classified as semiconductors, they have no PN-junctions. Vintage varistors were made from silicon carbide mixed with a ceramic binder. The mix was then pressed or extruded into the desired shape (disc, rod, washer) and sintered, resulting in hard ceramic-like material.

More modern zinc oxide or "MOV" (metal-oxide) varistors have sharper I-A characteristics. Polycrystalline zinc oxide is mixed with molten bismuth oxide and sintered. The bismuth oxide forms a rigid coating around the zinc oxide grains, so the varistor acts as an open circuit at low applied voltages, but when the voltage across it exceeds the value of its "clamping voltage" V_C the varistor suddenly changes its properties from a very high to very low resistance and conducts.

This makes them the simplest, cheapest, and most effective overvoltage protective device and transient suppressor. Due to their symmetrical I-V characteristic, varistors work in both DC and AC circuits in parallel with the load.

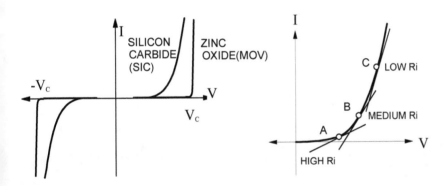

FAR LEFT: The I-V characteristics of two types of varistors

LEFT: The smoother, more curved I-V characteristics of SIC varistors show how as the voltage across them is increased, their internal resistance drops from very high at point A, to very low at point C. That fact was exploited in the Magnatone's vibrato circuit. The MOV type is not suitable for such a vibrato circuit.

Magnatone's vibrato circuit: AC bridge with nonlinear varistors in one arm

Since the patent mentioned above is available online, we will not analyze the circuit in detail. The four branches of the bridge are defined as between the four points, A, B, C, and D. AB and BC are audio signals of equal amplitudes but an opposite (inverted) phase.

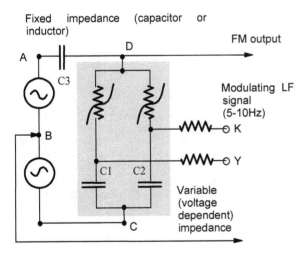

Two identical but out-of-phase audio signals

C3 is the fixed impedance in the third branch. The simple network of two varistors and two capacitors is the variable impedance, whose magnitude depends on the low frequency modulating voltage entering at points K and Y.

The push-pull audio signal (from tube guitar amp's preamplifier stages) between points C and D is modulated by a very low frequency signal from an LF oscillator, typically an RC phase-shift oscillator identical to those used to provide a tremolo or amplitude-modulated effect.

In the actual implementation of the bridge circuit, illustrated on the next page, tube V1 works as a split-load inverter, the out-of-phase identical signals are taken from its anode and cathode, points A and C. Tube V2 also works as a phase splitter, supplying out-of-phase modulating voltages to points K and Y.

Capacitor C4 practically short circuits the anode and the grid of tube V2 for signal frequencies so that the tube stage does not amplify the signal voltages from point C, tube V1's cathode.

Resistors R_K and R_Y isolate the vibrato stages from each other and the V2 phase splitter, so the varistors of each stage "float" with respect to the ground.

A single vibrato stage provides only a mild frequency modulation effect; for deeper, more pronounced frequency, a two-stage "wobbler" is needed, which is more than sufficient for a guitar. In some organ designs, a triple-stage vibrato design is used.

RIGHT: The practical implementation of Bonham's vibrato bridge circuit - two cathodyne (split load) phase splitters driving the bridge with push-pull signals, one with signal (audio frequency), the other with the low modulating frequency from the LF oscillator.

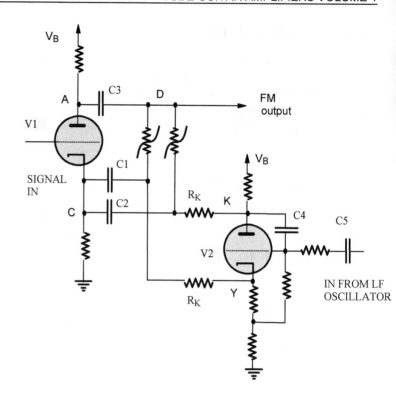

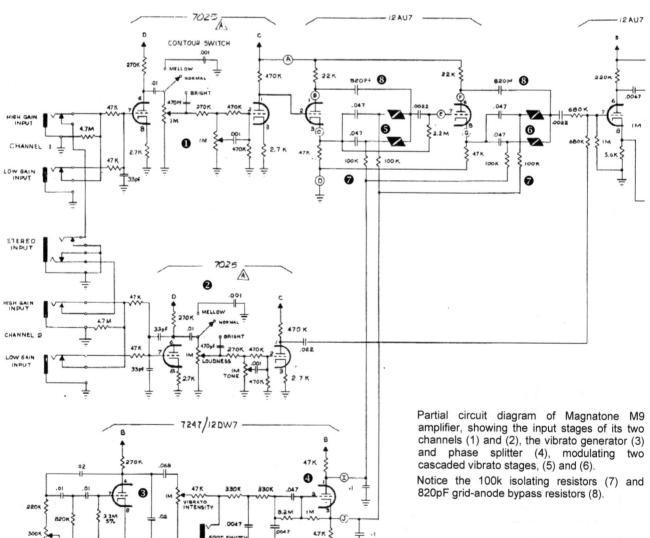

Partial circuit diagram of Magnatone M9 amplifier, showing the input stages of its two channels (1) and (2), the vibrato generator (3) and phase splitter (4), modulating two cascaded vibrato stages, (5) and (6).

Notice the 100k isolating resistors (7) and 820pF grid-anode bypass resistors (8).

Partial circuit diagram of Magnatone MP-1 amplifier, showing the input stage (1), tone controls (2), LDR-base tremolo (3), and reverb tank driver using a phase splitter (4) and a push-pull driver transformer (5). The wet and dry signals are mixed in the cathode circuit of the mic./reverb mixer tube (6), and that combined signal is mixed with the microphone input signal on its grid (7).

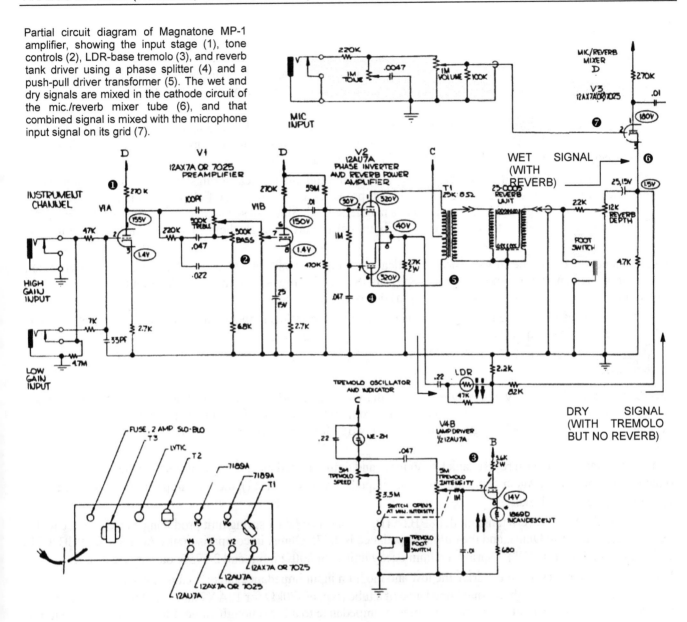

Estey T-32 and Magnatone models 431 and 432 have both vibrato and tremolo, which share the same LF tremolo oscillator. The player can select one or the other effect manually (8) or via a footswitch (9) but not both simultaneously. The *Intensity* and *Speed* controls thus work with both effects!

LEFT: Estey model T-32 is, for intents and purposes, Magnatone Starlight 431. Model 432 is an electrically identical combo but with a 12" speaker.

REVERB

Reverb or echo is an acoustic effect that happens naturally in large rooms due to various reflections of the sound waves from hard surfaces, mostly walls, floors, and ceilings. While too much of such effect is unwanted (the room is too "live"), the absence of echo is equally unpleasant; when the surfaces are too absorptive, the room is acoustically "dead."

The same effect can be created by various electromagnetic means. One of the options was a tape-based unit, which would record the original, non-delayed signal on a piece of continuously running tape and play it some short time later, and then combining such a delayed signal with the original sound.

A more popular way of creating a time-delayed signal was through mechanical springs or "reverb tanks." Signals are brought into an input transducer, whose magnetic coil produces an AC magnetic field that exalts mechanical force and vibrates the magnets placed in the transducer's magnetic field. Since these magnets are mechanically connected to transmission springs, the mechanical wave (vibrations) travels through them in finite time, determining the delay.

Once it reaches the end of the springs, the mechanical wave bounces back and travels in the opposite direction, albeit with a diminished amplitude, and this process repeats itself a few times. The mechanical parameters of the spring (coil diameter, wire thickness, and spring's length) determine the delay.

The reflected waves vibrate the moving magnets of the output transducer, an AC magnetic field is recreated and an electrical signal is induced in its coil.

Reverb tanks

Reverb tanks come in a variety of configurations, with two or three springs. In terms of delay time, they are grouped into short-, medium- and long-delay designs, ranging from 30 to 45 milliseconds (ms) per spring. Decay times (until the sound dies down) are much longer and range from two to four seconds for guitar and much shorter 1-1.5 sec for voice.

As a passive transducer, each reverb tank has an input and output impedance; it is important to match the driver stage to the input impedances. Again, there are low input impedance tanks (8-10Ω), medium impedance (150-600Ω), and high impedance tanks, 1-2kΩ!

For instance, the MOD reverb tank, model 8BB2A1B, is a short (9 ¾") 3 spring unit producing a medium decay. Its input impedance is 190 Ohms, and the output impedance is 2,575 Ohms. in comparison, the Accutronics 9EB2C1B reverb tank is a long (16 3/4") 3 spring medium decay unit with 800Ω input and 2,575Ω output impedance.

A transformer-coupled stage must drive the low and medium input impedance tanks because most of the drive signal would be lost on the very high internal impedance of a tube (typ. 60-70kΩ for 12AX7 and 30kΩ for 12AT7), so even paralleling two triodes would not reduce the internal impedance to a low-enough value. Luckily, since the required drive power is relatively low and bass frequencies below 200Hz do not contribute to the echo effect, such driver transformers can be very small and thus cheap.

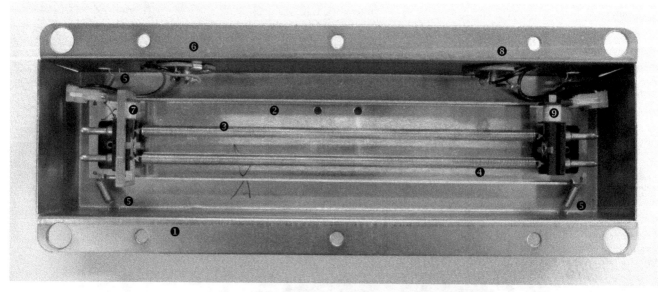

ABOVE: Spring reverb tank form Laney TF300 combo amp. Outer steel case (1), inner aluminium case (2), transmission springs (3 and 4), support (suspension) springs (5), input RCA socket (6), input transducer (7), output RCA socket (8), output transducer (9).

Two stage reverb driver: Teisco Checkmate 20 amp

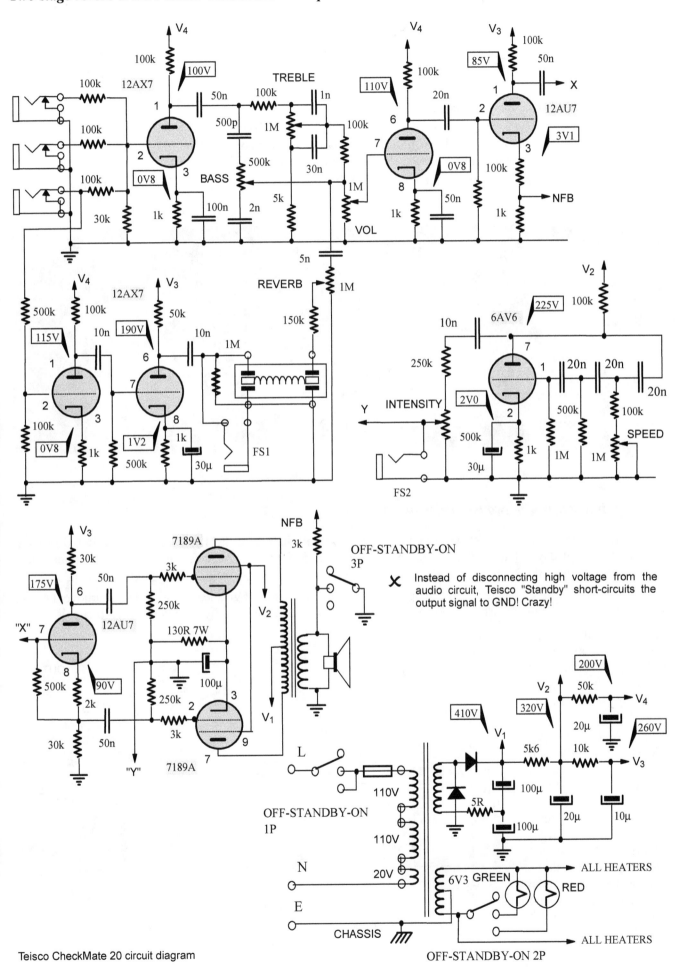

Teisco CheckMate 20 circuit diagram

The myth that Japanese guitar amps from the 1960s were simply copies of American amps is just that - a myth. The topology of this amp is very different from most American amps of the time.

All three paralleled inputs attenuate the input signal down to 23% of its value due to the 100k-30k voltage divider (30/130=0.23). Input #3 also feeds a two-stage 12AX7 reverb driver, meaning the reverb effect is only available if a guitar is plugged into input #3. Thus, the downstream bass and treble tone controls do not affect the reverb signal, which is returned to the volume control without any reverb recovery stage.

Instead of the usual "driver-reverb-recovery stage" layout, this amp has a "driver-driver-reverb" topology. Since the reverb tank is driven directly by the 12AX7 triode, it is obviously of a high input impedance variety.

The tremolo oscillator is more conventional, although it's based on a less commonly used 6AV6 duo-diode + triode tube. The two diodes are not used. Why was such an obscure tube chosen? Heaven only knows.

The choice of 12AU7 for the driver and concertina (split load) phase inverter was wise, the right tube for such application.

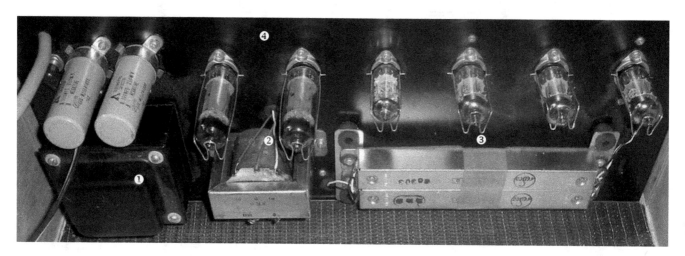

ABOVE: The internal layout of the Teisco amp is neat and logical. Tube shields were not used.

Reverb driver stages with two paralleled preamp triodes

The most common reverb drive stage in Fender amps used both halves of the 12AT7 tube in parallel, driving a transformer-coupled reverb tank. The current through the stage is 6/0.68 = 8.8mA. Its tiny driver transformer, rated at 2.5 Watts, reflects the 8Ω input impedance of the reverb tank back to the primary as a 25 kΩ impedance.

The high pass input filter, formed by the 500pF coupling capacitor and 1M grid leak resistor, has a turnover (-3dB) frequency of 318Hz, thus ensuring the lower bass frequencies don't unnecessarily muddle the reverb circuit.

To achieve the largest possible signal swing and clean headroom, the anode supply voltage of this stage is relatively high, +400V.

The reverb recovery stage is a standard common cathode amplifier with a 12AX7 triode, with a voltage gain of 45-50 times. Closing the reverb pedal's contact (normally open) short-circuits the grid of the recovery stage to ground and thus kills any reverb signal.

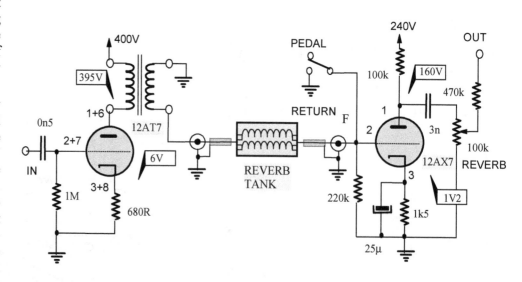

ABOVE: The reverb circuit of Fender Princeton Reverb amplifier

Reverb driver stages with a power tube

In VHT Special 12/20 RT amp (next page), the reverb tank is driven by a transformer-coupled single-ended 6V6 stage. The time constant of its grid RC network (2n2&1M) is 72 Hz. Should you wish to "clean" the reverb tone by preventing the bass frequencies from "clogging" its reverb tank (I've said that well, haven't I, in highly technical terms), decrease the coupling cap to 1n or even 470pF.

Bass frequencies do not contribute to the reverb effect anyway, so experiment, and you may be pleasantly surprised. This applies to all amps with reverb, not just this one.

Alternatively, reduce the cathode bypass cap from 22µF (-3dB frequency of 15Hz) to 1µF. The cathode circuit's -3dB frequency will increase to around 338 Hz.

The output of the reverb recovery stage (V2a) has a 3n3 coupling cap, but a 4n7 cap can be switched in parallel, which shifts the lower -3dB frequency of the coupling RC circuit from 482 Hz with only 3n3 in the circuit to 199 Hz when both are in parallel (deep reverb).

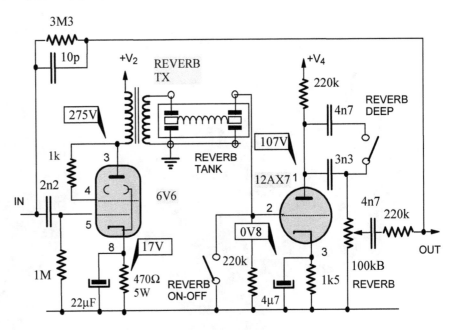

ABOVE: The reverb circuitry of VHT Special 12/20 RT, © VHT

EFFECT LOOPS

Since the "Send" signals for external effect pedals are usually taken after the first or second preamp stages, their amplitudes are more than large enough. Thus, the inclusion of a cathode follower, which reduces the signal level a bit (10-20%), does not affect the loop's operation. It does, however, lower the output impedance of the "Send" circuit and better matches the often low input impedance of solid-state effect pedals. This eliminates the loading effect on the previous tube amp stages and unwanted distortion.

Remember, a cathode follower is an impedance transformer due to its very high input and low output impedance.

The VHT Special 6 circuit is a typical example of a cathode follower driver stage and a common cathode "recovery" amplifier. Notice its very high anode resistor (200k), indicating a high voltage gain.

Strangely, the "send" and "return" levels are not adjustable, but such a modification isn't difficult; replace the 100k resistor at the "send" output and the 1M resistor at the "return" jack with potentiometers of the same values.

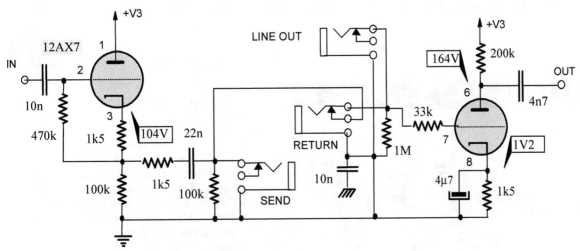

The cathode follower-driven effects loop of VHT Special 6 Ultra, © VHT

CATHODE FOLLOWER-DRIVEN EFFECTS LOOP: VHT SPECIAL 6 ULTRA

This feature-rich SE 6V6 combo amp has to be the best value of all. An eBay business seller had a 240V mains version for US$249, but after our low ball offer countered with US$199, which we duly accepted. Heck, you cannot buy an unloaded cabinet for that money, let alone a whole amp.

Its list of features makes a tube amp designer/builder salivate like a dog in Pavlov's experiments. There are three speaker impedances (for external speakers, the internal speaker is 16Ω), a *Standby-Pentode-Triode* switch, two inputs and channels (*Clean* and *Boost* or *Ultra*), a cathode follower-driven effects loop, and an extremely handsome chrome-plated twelve-inch speaker! What's not to like?

There is also a continuously variable power control (aptly named *Watts*), achieved via a potentiometer that varies the output of a solid-state high voltage regulator and thus changes the V+ supply voltages to the anode and screen grid of the 6V6 tube. However, changing the anode supply voltage also changes the operating points of that stage and its tone and distortion levels as well, not just the volume.

Although there is only a single *Tone* control knob, voicing is also varied through the 3-position *Texture* switch and an 11-position *Depth* switch.

Finally, what makes this large and heavy (for a 6 Watter) amp such an irresistible bargain is its terminal board construction. The terminal board (photo below) is the thickest we have ever seen. The components are well laid out and marked in terms of their part numbers (R1, C8, etc.) and their values and wattages! Thus, you can find each in a matter of seconds.

All the soldered connections are accessible from the top of the board, so you don't have to take it out to change any components.

The output tube is AC heated, while the preamp tubes are heated using a regulated DC voltage.

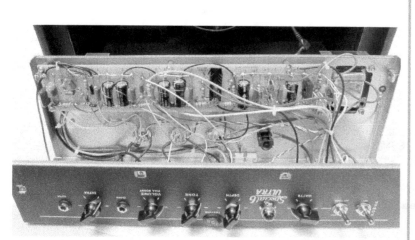

VHT Special 6 Ultra

- Controls: Clean Volume, Ultra Volume, Tone, Texture, Depth
- Inputs: Clean, Ultra
- 2x12AX7 + 1x6V6 tube
- Solid state rectification
- Standby-Pentode-Triode switch
- Variable power control ("Watts")
- Foot-switchable boost mode
- Cathode follower-driven effects loop
- Terminal board construction
- Output power: 6 Watts
- Speaker impedance: 4-8-16Ω (Switchable)
- Internal speaker: 12" VHT ChromeBack
- External Speaker jack, Line Out jack

The output transformer

The rule-of-thumb that applies to single-ended hi-fi output transformers says that the power rating of the transformer's magnetic core should be at least ten times the maximum output power. Special 6's OT complies with that rule; its lamination stack is rated at more than 65 Watts!

This is very rare for a guitar amp; most EL84 or 6V6 amps rated at 5-7 Watts use tiny transformers rated at ten or less Watts. Such transformers will saturate very early, especially at low (bass) frequencies.

VHT Special 6 Ultra
OUTPUT TRANSFORMER:

- EI76.2, A=25.4 mm, S=32mm
- $A=a*S = 8.1 cm^2$
- $P=A^2= 8.1^2= 65.6W$
- Primary DCR: $R_P=189\Omega$
- $L_P= 11.2H@120Hz$
- $L_P= 11.5H@1kHz$
- $L_L=36$ mH
- $QF = 11.5/0.036 = 319$

Based on the OPT size, Special 6 should be capable of producing a thundering bass, of course, providing that high pass filters inside the amp do not artificially attenuate those frequencies. The primary DC resistance is relatively low, under 200 ohms, indicating that a reasonably high diameter wire is used to wind its primary winding. The quality factor isn't great but is certainly above average, so the amp should reproduce treble frequencies quite well.

The power supply

ABOVE: Special 6 Ultra power supply, © VHT

The audio circuit

Whenever I see separate guitar inputs for high and low gain "channels" (Ultra In, Clean In), I am puzzled. You are merrily playing along, and at some point in a song, you want to change the tone to a screaming fuzz or crunch. You need to stop playing, unplug your guitar from one and plug it into the other input. It doesn't quite work, does it? It is so easy to switch inputs or channels with a foot pedal, so why would anyone use this topology is mind-boggling.

As can be seen more clearly from the block diagram of this design, most controls are implemented in the output stage. Apart from the relatively standard *Impedance Selector* and *Triode-Pentode* switch, *Depth* is an 11-position switch that changes the cathode resistor's bypass capacitor and the amp's lower -3dB frequency limit. The *Texture* switch is about the high-end of the spectrum, providing three different treble roll-off options.

Referring to the detailed circuit diagram on the next page, apart from the single-knob *Tone* control, there are many other voicing components in the preamp stages, the most obvious being a fixed tone stack (1).

The others are HF bypassing capacitors across various electrodes of preamp stages' tubes. For instance, both the *Ultra* and *Clean* stage have 100p caps between their anodes and cathodes (2), while the *Ultra* input stage has no less than three more HF compensating/shunting caps (3).

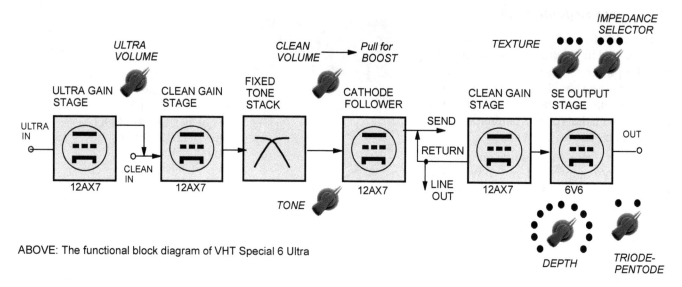

ABOVE: The functional block diagram of VHT Special 6 Ultra

It seems the designer of this amp spent some significant time trying to tame its high-frequency response. Each input also features a couple of HF shunting caps to the chassis ground (4).

The *Boost* function is activated by inserting a 68k resistor between the bottom of the tone stack (6k8 resistor) and the ground. The 68k resistor is normally bypassed by the NC (Normally Closed) manual switch on the *Clean* volume pot. When that pot is pulled, the switch opens, and the higher shunting resistance of the stack increases gain by reducing the voltage attenuation of the stack. Once the manual switch is pulled (open), the *Boost* footswitch FS is enabled, so its closing reduces the gain, and its opening activates the *Boost*. When the manual switch is closed (Boost not activated), the opening and closing of the FS have no effect.

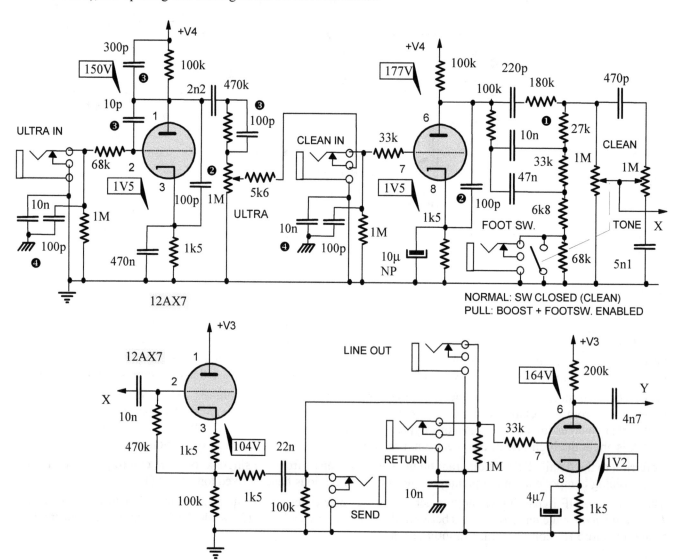

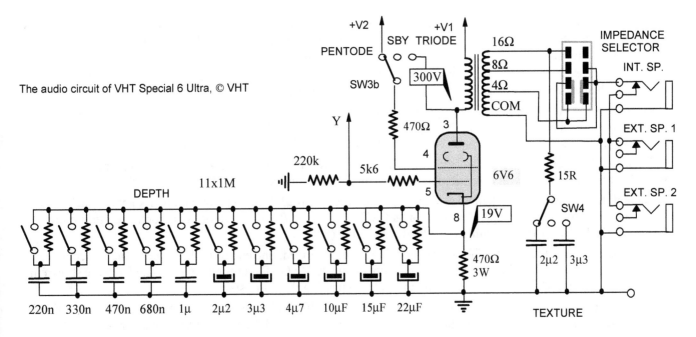

The audio circuit of VHT Special 6 Ultra, © VHT

Modified audio section

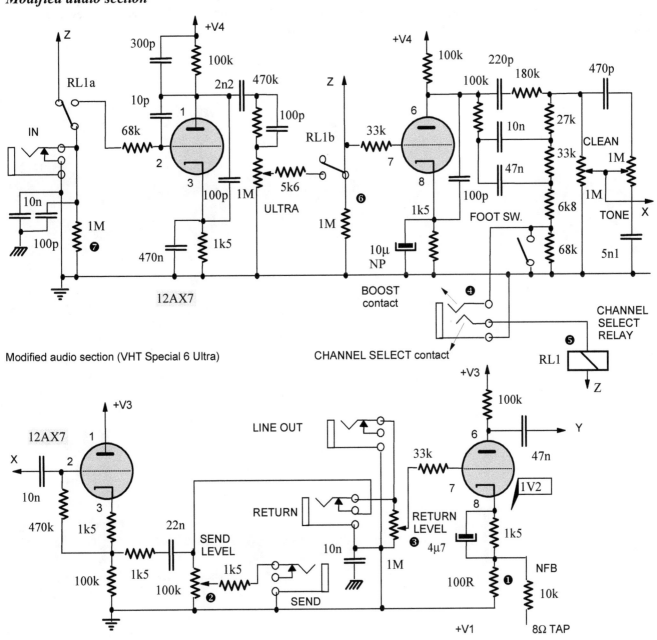

Modified audio section (VHT Special 6 Ultra)

Since there are two high gain stages in *Clean* mode and three in *Ultra*, there is plenty of gain in this amp. Should you wish to add negative feedback to firm up the bass and clean up the top end, that is easily done by adding a voltage divider, two resistors, 100Ω in the cathode of the driver stage (1), and the 10kΩ feedback resistor from the output tap you use most often (4Ω shown). The 10k would provide very mild feedback; for more feedback, less power and cleaner sound decrease its value to 6k8, 4k7, or even 2k2.

If you aren't using the *Clean* input jack anymore (assuming you implement our next mod, pedal channel switching), use its hole in the chassis to install an NFB pot. That would vary the NFB strength by making that 10k feedback resistor variable! Such a modification is easily reversible, and no chassis drilling is involved.

Adding *Send Level* or *Return Level* control (or both!) in the effects loop is also relatively easy. Replace the 100k (2) and 1M (3) fixed resistors with potentiometers, and you are in business.

Implementing foot-switchable channel selection is a bit more involved. First, replace a mono footswitch jack socket with a stereo or two contact type (4). One contact will remain the *Boost* activator, as before, the other one will change channels. How? By activating the added 6.3V relay (5).

When *Channel Select* footswitch FS is open, the relay is de-energized, one end of its coil is permanently connected point "Z," 6.3V$_{DC}$, but the other end is "in the air," not connected to GND. Its contacts Rl1a and Rl1b are shown in a de-energized state, which selects the *Clean* option by bypassing the Ultra stage.

Once the FS closes, it shorts the Channel Select contact to GND, and relay RL1 is energized. Contacts a and b change state and insert the *Ultra* gain stage between the amp's input and the input of the *Clean* stage.

Had RL1 been a "make before break" type, there would be no need for the added 1M grid leak resistor (6) since the grid at pin seven would always be referenced to GND through the 1M resistor of the *Ultra* stage (7), even during changeover times. Otherwise, an audible annoying "pop" would result.

POWER SUPPLIES FOR TUBE AMPLIFIERS

10

- EVALUATING POWER TRANSFORMERS
- TESTING POWER TRANSFORMERS
- RECTIFIER AND FILTERING CONFIGURATIONS
- A SIMPLE "HI-LO" POWER CIRCUIT: EMERY SOUND SUPERBABY
- VARIABLE HV ANODE SUPPLY: VHT SPECIAL 12/20 RT
- REGULATED POWER SUPPLIES: IBANEZ TUBESCREAMER TSA5TVR-S

There are at least two and often three types of supplies needed in a typical tube guitar amp. Tubes need to have their heaters energized and glowing happily. Although some choose DC heating, most amps use AC voltages for tube heaters. Audio circuits, preamp, and output (power) stages need steady and filtered DC anode supplies. Finally, if fixed (external) biasing is used, power tubes need a source of negative (with respect to ground or common terminal) bias voltages brought to their control grids.

In almost all cases, all three supplies share the same power (mains) transformer, whose three secondary windings provide step-down voltages for heater and bias circuits and step-up (higher) voltages for anode supplies. These are then rectified and filtered using chokes, resistors, and capacitors ("filters') to eliminate the AC or pulsating component ("ripple") of the DC voltage.

Some amp designers choose to regulate DC voltages using active elements (tubes, transistors, or integrated circuits). These act as servo amplifiers (in the same fashion as cruise control works in a car), constantly monitoring the DC levels at their outputs and adjusting it as the input voltage varies or as the load changes.

The five aspects of a power supply: voltage, current, energy storage, voltage regulation and ripple

A power supply inside audio equipment is invariably a DC (direct current) supply. Just as with amplifiers, we can define the attributes of an ideal DC power supply. First & foremost, it performs AC to DC conversion, converting mains AC voltage of a particular frequency (50 or 60Hz) to a steady DC voltage of the required level. This could be a low voltage used for tube heaters, typically 6.3 or 12.6V, a negative bias voltage, typ. -10 to -60V, or a high anode voltage, typ. 200-500V.

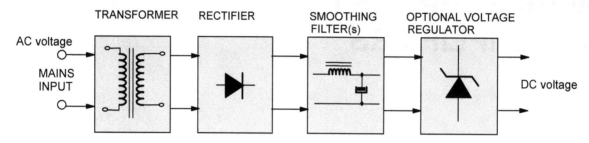

ABOVE: Block diagram of a typical power supply in a tube guitar amplifier

Apart from the desired voltage, every power supply must have a required current capacity; for instance, it must be able to supply $400V_{DC}$ anode voltage at 160mA, which would be the current drawn by the load (the audio section of the amplifier). Since we have the voltage and the current drawn by the audio section, the whole audio section (amplifying section) of an amp behaves like a resistor from the power supply's point of view! In Class A amplifiers, the current draw of the audio section is almost constant, so the audio section behaves like a fixed resistor; in this case, its load resistance is R_L=400V/0.160A= 2,500 ohms.

A power supply must also provide adequate energy storage. Energy can only be stored in the filtering chokes and capacitors, not in resistors, so "stiff" superior power supplies use high inductance chokes and large capacitor banks.

A power supply must be efficient in its energy conversion from an AC to DC supply, i.e., its internal losses should be as low as possible. Low losses mean less heat dissipation, less vibration, no buzzing of transformers, and a cooler and quieter running amplifier. Internal losses are modeled by the internal resistance of a power supply, which would be zero in an ideal case.

One consequence of the internal resistance of a power supply is the voltage drop that load current will cause on such a resistance, thus reducing the output voltage (voltage divider effect). So, the output of an actual power supply will drop (sometimes called a voltage "droop" or "sag") with increasing load current. An ideal DC supply provides a constant output voltage regardless of the load and its current demands (current draw). The factor that describes how close an actual power supply comes to an ideal one is called voltage regulation.

Since the output DC voltage is obtained by rectification and filtration from an AC waveform, some remnants of that AC, called ripple, remain. While an ideal DC supply would provide a perfectly horizontal line, the output voltage of real power supplies fluctuates above and below some steady DC value V_0. Thus, the smaller that "ripple" or AC component, the lower the hum on the amplifier's output.

For instance, if the voltage of a push-pull guitar amp drops from 350V at no load to 300V at full load, the regulation factor of that power supply is 16.67%.

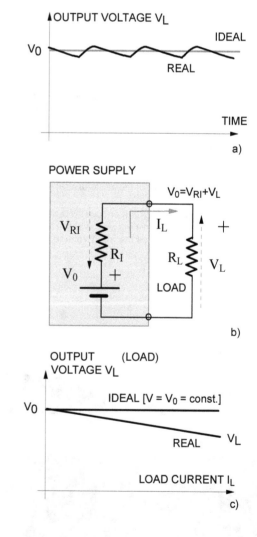

a) The output voltage waveform of an ideal and real DC power supply

b) The model of a real DC voltage source with internal resistance R_I

c) The output voltage versus load current characteristics for the ideal power supply, with (constant output regardless of the load ($V_L=V_0$) and a real, unregulated power supply (drooping voltage V_L)

EVALUATING POWER TRANSFORMERS

Transformer theory and operating principles are a complex subject, and most tube amp builders find it too difficult and time-consuming to master this specialist topic. So, we will limit ourselves here only to the practical aspects of power and output transformers used in guitar amps. For detailed coverage of the subject, please refer to my book "Audiophile Vacuum Tube Amplifiers," Volume 2.

As with guitar amps, we have three choices, buying a finished transformer, reusing a transformer from another device, or designing & winding one from scratch. It is important to know if a particular transformer is suitable for use in your design, so we will focus our analysis on that issue. This will ensure a transformer is appropriately sized for the load, both in magnetic (the core size) and electric terms (the wire sizes), so it can deliver the voltages and currents needed by the audio section of the amplifier.

How to identify windings of an unknown power transformer

The rules for identifying transformer windings are "the lower the voltage, the higher the current" and "the thicker the wire, the lower the resistance of the winding".

The heater winding will have the thickest wire and the lowest winding resistance (say 0.3-3Ω). The primary winding will have an order of magnitude higher resistance (say 10-30Ω), and the high voltage winding an order of magnitude higher again (1:10:100Ω).

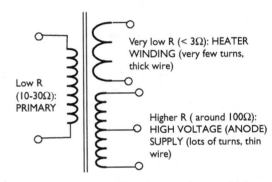

Typical DC resistances of mains transformer windings

Quick estimation of EI transformer's power rating

EI power transformers are still the most common in tube guitar amplifiers. The following discussion applies only to transformers that use standard (wasteless) EI lamination, not toroidal or C-core transformers. The power rating of the core is the area of the center-leg's cross-section in cm^2 (CL times the thickness of the lamination stack S) squared. The width of the Center Leg is double the width of the I-section. From the drawing, you can see that you can determine U (one unit or the width of I) by measuring the outside dimensions. Ensure you measure the laminations, not the overall dimensions, including the thickness of end-bells or covers.

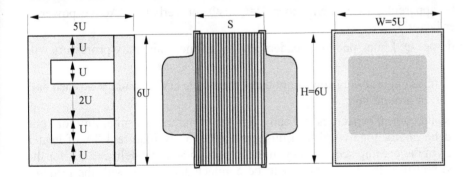

By measuring external dimensions of wasteless EI-laminations on finished transformers, their type and internal dimensions can easily be determined.

S is the thickness of the lamination stack, H is the height (the larger of the two outside dimensions).

For example, transformers using EI96 laminations will have H= 96mm, so their I=96/6 = 1.6 cm. To get the cross-sectional area, divide the height by three and multiply by the stack thickness S. H/3 = 3.2 cm, and if the stack thickness is 4.5 cm (one of the standard bobbin sizes), A is 14.5 cm^2 and the power rating of that transformer core is $P=A^2 = 14.4^2 = 207$ VA

P [VA] = A^2 = $(CL \times S)^2$ and since CL=2U and H=6U, we get P [VA] = $(HS/3)^2$

EI TRANSFORMER'S POWER RATING

P [VA] = A^2 = $(HS/3)^2$
H and S must be in cm!

MAINS TRANSFORMER CORE SIZING

For 50 Hz $A_{EF} \approx \sqrt{P}$ For 60 Hz $A_{EF} \approx 0.9\sqrt{P}$
P in VA, A_{EF} in cm^2
For C-cores halve these A_{EF} figures.

Case study: power transformers Sherwood S-1000 amplifiers

You are building a guitar amp with two 6L6 in push-pull and are considering buying a vintage power transformer on eBay, salvaged from Sherwood S-1000 mono hi-fi amplifier. It has a 115V primary, and the secondaries have most of the required information marked on the S-1000 circuit diagram available on the web (shown on the diagram). How to assess these transformers' suitability for your project?

1. The rectifier heater winding specs are not given, but you know that the GZ34 rectifier in S-1000 amp needs 5V at 1.9A, so you can conclude that you have a 5V@2A winding.

2. The heater winding for output tubes is also unmarked, but the amp uses four 6BQ5 pentodes in push-pull; these need 6.3V@ 0.8A each, or 6.3V@3.2A total, so the winding is most likely rated at 6.3V@4A!

3. The 21V CT winding serves a dual purpose; once rectified it provides a -30V DC bias supply and also heats three 0.15A preamp tubes' heaters (two at 12.6V and one at 6.3V). The 150 mA current rating is marked on the schematics.

4. Finally, the HV winding is 370V CT, rectified into $480V_{DC}$ @ 80 mA. Now, 80 mA for four EL84 output tubes seems very low, most likely the idle current, not the maximum current the winding can provide.

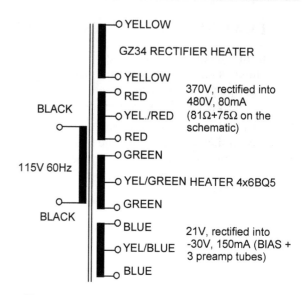

There are two ways to estimate the maximum current this winding can provide without testing the transformer. The 6BQ5s in PP class AB1 can draw up to 46 mA each, plus four mA each of the screen current, so four tubes x 50mA can draw up to 200 mA in total.

The more information the manufacturer provides the easier and quicker it is to evaluate a power transformer without actually testing it.

Each half of the center-tapped winding must be able to provide at least 100 mA since their currents add up.

The second method uses power draws. You need to estimate the power rating of the transformer core from its dimensions, calculate power draws for all secondary windings except the HV winding, then subtract the secondary power draws from the power rating of the core to get the rough power rating of the HV winding.

The longer dimension of the core is 10.4 cm, the shorter one is 8.7 cm, and the stack is 3.0 cm. From imperial lamination sizes (see the reference data at the back of the book), we find that these are EI104.8 (or L137) laminations, 10.48 cm long and 8.73 cm wide. The center leg is 3.492 cm wide, so A=aS= 10.47 cm^2 so we estimate A_{EF} =0.96A=10.05cm^2 and get $P_{TOT} = A_{EF}^2 = 101$ VA.

$P_S = 5V*2A + 30V*0.15A + 6.3V*3.2A = 10+4.5+20.2 = 34.7$ VA, so $P_{HV} = P_{TOT}-P_S = 101 - 34.7 = 66.3$ VA

Instead of W (Watts), the unit for real power, we use VA (volt-amperes) here. Since we don't know the cosΦ (the power factor), we are talking about the apparent power. Remember, P=QcosΦ, or verbally, "Active power = Apparent power multiplied by the power factor"!

Strictly speaking, instead of P, we should use Q; I hope pedantic readers and my former university professors will forgive me this transgression.

Finally, $I_{MAX} = P_{HV}/V_0 = 66.3/386 = 172$ mA. This is the current supplied by the whole HV secondary, so each half of the CT winding can provide half of that, or around 86 mA.

Now you have all the secondary voltages and current capacities and can judge the transformer's suitability for your project. 200mA of anode supply should be plenty for two 6L6 in push-pull. They only require 1.8A of heater current, plus 600mA for, say, two 12AX7 in the preamp section, so your heater secondary can deliver 2.5A for the heaters. If you don't want to use a tube rectifier, the 5V winding can be connected in series with the primary winding, as per the trick explained below.

Power transformer buck or boost connection

If your mains voltage is around 120V or even higher, but your mains transformer has a 110V or 115V primary, you can connect an unused 5V or 6.3V secondary winding in series with it to get a higher voltage rating.

Likewise, if the primary was designed for 230V and your mains voltage is nominally 240V and in reality closer to 250V, as ours is, you can fix that problem with a 12.6V secondary winding or make it less troublesome with a 5.0 or 6.3V winding!

BOOST: Connect the primary and the rectifier heater winding in phase, so their voltages add up.

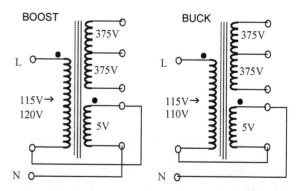

The buck and boost connections of a low-voltage secondary winding (5V in this case), in series with the primary winding!

Instead of 115V primary, you now have a 120V primary. All secondary voltages will be 120/115 = 4.4% higher. Instead of 2x375V HV secondary you will get 2*375*1.044 = 2x391V!

BUCK: The windings are in series, but this time out-of-phase. The winding ends without a dot are connected together. Instead of 115V primary, you now have a 110V primary. 110/115 = 0.9565, so all secondary voltages will be 4.4% lower. Instead of 2x375V you will get 2*375*0.9565 = 2 x 359V!

CAN POWER TRANSFORMERS DESIGNED FOR 60Hz BE USED ON 50Hz MAINS, AND VICE VERSA?

From the fundamental TX equation we can express turns per volt $TPV = N_1/V_1 = 104/(4.44fBA_{EF})$

Since TPV and A_{EF} are constant for a certain transformer, when a 60Hz unit works on 50Hz mains, the frequency will go down 16.67%, so B (magnetic flux density) will have to go up 20% (1.2 times) to make up for it. Thus, it all depends on the level of B the transformer was designed to operate at 60Hz.

If the transformer was operating at high levels of B, 1.5-1.6T, it may saturate at 50Hz (B up to 1.8T)! Many American mains transformers from vintage tube amps were made to a strict budget with a penny-pinching attitude and would be marginal at 50Hz at best or totally in saturation at worst.

For 50Hz transformers working on 60Hz mains, the frequency f will go up, the flux density B will go down, so such transformers will run cooler and quieter.

Case study: Triad isolation transformers

Many of the amps we bought for testing, analysis, and improvement, to be used as case studies in this book, we got from the USA, where the mains voltage is 115/120V, while our mains voltage in Australia is 240V. Instead of using an outside step-down transformer, it is neater to either replace the amp's power transformer with one that has the same secondary voltages but a 240V primary winding or to install a step-down transformer inside the amplifier.

Triad N-68X is made in China and sold by Allied Electronics (US$13.30 in 2017). Its claimed power rating is 50VA, which should be enough for small single-ended guitar amps.

Its bigger brother, model N-77U, is rated at 100VA and in 2017 was being sold for US$33.80. The shipping was swift (by DHL air courier) and cost US$35 for four small and one large transformer.

The idle current draw of a transformer is a good indication of its quality. While 64mA for a 100VA unit was relatively high, by coincidence, the idle draw of the smaller transformer was also 64mA! That was the first warning sign, indicating that poor quality (high loss) laminations were used.

The magnetic core of N-68X is only rated at around 30VA, so proclaiming it to be capable of 50VA is fraudulent. When loaded with a 50-watt load, that transformer started buzzing, and after 5 minutes, it got so hot that one could small the melting varnish and the metal frame was far too hot to touch! The verdict: Don't waste your money!

The 100VA N-77U has a 118VA core so that rating is accurate once losses of 18VA are deducted.

TRIAD N-68X MEASURED RESULTS:

- No load primary current: 64mA
- EI66 laminations (a=22mm)
- S=25mm stack thickness
- Core power rating: $P=S^2=(a*S)^2 = (2.2*2.5)^2 = 30.25$ VA

TRIAD N-77U MEASURED RESULTS:

- No load primary current: 64mA
- EI85.8 laminations (a=28.6mm)
- S=38mm stack
- Core power rating: $P=S^2=(a*S)^2 = 118$ VA

BELOW: Triad N-68X (left) and N-77U isolation transformers.

Case study: Vintage Nordmende power transformer

Vintage (and modern) USA- and UK-made power transformers usually only specify AC voltages their various windings produce. Even the crucial info, such as their current ratings, is often omitted.

In contrast, vintage German-made power and output transformers offer more detailed specifications. This salvaged Nordmende power transformer includes the allowable variations from the nominal secondary voltages, in this case, +/-5%.

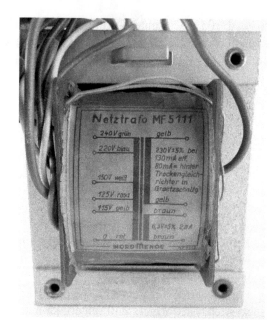

So the high voltage of 230V can vary between 0.95*230V and 1.05*230V, or 218.5V - 241.5V, and the heater voltage between 0.95*6.3V and 1.05*6.3V or 5.985V - 6.615V.

More importantly, it warns the user about the relationship between the AC current a secondary winding can supply and the maximum possible DC current after full-wave rectification; "Trockengleichrichter" means dry rectifier and "Graetzschaltung" means a full-wave rectifier with four diodes (a "bridge" rectifier).

Notice that although the transformer's HV secondary can supply 130mA$_{eff}$ the DC current is only 80mA! The subscript "eff" means "effective" or "Root Mean Square" (RMS).

So, the Rule-of-Thumb says that only about 62% of the windings AC current rating is available as a DC current after full-wave rectification. For instance, if the high voltage power supply of your amp must be able to provide 100mA$_{DC}$, the HV secondary winding of the power transformer must be rated at a minimum of 100/0.62 = 160mA$_{DC}$!

TESTING POWER TRANSFORMERS

Expensive test instruments aren't needed to perform the basic transformer tests and measurements. A good quality multimeter is a must-have, preferable a True-RMS type. It is also essential that the multimeter be sensitive enough to measure AC and DC millivolts (mV), not just Volts.

Measuring transformer's no-load current and power losses

BOB'S BLACK BOX (A SIMPLE POWER METER)

DIY PROJECT

Bob's Black Box (or BBB for short) is a simple wattmeter, or, more precisely, a VA-meter (apparent power, or the products of volts and amps).

The load current flows through the 10Ω, 10 Watt-rated series resistor. Connect a voltmeter or multimeter on the V$_{AC}$ range to the two test points TP1 and TP2. The meter will measure the AC voltage drop on the 10Ω resistor; that voltage is proportional to the load current. The load current is I$_L$=V/10, and by multiplying this current by the mains voltage, we get the power draw in VA.

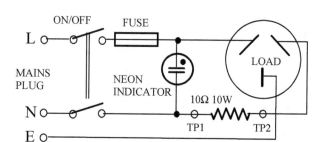

The diagram shows an Australian power outlet, but obviously, any international type can be used. There is a fuse in the active (line) feed for safety reasons, the test points are on the neutral side, and a double-pole switch is used for both line and neutral connections. We housed or unit a plastic ABS box (see the photo).

For safety reasons, use a mains-rated switch with inbuilt neon light, so you can easily see when the box is energized!

BBB is not useful just for powering up and testing amplifiers; by connecting unloaded mains or output transformers to it, we can measure their no-load or magnetizing currents and voltage ratios.

ABOVE: The circuit diagram of Bob's Black Box

RIGHT: A finished power transformer, would by us for one of the projects in this book, is tested using BBB. Its primary winding is connected as load and the secondary voltages are measured by a multimeter (not shown inn the photo) using crocodile clips.

WARNING: This is a dangerous test, since high and lethal voltages are present, so exercise extreme caution!

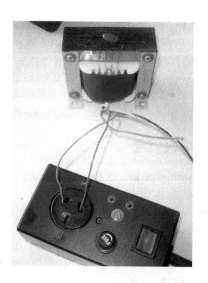

EXAMPLE: What is the power consumption in VA of a power transformer without load (open secondaries), if the measured voltage drop on Bob's Black Box is 0.527 V_{AC} and the measured mains voltage is 247 V_{AC}? The current is 0.527 V / 10 ohms = 0.0527 Amperes or 52.7 mA. The power draw (losses) are 247V*0.0527A = 13VA!

Power transformer phasing check

In most applications, it is crucial to connect two or more secondaries of a transformer the right way so that their voltages are in phase. Say you have a power transformer with two primaries, and you either have to connect them in parallel for 115V operation or in series for 230V mains. How to determine which terminals are "in phase"?You'll need an LCR meter set on "L" or inductance. There are a few possible test combinations, but the principle is the same.

For instance, join wires Y & U and measure the inductance of this series connection at the end terminals X & V.

Say you got a very low reading in mH (milliHenrys). That means the two windings now oppose each other; they are "out-of-phase. Terminals with the "dot" are in phase. So, if you need a series connection for 230V mains, you should join Y & V and use X &U as mains terminals, or join U & X and use Y &V as mains terminals.

For a parallel or 115V connection, join X&V as one input and Y&U as the other!

If your reading was high, a few Henrys, typically 5-10H, that means points X and U are in phase (as are Y and V), so the two windings "support" one another and send their magnetic fluxes the same way. Thus, bridge U &Y and use X & V for mains connection in series.

For a parallel 115V connection, join X&U as one input and Y&V as the other!

The same testing method can be used for secondary windings and for audio (output) transformers.

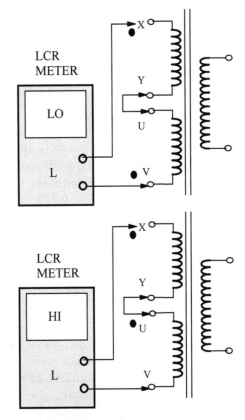

ABOVE: The phasing check using an LCR meter to measure inductance

Case study: Power transformer

Let's practice the basic transformer deciphering skills. Please study the sticker on this power transformer and:

1. Draw its wiring diagram.

2. Add up the total nominal secondary power draws.

3. Compare the total secondary power draw with the rule-of-thumb estimate of the magnetic core's power rating.

EI76x40 means that EI76 laminations were used and that the lamination stack is 40mm wide (S=40mm). EI76 laminations have a=25.3 mm, the width of their center leg.

Using the Rule-Of-Thumb estimation, the center leg cross section is A=aS = 2.53*4= 10.1 cm^2, and its power rating is P=A^2= 10.1^2= 102 VA.

Since P=V*I (voltage multiplied by current), the total power rating of the three secondary windings is P=P1+P2+P3 = V1*I1+V2*I2+V3*I3 = 6.3*3.2 + 40*0.18 +350*0.25 = 20.16 + 7.2 + 87.5 = 114.86 VA!

This agrees with the declared power rating of 115 Watts but exceeds our estimation by 115/102 = 1.127 or 12.7%. Unless very high quality (low loss) laminations were used, this transformer would get very hot if fully loaded.

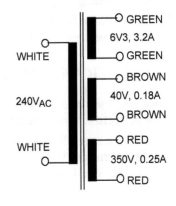

Case study: Power transformer

Bought on eBay for US$20.50 and US$37 airmail (2017). Made in Taiwan by SanLang Co., the primary of this transformer has three taps, 120V, 230V, and 240V.

A 130VA magnetic core can supply 230V@150mA and 6.3V@5A, or a total of P=230*0.15+6.3*5 = 66VA. So, its (electrical) power rating is very conservative for such a large core (only 50%), always a good sign, meaning it will run quietly in an amplifier and stay cool even under full load.

With its HV secondary loaded with a 1,500Ω 50W rheostat (approx. 150 mA of AC current) for 15 minutes, it stayed cool; that winding could most likely supply 200 mA or more!

The voltage regulation VR of the HV winding can be calculated from the voltage drop under full rated load (150mA) as VR= $(V_L-V_0)/V_0$ = (219-236)/236 = -0.072, or -7.2%, a good result. The very low idle current draw of only 22.2mA indicates that high-quality (low loss) laminations were used.

"What were they thinking?!?" case study: A Taiwanese power transformer

We considered this power transformer, made in Taiwan by SanLang Co., for one of the DIY projects in this book. Three primary taps, 120V, 230V, and 240V, are great news. Most similar China- and Taiwan-made units have only a 115/230V primary, unsuitable for use with 240-250V mains voltage.

MEASURED RESULTS:

- Idle primary current (no load): 22.2mA
- EI85.8 (a=28.6mm), S=40mm stack
- Core power rating: $P=S^2=(a*S)^2$ = 130 VA
- HV sec: 236V (no load), 219V (150 mA)
- Heater sec: 6.66V (no load)
- Weight: 1.90 kg

EI96 laminations were used, but the seller didn't mention the stack thickness. The power rating at 50Hz was specified as "79.5VA max". The transformer weighs just under 3kg.

There are three secondary windings, 0-6.3V@5A, for audio tubes' heaters, 0-5V@3A, presumably for the heater of a rectifier tube, and an HV winding 0-275V@120mA.

The winding is neat, and winding wires protrude between the bobbin and the terminal lugs onto which they were soldered. Thus, their thickness can easily be measured with a micrometer.

The current and power ratings for 60Hz operation were also given: 6.3V @6A, 5V@3.6Aand 275V@144mA, for a total of 95.4VA. A quick check: 60Hz/50Hz=1.2, $P_{60Hz}=P_{50Hz}*1.2$ = 79.5*1.2 = 95.4VA! So, the voltages stay the same, but current ratings are 20% higher when a transformer designed for 50Hz operation works on 60Hz mains.

So far so good. However, go back and have a closer look at the secondaries and their current ratings. Firstly, a 5A heater winding would make this transformer suitable to power the heaters of four 6L6 and three 12AX7, for example. That would be a 100 Watt amplifier. Can this transformer power such an amp? No way.

Notice the very low HV current rating of 120mA. Remember, that is an AC current rating, meaning that when rectified we will not get more than 80-90mA of DC current! That isn't enough even for two 6L6 in push-pull.

So, the heater winding is oversized (the wire used is too thick), while the HV winding is undersized, its wire is too thin, and as a result, its current capacity is too low! This transformer is only suitable for a single-ended amplifier with one 6L6, EL34, etc. It will be marginal even in a small PP amp with two 6V6, EL84, or similar low power pentodes or beam tubes.

Worst of all, the HV winding is not center-tapped, so a rectifier tube cannot be used, it requires a solid-state rectifier bridge or a voltage doubler, yet there is a 5V/3A winding for a tube rectifier?

RECTIFIER AND FILTERING CONFIGURATIONS

Once suitable AC voltages are present at the power transformer's secondary windings, all except heater voltages must be rectified and filtered into DC voltages needed for various amplification stages.

As we have seen already, a diode (a solid-state or vacuum diode) is a rectifier due to its unidirectionality. Since it allows the current to flow in one direction only, it "rectifies" an AC signal at its input so only positive peaks are passed through, thus making it a pulsating signal with a DC component.

There are three basic rectifier circuits, a half-wave rectifier, a full-wave rectifier using transformer's center-tap, and a full-wave bridge rectifier. Once we understand how they work, we'll discuss the fourth most commonly used circuit, the voltage doubler.

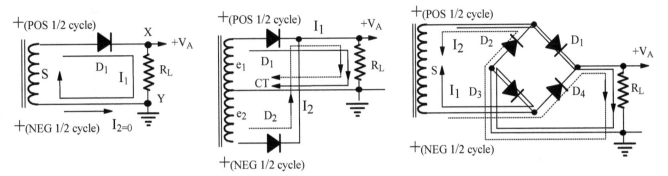

Three basic rectifier circuits (L-R): half-wave, full wave (center-tap) and full-wave bridge. Current flows through diodes for the positive (I_1, full line) and negative (I_2, dotted line) halves of the secondary voltages are indicated.

Half-wave rectifier

Half-wave rectification suffers from three main problems: 1. high ripple 2. unbalanced DC current flowing through the secondary can saturate the transformer's core, and 3. the low efficiency of 40% (DC power out/AC power in). Therefore, it is seldom used, except for negative bias supplies in Class A_1 and AB_1 amplifiers, where it supplies a practically infinite load impedance (no grid current flowing). To get such negative bias voltage, point X should be grounded, and the negative voltage taken from point Y (diagram above left).

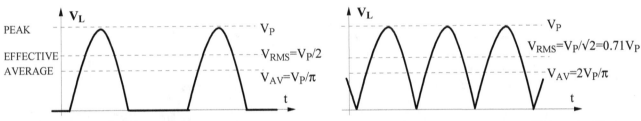

Half-wave rectified voltage (resistive load without any filtering)

Full-wave rectified voltage (resistive load, no filtering)

Full-wave center-tap rectifier

This was the most common circuit with tube rectifiers in the 1950s and 60s due to the large size and high cost of tubes. Once silicon rectifiers took over, this topology was abandoned in favor of the bridge rectifier. The bridge rectifier with tubes would need two separate heater power supplies (or secondary windings), while only one is required for the center-tap arrangement.

The two DC currents (I_1 and I_2) through the secondary winding cancel out, so there is no DC magnetization of the core. Higher efficiency than the half-wave rectifier; however, two secondary windings are needed, doubling the number of secondary turns.

Full-wave bridge rectifier

As with the center-tapped circuit, there is no DC magnetization of the core. Higher efficiency than the center-tapped arrangement, plus half the number of secondary turns required. The price to pay is the need for four rectifiers instead of two, but with small and very cheap silicon diodes, that is not an issue.

In this case, the effective value of the output (load) voltage V_{RMS} is 41% higher than the half-wave circuit ($0.71V_P$ versus $0.5V_P$), the same as for the center-tapped circuit.

RIPPLE VOLTAGE FOR FULL-WAVE RECTIFIERS

$V_{RP} \approx 2.1 I_{DC}/C$ [V_{RMS}]
where I_{DC} is in mA and C in μF

PEAK, AVERAGE AND RMS RECTIFIER VOLTAGES

HALF-WAVE:	FULL-WAVE:
$V_{AV}=V_P/\pi$	$V_{AV}=2V_P/\pi$
$V_{RMS}=V_P/2=0.5V_P$	$V_{RMS}=V_P/\sqrt{2}=0.71V_P$
RIPPLE FACTOR $\gamma=1.21$	RIPPLE FACTOR $\gamma=0.48$

Voltage doubler

If the window area in the power transformer is at a premium and you cannot fit in enough winding turns to get the voltage high enough for a full-wave center-tapped arrangement, and there isn't enough space even for a bridge topology, all is not lost. Providing you can fit half of the required turns, you can always use the voltage-doubling circuit where $V_{DC}=2V_P$, where V_P is the peak value of the secondary voltage.

There is no DC magnetization of the core. However, notice how the secondary winding must be floating; you aren't allowed to ground it due to the specific topology of the circuit. That point can be used as an additional DC voltage source at half of the main V_{DC}.

When point X is positive, current I_1 flows through D_1 and charges capacitor C_1. D_2 does not conduct. When point Y is positive, during the next half-period of the sinusoidal secondary voltage, I_2 flows through D_2 and charges C_2. The voltages on C_1 and C_2 are in series and add up!

The doubler is also less affected by the variations in the AC input than the other full-wave supply circuits. Its topology prevents spikes and interference from being fed from the amplifier back to the mains transformer's secondary and primary.

The output voltage V_{DC} drops as the load current increases, meaning that voltage doublers have a poorer regulation. The ripple factor also worsens in that case. The maximum reverse voltage on diodes is approximately equal to double the AC voltage amplitude on the secondary.

Capacitive filtering

We have already seen capacitive filtering in action in the voltage doubling circuit. A capacitive filter is inherent in this circuit; voltage doubling would not be possible without the capacitive action!

A full-wave rectifier using a vacuum rectifier tube is illustrated on the right. For a moment, disregard the choke L and the second capacitor C_2 and assume that only C1 is connected, a simple capacitive filter.

At point A the output voltage (on the capacitor) follows the sinusoidal waveform. The capacitor is charged by a current supplied by the transformer's secondary winding and the rectifier tube. As soon as the voltage reaches point B, the peak, the voltage on the capacitor becomes higher than the input voltage pulse from the rectifier tube, and the charging current i_S stops flowing. The capacitor now provides the load current, and due to the discharging effect of the load current, the voltage on the capacitor starts to drop.

RIGHT: The price that needs to be paid to achieve low ripple in capacitor filters is the high and sharp current peaks through the rectifiers and mains transformer's secondary winding. Half-wave rectification is used to better illustrate the more pronounced voltage drop between the pulses, but the discussion also applies to full-wave rectification.

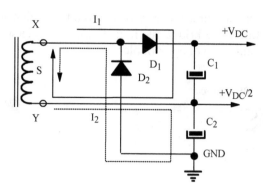

ABOVE: Typical HV power supply using a voltage doubling arrangement with solid state diodes D_1 and D_2

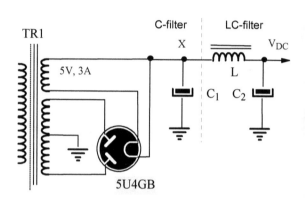

ABOVE: Typical HV power supply using a rectifier tube and a CLC-filter

BELOW: The voltage ripple (AC component superimposed on the DC voltage) and sharp current pulses through the diodes and the power transformer

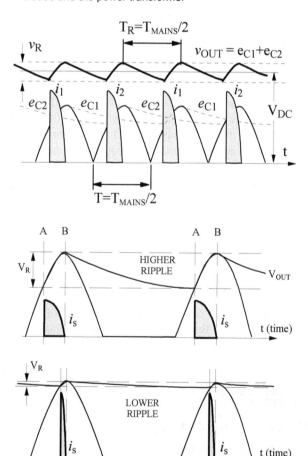

With capacitive filtering, the current through the transformer secondary flows in very short high current pulses, whose spectrum contains many high-order harmonics. These high-frequency components heat both the transformer core and the winding (hysteresis and copper losses).

In some tube amps we had on our test bench, the power transformers got so hot that one could barely touch them after an hour of operation. These transformers will have a short and stressful life and their owners an unavoidable significant repair expense!

Inductive (LC) filters

L-filter (also called LC-filter) features a choke connected to the rectifier output without any filtering capacitor between that point (marked X) and ground (unlike CLC or PI-filter).

The currents through two halves of the high voltage secondary winding are almost constant (the waveform on the next page), again in contrast to currents in transformers feeding CLC filters, which are short-duration high amplitude spikes. The inductor represents a high reactance to the ripple voltage, while the shunt capacitor has low reactance and shunts or bypasses this AC ripple to the ground.

The ripple factor γ (Greek letter *gamma*) is the ratio of the AC (ripple) current and DC load current, or the RMS value of the ripple voltage divided by the DC component of the load voltage:

$\gamma = I_{RMS}/I_{DC} = V_{RMS}/V_{DC}$ and for LC filter $\gamma = \sqrt{2/3} * X_C/X_L = \sqrt{2/3} * 1/2\omega C * 1/2\omega L$. With L in [H] and C in [μF], $\gamma = 0.83/LC$ for 60Hz mains and $\gamma = 1.2/LC$ for 50Hz mains!

Ideally, above I_X, the minimum load current, the output voltage V_L voltage stays constant ($V_L = V_0$) as per curve A). The load current is constant and is comprised of two almost square-wave components i_1 and i_2. In reality, due to DC resistance and AC reactance of transformer's secondary winding, the rectifier tube's internal resistance and choke's resistance, the voltage drops with increased load current (curve B), since the voltage drop on all those resistances increases and subtracts from the ideal level V_0

Secondary transformer voltage is $V_S = V_M\sin\omega t$ and the output DC voltage is $V_L = 2V_M/\pi = 0.637V_M = 0.9V_{RMS}$

By relatively complex mathematical analysis it can be shown that the minimum or critical value of choke's inductance L_{MIN} needed to achieve I_X and proper operation of the filter depends on the load resistance: $L_{MIN} = R_L/(3\omega)$, which is approx. $R_L/945$ for 50 Hz mains frequency or $R_L/1131$ for 60 Hz.

LC filters are seldom used in guitar amplifiers due to two main reasons. A large and expensive choke is required, and even then, the AC ripple is larger than for capacitive filters. The only advantage of LC filters is in improved sonics; that is why many audiophiles prefer hi-fi amps with LC filtering.

Nobody really knows why amplifiers with LC filtered power supplies sound better. The most likely reason is the elimination of sharp current pulses through the power transformer. These pulses contain high energy harmonics which propagate through both the mains circuit and the audio circuitry of amplifiers and "smudge" or defocus their sonic presentation!

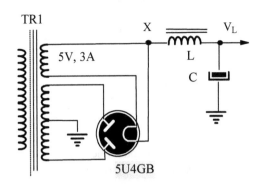

Typical HV power supply using a directly heated rectifier tube (5U4GB) and LC-filter

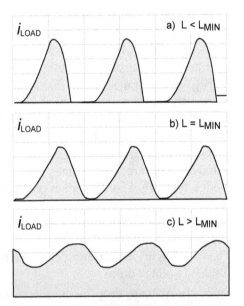

Load current in a full-wave rectifier with an LC filter for a) L < L$_{MIN}$ b) L = L$_{MIN}$ and c) L > L$_{MIN}$, where L$_{MIN}$ is the critical inductance

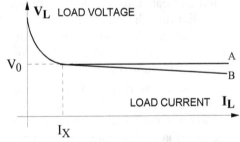

ABOVE: The regulation curve of an L-filter, ideal (A) and real (B).

LC FILTER

$V_0 = 2V_M/\pi = 0.637V_M = 0.9V_{RMS}$

$L_{MIN} = R_L/(3\omega)$

V_{RMS} = input AC voltage

R_L = load resistance

Ripple factor (L in H, C in μF):

γ (60Hz mains) = 0.83/LC

γ (50Hz mains) = 1.2/LC

Tube- versus solid state rectifiers

Before selenium, germanium and silicon rectifiers were invented, tube rectifiers ruled the field. Compared to their solid-state descendants, tube rectifiers and rectifying circuits suffer from at least seven drawbacks:

The mains transformer requires an additional winding (1) to provide heating voltage for the rectifier tube, typically 5V, 2A (10 Watts) or 5V, 3A (15 watts). Furthermore, a high voltage winding with a center tap (CT) is needed (2), and such secondary has double the number of turns compared to the secondary designed for bridge rectification.

The value of the first filtering capacitor is limited, typically from 10 to 47 µF (3). Each tube rectifier has the maximum capacitance value it can tolerate; otherwise, it will start arcing when the amp is powered on due to high capacitor charging currents. Its life will be drastically shortened, and it will eventually fail.

The total energy reserve of a tube power supply is way lower than the energy reserve of the same voltage solid-state supply. $E=C*V^2$, so the energy stored in 32 µF filtering cap used with a tube rectifier producing $500V_{DC}$ is 8.0 Joules, while that of the 680 µF with SS rectifier at the same voltage is 170 Joules. That is 21.25 times more energy, and that means better bass response and better dynamics.

Tube rectifiers have high internal resistance, so their voltage drops are 20-50 times higher than silicon diodes, typically 15-40V, versus only 0.6V for silicon rectifiers. For the same transformer secondary, the voltage at point X on the diagram will be much lower than the same circuit using SS diodes.

The table (right) lists the most common rectifiers with full load voltage drops.

Since the voltage drop on tube rectifiers increases with load current, their voltage regulation is worse than that of SS power supplies. With the increased amplifier power draw, more and more power is wasted as heat, plus the anode voltage will sag or droop, limiting the output power. All other aspects being equal, an identical amplifier with a SS rectifier can produce higher output power than the same amplifier with tube rectification.

Tube rectifiers work hard and continuously, so they are usually the first tubes to fail in an amp. Tube rectifiers are much bulkier and 500-1,000 times more expensive than silicon alternatives, typ. 5 cents versus $25-$50!

Another limitation in the table is the minimum resistance R_{MIN} on the AC or anode side tube rectifiers need to be connected to. If a transformer primary and secondary were wound with a large diameter wire, the total secondary transformer winding resistance might be below the minimum needed, so series resistors may need to be added (4).

Both audiophiles and guitar players claim that amps with tube rectifiers sound better than solid-state diodes despite the limitations and drawbacks of tube rectifiers. By better, they mean smoother, softer, more "natural," and "emotionally engaging."

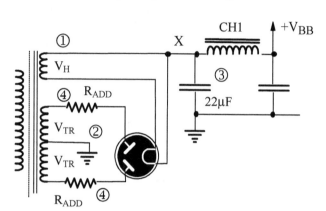

ABOVE: A typical high voltage power supply using a "full wave rectifier", a vintage name for a twin-diode vacuum tube. The heater-cum-cathode is obviously common for both diodes. A directly-heated type is illustrated, but the external connections are identical for indirectly heated types.

TUBE	V_H (V)	I_H (A)	V_{TR} (V)	R_{MIN} (Ω)	C_{MAX} (µF)	I_{MAX} (mA)
5R4GY	5	2	750	250	4	250
5R4WGY	5	2	850	500	4	150
5U4G	5	3	300	170	32	245
5U4WG	5	3	450	75	32	225
5U4GB	5	3	300	21	40	300
5U4GB	5	3	450	67	40	275
5V4	5	2	375	100	10	175
5X4G	5	3	300	170	32	245
6X4	6.3	0.6	325	525	10	70
EZ80	6.3	0.6	350	175	50	90
6CA4	6.3	1	350	230	50	150
5Y3	5	2	350	50	20	125

ABOVE: Common tube rectifiers and their main operating parameters, heater voltage/current, maximum secondary AC voltage on their anodes (plates), minimum required transformer secondary resistance, maximum capacitance of the 1st filtering capacitor (3), and maximum DC current they can provide

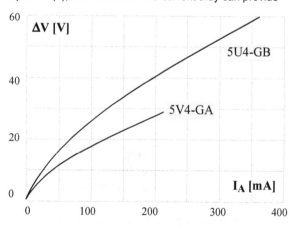

ABOVE: Voltage drop across 5U4-GB and 5V4-GA tube rectifiers for various levels of their DC output current.

A SIMPLE "HI-LO" POWER CIRCUIT: EMERY SOUND SUPERBABY

Many moons ago, this was the first modern "boutique" amp on our test bench. While the audio circuit is plain vanilla, the Hi-Lo switch changes DC voltages in the power supply and the whole amp and thus results in higher and lower output power levels, respectively. It could almost be called a "half-power feature"; the maximum power was measured as 6.3 Watts in "Hi" mode and 3.8 Watts in "Lo" mode. DC voltages in the "Lo" mode are marked in brackets on the diagram.

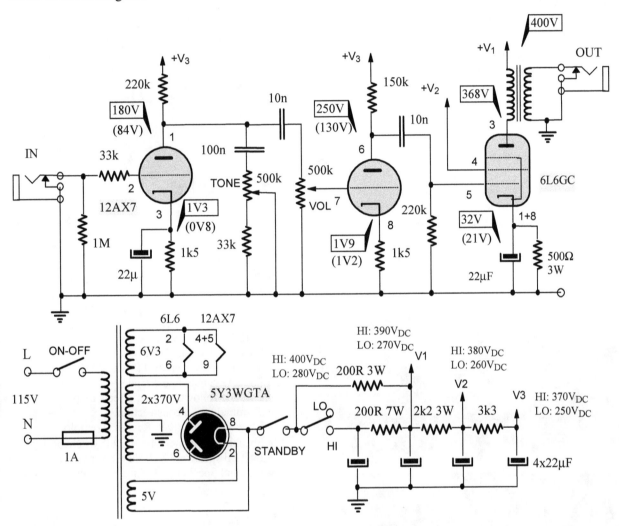

Emery Sound Superbaby (2001) circuit diagram, © Emery Sound

NOTE: DC voltages in rectangular frames were measured in the "Hi" mode, those in brackets below were DC voltages in the "Lo" operating regime

In "Hi" mode, with 64 mA cathode current flowing through the cathode resistor of the 6L6 output tube, the anode current is approx. 60mA (about 4mA of screen current), so the power dissipation of that tube is $P = V_{AK} * I_A = (368-32)*0.06 = 20.2$ Watts, which is fine (below the maximum for 6L6).

Notice a very high DC voltage drop on the primary winding of the output transformer; 32V are lost. That is due to its high DC resistance of 500Ω!

The power dissipation on the 500R cathode resistor is slightly above 2 Watts; that 3-Watt rated resistor is marginal, it will get very hot. A 5- or even a 10-Watter should have been used.

All these measurements are with a 5Y3 tube rectifier, which is known for its high internal resistance and a significant voltage drop. Using 5V4 or especially 5AR4 rectifiers, with much lower internal voltage drops, would raise all DC voltages by 15-20V.

MEASURED RESULTS (LO):
- -3dB BW @ 1 Watt: 123Hz - 4,960kHz
- P_{MAX}= 3.8W (1 kHz)

MEASURED RESULTS (HI):
- -3dB BW @ 1 Watt: 137Hz - 4,680kHz
- P_{MAX}= 6.3W (1 kHz)

VARIABLE HV ANODE SUPPLY: VHT SPECIAL 12/20 RT

Like its baby brother, the Special 6 Ultra, this combo has a list of inclusions longer than Melania Trump's shopping list.

There are two switch-selectable power ranges, one declared as 12 Watts (using 6V6 output tubes), the other "20 Watts", obtainable if 6L6 or EL34 are substituted into the same octal sockets. 20 Watts from a pair of 6L6 is very low output power, indicating Class A operation of the output stage.

No matter how many controls and features an amp has (and this one has quite a few!), there will always be those who want more features or a different sound. However, modifications and additions would not be easy despite this amp's terminal board construction, and one look under its chassis (next page) tells you why.

The internals are a spaghetti-like mess of interconnecting wires, crisscrossing the space between two terminal boards, chassis-mounted tube sockets, and top controls. Access to tube socket lugs, test points, and many components is difficult, if not wholly impossible.

Likewise, there is no room on the top control panel to add any knobs or switches. Even the back of the chassis is chock-a-block!

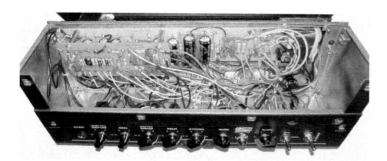

ABOVE: The amp's top-facing control panel and the back of its chassis are completely filled with knobs, jacks and switches so adding anything wouldn't be easy!

SPECS

VHT Special 12/20 RT

- Controls: Volume w/ Pull Boost, Tone, Reverb w/ Pull Deep, Speed, Tremolo, Depth & Watts, Effects Send, Effects Return (Back Panel)
- 3x12AX7 + 2x6V6 tube
- Standby-Pentode-Triode switch
- Variable power control ("Watts")
- Footswitchable boost mode
- Cathode follower-driven effects loop
- Terminal board construction
- Effects: Tube Driven Reverb, Tremolo
- Speaker impedance: 4-8-16Ω (Switchable)
- Internal speaker: 12" VHT ChromeBack
- External Speaker jack, Line Out jack

Power supply

Five power supply sections are apparent from its circuit diagram (next page).

The AC side of the mains transformer (1) includes a dual voltage primary (not shown due to space limitations), a fuse, on-off switch, and an NTC thermistor or "thermal resistor".

When a tube amp is switched on, assuming its filtering capacitors are discharged, there is a current spike through the heaters of all tubes, which are cold, and through the high voltage secondary winding. This spike is reflected on the primary side, so there is also a primary inrush current. Sudden voltage and current spikes can also damage or destroy the transformer's insulation.

The high resistance of NTC thermistors in a cold state and high resistance in a hot state makes them ideal as limiters for inrush currents, which is why VHT used them here.

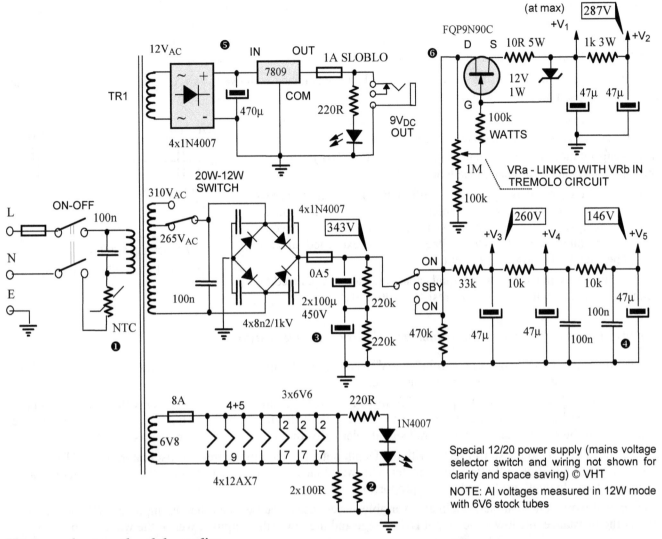

Special 12/20 power supply (mains voltage selector switch and wiring not shown for clarity and space saving) © VHT

NOTE: Al voltages measured in 12W mode with 6V6 stock tubes

Heater and external pedal supplies

All tubes are heated with 6.3V AC voltage (6.8V measured). Instead of grounding one end of the heater secondary winding, two 100R resistors were used (2), one between each winding end and GND. This is a standard way to minimize hum. The "Power On" LED is also connected across the heater supply, a better option than placing the "Power On" indicator on the primary side of the mains transformer. At least here, its glow indicates that the tube's heaters are getting the power.

The $9V_{DC}$ power supply for external pedals is regulated by a 7809 3-terminal IC voltage regulator integrated circuit and fused.

The high voltage supply

The HV secondary winding has two switch-selectable taps. The $265V_{AC}$ tap is used in the 12W mode with 6V6 power tubes. The other tap provides a higher voltage ($310V_{AC}$) which was supposed to be used with EL34 and 6L6 power tubes. The tubes themselves are not switchable; there are only two octal sockets, so the amp needs to be turned off, 6V6 tubes unplugged, and either 6L6 or EL34 substituted. 6K6 and 6F6 can also be used instead of 6V6, and KT66, KT77, 5881, and other 6L6/EL34 relatives can be used in the 20 Watt mode.

Lower capacitance value film capacitors are used to bypassed the last two elcos in the filtering chain (4). Relatively unusual in guitar amps, that is a common way to improve the sound of preamp stages in hi-fi amplifiers, especially at higher frequencies where the filtering action of elcos diminishes and the sonics suffer as a result.

Series connection of electrolytic capacitors

Notice two 100μF electrolytic capacitors connected in series after the HV rectifier, each bypassed by a 220kΩ resistor (3). Why go to all trouble? In the 20W mode, the DC voltage after the rectifier will be around or just under 450V, but before tubes warm up and start drawing current, the voltage may shoot up to 500V or even higher! So, even a $450V_{DC}$ rated elco would be very stressed, and its failure (explosion!) would be only a matter of time.

When two identical elcos are strung in series, their voltage rating doubles, so a high safety margin is thus achieved. In a series connection, each capacitor will carry an equal charge Q. If potential drops across the capacitors are V_1, V_2, ...V_N, then $V_{TOT} = V_1 + V_2 + ... + V_N = Q/C_1 + Q/C_2 + ... + Q/C_N = Q(1/C_1 + 1/C_2 + ... + 1/C_N)$ The factor in brackets can be considered an equivalent capacitor, so $1/C_{TOT} = 1/C_1 + 1/C_2 + ... + 1/C_N$

However, there is a price to be paid. The equivalent capacitance is halved; two 100μ elcos in series have an equivalent capacitance of only 50μF.

Also, since capacitance tolerances of these capacitors are wide, typically +/-20% or more, in the worst-case scenario, one 100μF capacitor may have a capacitance 20% lower than this nominal value (80μF), while the other may have a 20% higher capacitance, or 120μF!

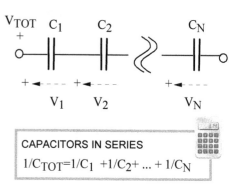

Connected in series, they would have different voltage drops across them. The capacitor with a larger capacitance will have a smaller voltage drop across it, while the cap with a lower capacitance will be subjected to a higher voltage drop. Its rated voltage could be exceeded, and we would be back to the original problem.

Luckily, the DC voltages across unequal capacitors can be equalized by shunting each elco with a high-value resistor (200-470k).

CAPACITORS IN SERIES

$$1/C_{TOT} = 1/C_1 + 1/C_2 + ... + 1/C_N$$

"Watts" power control: series pass tube or FET as a variable resistor

The continuously variable power control ("Watts") varies the V+ supply voltages to the anode and screen grids of the output tubes, thus determining the output power.

The 6V6 reverb driving stage is also powered from +V2. Its voltages are also scaled together with those of the output tubes (lower voltage, lower output power, and lower reverb driving voltage). This was done to maintain the proportion of the reverb signal in the overall signal relatively constant as the output power is adjusted.

Also, instead of a fixed resistor in series with the Depth potentiometer in the tremolo oscillator, the VHT designer used one gang of dual-gang Watts potentiometer. This was done for the same reason, so as the output power is scaled down, the tremolo depth is also reduced in proportion.

The control works on a simple principle of variable series resistance between the DC input and DC output. The higher the resistance, the lower the output DC voltage and the lower the output power of the whole amp.

Instead of a large and primitive rheostat, which would have to pass the total load current and would thus get very hot, a tube or FET transistor is usually used instead. We know that the internal resistance of such active elements (between drain D and source S in this case) varies with voltages between their electrodes.

Notice that the gate-source voltage (V_{GS}) is kept constant by a 12V Zener diode, and the gate-drain voltage V_{GS} is adjusted by the "Watts" potentiometer. In the upper position, the gate is at the input voltage level; in the bottom position of the wiper, the gate is at 100k/(100k + 1,000k) = 9% of the input voltage (the 1M - 100k voltage divider).

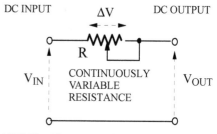

ABOVE: The operating principle behind continuously variable power control

The audio section

Although there is only a single Tone control knob, voicing is also varied through the Texture and Depth switches (next page). Also, keep in mind that the Triode/Pentode switch changes not only the output power but also the tone or the spectral signature of the amp.

The amp is quite noisy, especially in the middle position of the Texture switch, which increases the high-frequency hiss in the speaker. Increasing the Watts control (turning it CW) increases the hum and hiss.

A transformer-coupled single-ended 6V6 stage drives the reverb tank. The time constant of its grid RC network (2n2 & 1M) is 72 Hz. Should you wish to "clear" the reverb tone by preventing the bass frequencies from "clogging" its reverb tank, decrease the coupling cap to 1n and even 470pF.

The output of the reverb recovery stage (V2a) has a 3n3 coupling cap but a 4n7 cap can be switched in parallel, which shifts the lower -3dB frequency of the coupling RC circuit from 482 Hz with only 3n3 in the circuit to 199 Hz when both are in parallel (deep reverb).

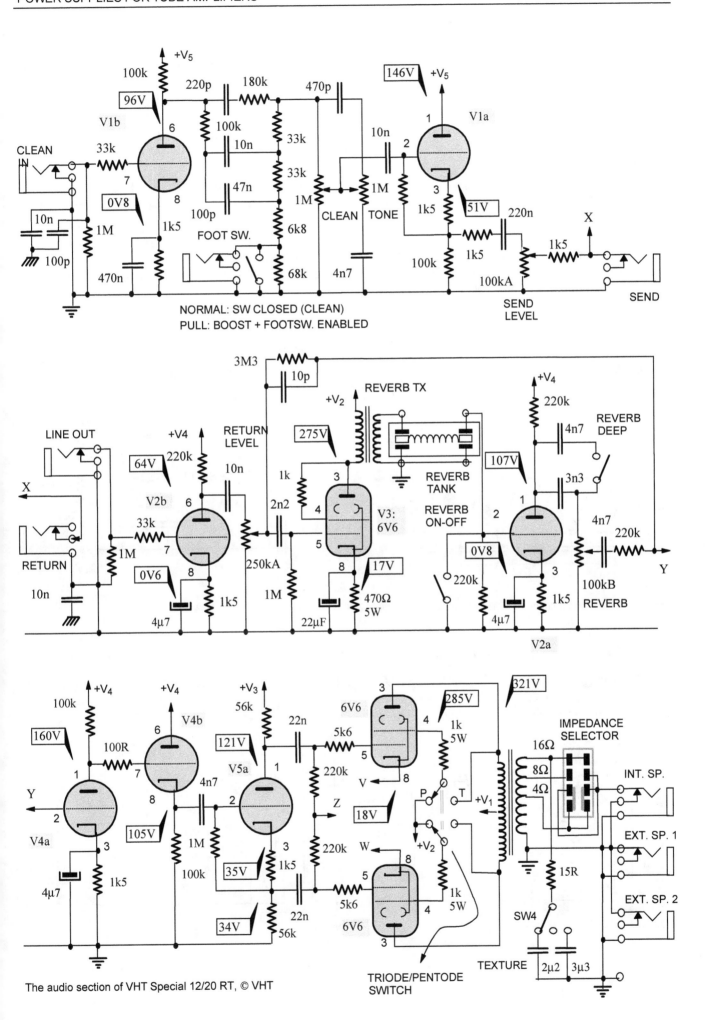

The audio section of VHT Special 12/20 RT, © VHT

Tremolo oscillator and cathode "DEPTH" circuit

A standard phase-shift oscillator achieves tremolo with Speed and Depth controls. There is also a less common Fast/Slow speed switch. The Depth of amplitude modulation is also changed when Watts control is adjusted via potentiometer V3b. Power changes mean signal level changes, and thus tremolo depth must be scaled accordingly to keep it at the same percentage regardless of the Watts control level.

The Depth switch makes one of six RC networks in the cathode circuit of the output tube "active" by short-circuiting one of the six 1M series resistors. That effectively connects its corresponding capacitor in parallel with the 470R cathode resistor.

On the circuit diagram, the contact in series with 22µF cathode bypass cap is closed (one for each output tube), so the lower -3dB frequency of the cathode circuit is $f_L = 1/(2\pi RC) = 15.4$Hz, which means full bass reproduction akin to hi-fi amps. This is the "deep end."

On the other end of the adjustable range is the 470 nF capacitor. When its 1M resistor is shorted, the lower -3dB frequency of the cathode circuit is much higher, 720 Hz. This means that the lowest decade of the guitar's fundamental tone range, from 80 Hz to 720 Hz (or three bottom octaves, 80-160-320-640Hz), will be progressively attenuated, resulting in "bright" sound and less bass.

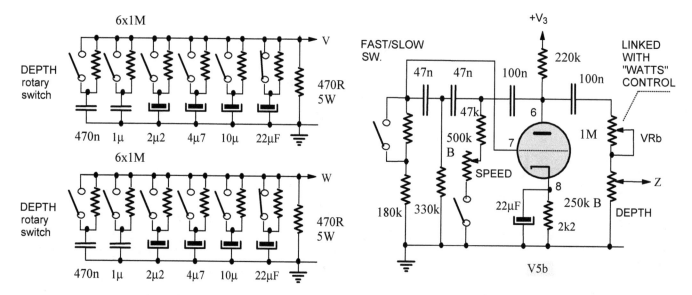

Tremolo oscillator and cathode "DEPTH" circuit of VHT Special 12/20 RT, © VHT

Using a CT high voltage secondary winding with a diode bridge

The high voltage secondary winding of VHT 12/20 amp's transformer had two taps, $265V_{AC}$ and $310V_{AC}$. If you have a power transformer without such power taps but with a center-tapped HV secondary winding, you can also achieve two power levels, just as Kustom V15 designers did in the circuit below.

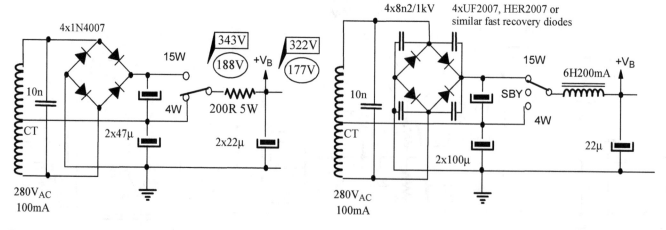

The original HV power supply section of Kustom V15
© Kustom Musical Amplification Inc

The improved HV power supply

Instead of grounding the CT (as is usually done when two diodes are used), it is taken to a junction of two 47mF filtering elcos. In the "15W" position, the filtering chain is supplied from the top elco, and in the "4W" position from the bottom elco, where the DC voltage is just over half of the "15W" value and the power is almost four times lower (since power raises with the square of voltage).

The possible improvements to this circuit include:

1. Replace 1N4007 diodes with fast recovery diodes such as UF2007, HER2007 or similar

2. Bypass each diode with a 4n7-10n 1kV rated ceramic capacitor (for interference and noise suppression)

3. Increase the energy storage and filtration by replacing the first filtering elcos with100µF capacitors

4. Replace the SPDT "1/4 Power Switch" with SP3T; the middle position will leave the amp in standby mode

5. Replace the 200W 5 Watts series resistor with a 2-10H choke, rated at 150-200mA

REGULATED POWER SUPPLIES: IBANEZ TUBESCREAMER TSA5TVR-S

This hybrid combo amp uses simple series voltage regulators with bipolar transistors in its low voltage heater and high voltage anode supplies. Similar voltage regulators can be used in so-called "active" decoupling circuits, which achieve a much better hum reduction than the ordinary or "passive" RC decoupling filters used in most amps.

This is how the Ibanez website introduces this little amp: "The TSA5TVR is an all-tube, Class-A, 5-watt combo with a genuine Tube Screamer built into the front end of the amp."

Well, firstly, the "Tube Screamer" is not a tube module at all. Solid-state integrated circuits are used, the chips are soldered directly into the PCB.

Secondly, TSA5TVR is *not* an all-tube amplifier. The audio circuit is tubed, but 1) solid-state rectification is used, 2) solid-state devices drive the spring reverb, and 3) the input goes into the emitter of the A1015 PNP transistor, which is always in the signal path, so calling this amp "all tube" is misleading.

Solid-state rectification is used, yet there is no standby control, meaning the high anode voltage appears immediately, yet the tubes had no time to warm up. This will shorten the life of the tubes.

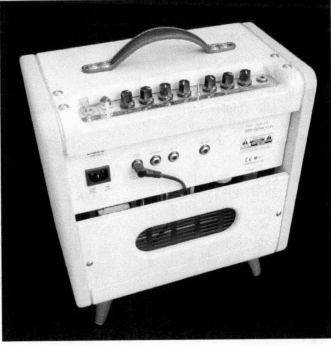

SPECS

IBANEZ TUBESCREAMER
AMPLIFIER TSA5TVR-S

- Controls: Master Volume, Reverb, Bass, Treble
- Tube Screamer on/off, Overdrive, Tone, Level
- 1x12AX7 preamp tube, 1x6v6 power tube
- Solid state rectification
- PCB construction
- Output power: 5 Watts into 8Ω, 15% THD
- Jensen C8R speaker
- Accutronics® spring reverb, IC-driven
- External speaker output, headphone out, line out
- Dimensions: 335 (W) x 210 (D) x 430 (H)
- Weight: 8.3 kg (18.2 lb.)

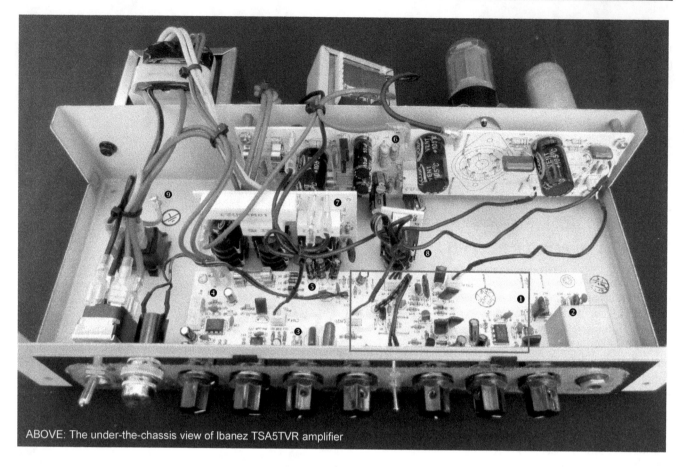

ABOVE: The under-the-chassis view of Ibanez TSA5TVR amplifier

Under-the-chassis peek and the tube audio circuit

The preamp PCB contains the "Tubescreamer" section (1), the input RC circuitry and a bipolar transistor which cannot be seen behind the input metal shield (2), the tone control components (3), and the IC-driven and amplified reverb circuit (4). These have their own low voltage DC power supply (5).

The power PCB contains two DC power supplies for the heater and the high voltage (6), plus the two tubes and their associated RC components. The output PCB carries three female jacks - for the speaker, headphones, and line out (7) -and the associated resistors and capacitors. The smallest PCB (8) has only the footswitch jack, so the "Tubescreamer" can be turned on or off.

The tube section on the power PCB is relatively easy to trace since its topology and values are very common. We didn't have the patience to trace the preamp section, but circuit diagrams of a few different versions of the solid-state "Tubescreamer" circuit are available online for download, so they won't be shown here.

A filtering choke could be added in the corner where the earth lug of the chassis is (9), but since the ripple filtering circuit works well (discussed on the next page), there is no need for it.

The other easy mods include a gain switch (DPDT) which would bypass both cathode resistors with elcos, a pentode-triode switch, and an NFB on-off switch. There is no room for additional switches or controls at the front panel, so they would have to be installed at the back.

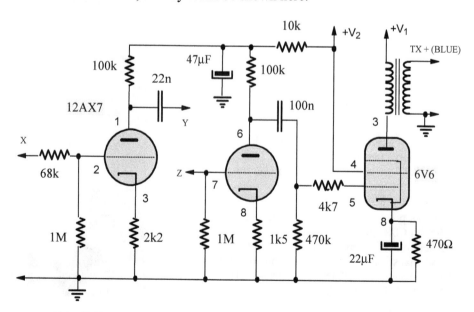

Note: X, Y and Z go to tone control circuitry on the preamp PCB via shielded cables

The heater and HV power supplies

The heater and HV power supply circuits of this amp use the same topology. It looks pretty similar to the basic textbook example of a series voltage regulator with a bipolar transistor, and it is. The only difference is that in voltage regulators, a Zener diode or another similar component provides a stable or "referent" voltage for the base of the transistor, and here we have an elco in parallel with a resistor.

So if this circuit does not regulate the DC voltage on the transistor's emitter (the output), how does it work, and what purpose does it serve? To understand that, we need to go back and refresh our knowledge of decoupling circuits.

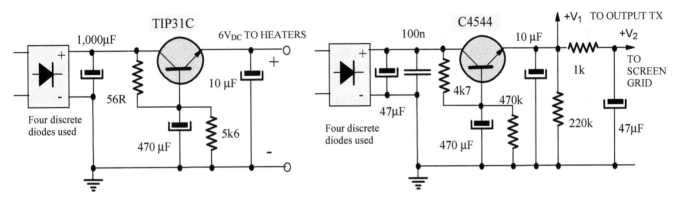

ABOVE: The heater supply circuit of Ibanez TSA5TVR amp   ABOVE: The HV supply circuitry of Ibanez TSA5TVR amp

Passive and active decoupling circuits

Ideally, the power supply's internal impedance would be zero, and decoupling circuits would not be needed. A decoupling circuit is an RC filter that reduces the feedback between different amplifier stages due to the finite common power supply impedance. It also reduces the ripple voltage (AC component) superimposed on the high voltage DC line. This reduction is in the ratio of V_{OUT}/V_{IN} due to its voltage divider effect. This voltage dividing action is only effective if the capacitive reactance X_C is small small compared to R_D at low frequencies. A typical frequency at which such effectiveness is evaluated is the lower -3dB frequency f_L.

Of course, the DC anode current(s) of the previous stage(s) must pass through R_D so the resistance of R_D cannot be too high; otherwise, too much of the supply voltage would be lost on such a series resistor. Circuits are usually drawn from left (input) to the right (output), but because decoupling filters are drawn in the opposite direction, their inputs are on the right and outputs on the left side (the power flow is right-to-left).

The effectiveness of the passive RC decoupling filter $V_{OUT}/V_{IN} = X_C/(X_C+R_D)$ relies on two factors. The first is a very low X_C, or shunting capacitor's reactance, which means a high value of the shunt capacitor's capacitance is needed. The other factor is the high value of the series resistor R_D. Since that causes a proportionally higher DC voltage drop on such a resistor, there is a limit to what can be tolerated.

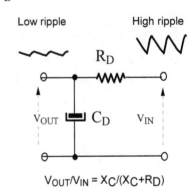

$$V_{OUT}/V_{IN} = X_C/(X_C+R_D)$$

HOW EFFECTIVE ARE DECOUPLING CIRCUITS AT REDUCING RIPPLE?

CALCULATION

Determine ripple voltage V_{ROUT} at the output of the typical decoupling circuit between the output stage and preamp stages of a guitar amplifier, if the ripple voltage at its input is $V_{RIN}=1V_{PP}$. Assume 50Hz mains frequency.

Since $X_C=1/(\omega C)$, at 50Hz mains frequency, the ripple frequency is double that or 100 Hz, so we have $X_C=1/(2*\pi*100*22*10^{-6}) = 72\Omega$

The attenuation of ripple is $A=V_{OUT}/V_{IN} = X_C/(X_C+R_D) = 72/(72+10,000) = 0.00715$

Or, in dB: $A= 20\log0.00715 = 20*-2.146 = -42.9$ dB

Finally, $V_{ROUT} = V_{RIN}A = 1*0.00715 = 7.15$ mV$_{PP}$, meaning the peak-to-peak value of the ripple will be reduced from 1 Volt to 7.15 mV!

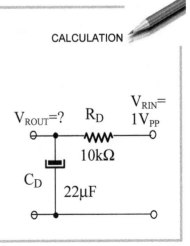

In this active decoupling filter ("active" because an active element such as a bipolar transistor is used), the transistor works in the common-collector (emitter follower) configuration, since the "load" comprising of the anode resistor RA and the internal tube resistance rI are in the collector's circuit.

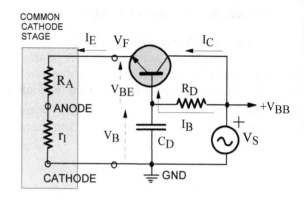

Active decoupling circuit with a bipolar transistor

V_S is the unwanted audio signal or the AC ripple riding on top of the DC voltage. The unwanted signal feedback voltage V_F equals the signal voltage V_B on the base, which is the voltage across the decoupling capacitor CD.

The first advantage of this circuit is that series resistor R_D can now be much larger than before without causing a significant DC voltage drop. This is because now only a small base current I_B flows through it, not the much larger collector current I_C!

The second advantage is that now capacitance C_D can be much smaller. The filter only deals with the base current, which is much smaller than the load or collector current (by the factor β of "current gain" of the transistor), so the circuit effectively multiplies the capacitor action by factor β, which is in the order of 100 or more!

DESIGNING AN ACTIVE DECOUPLING CIRCUIT   CALCULATION

The load current (also emitter current I_E in the model above) to an input stage of a tube amplifier is 4 mA. The transistor used has current amplification factor $\beta_0=100$. If the maximum allowable voltage drop across the series transistor is 10V and attenuation of 100Hz ripple signal of at least 100 (-40dB) is required, determine the values for R_D and C_D.

Since $I_E=4mA$ and $I_E= \beta_0IB$, $I_B=0.004/100 = 40\mu A$! Assuming a base-emitter voltage drop V_{BE} in the "on" state as 0.6V, the voltage across R_D is the maximum voltage drop minus V_{BE} or 10-0.6 = 9.4V. Thus, R_D must be $R_D=9.4V/40mA= 235 k\Omega$.

$X_C=1/(\omega C)$, at 50Hz mains frequency the ripple frequency is double that or 100 Hz, so we have $X_C=1/(2*\pi*100*22*10^{-6}) = 72\Omega$

The attenuation of ripple is $A=V_{OUT}/V_{IN} = X_C/(X_C+R_D) = 1/100$ so $100X_C=X_C+R_D$, so $99X_C=R_D$ or $X_C=R_D/99$. Finally, we have $C=99/(R_D*2*\pi*100) = 0.67\mu F$, so one would use a $1\mu F$ cap.

A $10\mu F$ capacitor would increase the attenuation to more than 0.001 or -60dB!

Going back to the Ibanez circuitry, the R_D in the heater supply is 56Ω and $C_D=470\mu F$ so at the 100Hz ripple frequency (120Hz in USA), $X_C=1/(\omega C) = 1/(2*\pi*100*470*10^{-6}) = 3.4\Omega$. The ripple attenuation is $A=V_{OUT}/V_{IN} = X_C/(X_C+R_D) = 3.4/(3.4+56) = 0.057$ or 5.7%

The DC voltage at the base of the transistor is determined by the voltage dividing effect of the 56Ω and 5k6 resistors, so it is 5,600/(5,600+56) = 0.99 or 99% of the input voltage. Since 5k6 is 100 times larger than 56, we could have determined that mentally without resorting to any calculations.

In the HV section, the DC voltage at the base is 470/(470+4.7) = 0.99 or 99% of the input voltage. Its R_D is 4,700Ω and C_D is the same as in the heater supply, 470μF, so its 100Hz reactance is the same, 3.4Ω.

The ripple attenuation is $A=V_{OUT}/V_{IN} = X_C/(X_C+R_D) = 3.4/(3.4+4,700) = 0.000723$ or -62.8 dB, much higher than in the heater circuit (24.9dB). Ripple isn't that critical in the heater circuit, but it is important in the HV supply, especially in single-ended amps like this one.

SINGLE-ENDED TRIODE, PENTODE AND ULTRALINEAR OUTPUT STAGES

11

- SINGLE-ENDED TRIODE POWER STAGE
- SINGLE-ENDED OUTPUT STAGES WITH PENTODES AND BEAM POWER TUBES
- DESIGNING AN EL84 SE PENTODE OUTPUT STAGE
- LOW-POWER OCTAL PENTODES AND BEAM TETRODES: 6V6, 6K6, 6F6, 6Y6
- DESIGNING A 6F6 SE PENTODE OUTPUT STAGE
- ULTRALINEAR CONNECTION
- 6L6 SINGLE-ENDED STAGE: TRIODE, ULTRALINEAR AND PENTODE COMPARISON
- ANALYZED: BELCAT TUBE-10 AMP

Although most guitar amps feature a pentode output stage, single-ended in smaller amps and push-pull arrangement of two or four tubes in bigger amps, many have a switchable mode that connects those pentodes or beam tubes as triodes. These significantly reduce the output power levels but also change the tone or the voicing of the amp.

Since triodes are simpler than pentodes, we will start with a triode power stage and move towards pentode stages.

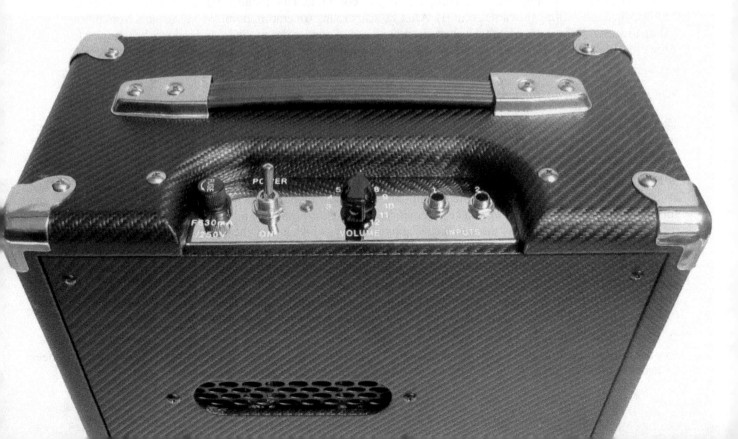

SINGLE-ENDED TRIODE POWER STAGE

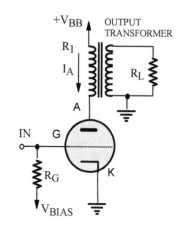

The most common single-ended output stage has the output transformer's primary winding in series with the output tube. The DC anode current I_A flows through the primary winding and magnetizes the magnetic core, reducing the incremental inductance significantly. That is why SE transformers must use larger cores and more windings than push-pull transformers, through which the net primary DC current is zero (ideally, with perfectly matched tubes).

Static and dynamic load lines

The primary winding of output transformers has a small DC resistance R_1, 70-300Ω. Anode current flowing through the primary winding creates a voltage drop on such resistance, so the quiescent voltage on the anode V_0 is lower than the anode supply voltage V_{BB}. Due to the large difference between output transformers' primary DC resistance and its much higher AC impedance, the static and dynamic load lines also differ significantly.

The static load line has slope $\tan\beta = V_{R1}/I_0$. It passes through the operating point Q_A and the point V_{BB} on the V_A axis. Usually we neglect this small voltage drop V_{R1} and analyze the output stage as V_0 would be equal to V_{BB}. For instance, typically, $V_{BB} = 400V$ and $V_{R1}=10V$, so $V_0=390V$, introducing the error of $(400-390)/390 = 2.5\%$

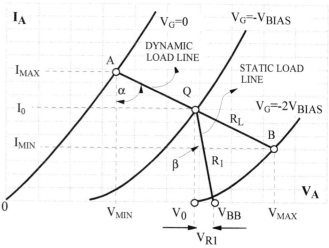

RIGHT: The static and dynamic load lines for a transformer-loaded single-ended stage

---

HOW CAN THE ANODE VOLTAGE EXCEED THE POWER SUPPLY VOLTAGE V_{BB}?

Theoretically, for ideal triodes and no losses of any kind, the anode voltage can swing between zero and $2V_{BB}$. This is possible due to the transformer action, more precisely, the self-induction in the primary winding of the output transformer.

With the grid signal falling from $V_G=0$ (point A) towards point Q, the voltage in the transformer's primary is induced to prevent or oppose such a drop (to keep I_A constant). That induced voltage keeps rising for as long as I_A keeps falling (towards point B). After I_A starts rising (operating point swinging back from point B up towards Q) the induced voltage changes, now trying to oppose the rise in I_A and reduces back towards lower anode voltages! Chokes operate on the same induction principle, opposing any change in current.

The model of SE transformer-coupled triode output stage

The model of the output stage for a series-fed single-ended output stage is quite simple. V_{BB} is the anode DC power source, supplying current I_A, which is also the DC current I_1 flowing through the transformer's primary. The AC (signal) voltage that the output valve as an AC generator provides is $v_s(t)=-\mu v_G(t)$, where $v_G(t)$ is the signal on the tube's grid and μ is the tube's amplification factor.

Points A and B are the primary transformer terminals. For AC signal point B is at ground potential since the ideal DC source V_{BB} has a zero internal impedance. In practice, that is not the case; the power supply always has some internal resistance. The signal $i_1(t)$ flows through the DC power supply, which is why the quality of a single-ended tube amp's power supply directly impacts its sound! The same applies to interstage transformers; the driver stage's plate current flows through the primary, but the secondary (grid) impedance is very high, so there is no secondary current.

Unbalanced DC current I_1 flows in transformer's primary. A transformer does not pass or "transform" DC currents or voltages, so there is no secondary DC current, only the AC signal current $i_2(t)$. The load is reflected onto the primary side as $(N_1/N_2)^2 Z_L$; this ratio $(N_1/N_2)^2$ is called an **impedance ratio of the output transformer**.

DC power supply is a short circuit for AC signal, so the model is a simple resistive voltage divider between the tube's internal resistance and the load. This assumes an ideal transformer and a purely resistive load, both of which are sheer fantasies but make it easier to understand the fundamental behavior of the output stage.

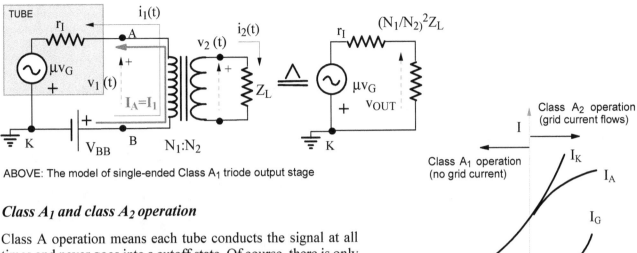

ABOVE: The model of single-ended Class A_1 triode output stage

Class A_1 and class A_2 operation

Class A operation means each tube conducts the signal at all times and never goes into a cutoff state. Of course, there is only one tube in SE stages, so it must reproduce the whole signal, but this definition also applies to push-pull amplifiers, where both tubes (or four if paralleled) conduct the signal at all times.

In the class A_1 operation, the grid is biased negatively with respect to the cathode, and no grid current flows, meaning the input resistance of the grid circuit is infinite. Usually, when we say Class A, we mean Class A_1.

However, if the grid of a power tube is driven into the positive region $+V_G$ (grid positive with respect to cathode), as in Classes A_2 and AB_2, the grid will attract electrons from the cathode and act as a mini-anode.

The grid current will start flowing, and the input impedance of the output stage (both single-ended and push-pull) will drop to a very low value, typically around 1 kΩ.

That would cause the loading of the driver stage (increased driver tube's anode current) and a large distortion unless the driver stage was properly designed for such operation. RC-coupled output stages cannot cross into Class A_2, but those driven by an interstage transformer specially designed to operate with secondary DC current (grid current) can.

Optimal anode load

The choice of the anode load for output tubes is a compromise between maximum power and minimum distortion, more so for triodes than for pentodes or beam power tubes. For maximum triode power, the load impedance should be 3-4 times the internal impedance of the tube. For lower distortion levels, higher load impedances should be used. Notice that at the load impedance producing maximum power Z_{AP} the distortion is higher than with larger load impedances, such as Z_{AD}. Power and distortion drop off at approximately the same rate with increasing load.

ABOVE: Cathode, anode and grid currents in a triode with fixed anode voltage, as a function of grid voltage: $I_K = I_G + I_A$

ABOVE: Typical triode's power and distortion curves as a function of anode resistance. D2 and D4 are the 2nd and 4th harmonics

OPTIMAL ANODE LOAD AND GRID BIAS FOR SE TRIODES

$Z_{AOPT} \approx 3r_I - 4r_I$,
r_I = internal resistance of a tube
$V_{G0} = 0.75 V_A/\mu$ [V]
V_A = DC anode voltage
μ = amplification factor

The slope of the load line $\Delta I_A/\Delta V_A$ (next page) is inversely proportional to the load impedance Z_A and since the output power is a product of I_A and V_A, the issue of maximizing the output power can be graphically interpreted as maximizing the area of the triangle whose hypotenuse is the load line and whose base is the I_0 horizontal line. Three examples of load lines are illustrated.

The quiescent DC anode voltage is the same in all three cases (V_{AQ}), while the bias voltages E_G vary, resulting in very different quiescent DC currents I_{01}, I_{02} and I_{03}.

Z_{L1} has the smallest slope (it's the most "horizontal" of all) and therefore the highest load impedance. The voltage swing, the difference between the V_{AMIN} and V_{AQ}, is very large, but its anode current swing (current in point A minus quiescent current I_{01}) is very small. Z_{L3} goes to the other extreme.

The efficiency of Class A stages with a transformer- coupled load (speaker)

You may recall a very low efficiency (a theoretical maximum of 25%) of a gain stage where the load was DC-coupled, in series with the anode. However, the efficiency of the preamplifier stages where such arrangements are used is generally of no concern to designers since we aim to amplify the voltage signal and not power.

Because output stages amplify power (both voltage and current), to reduce the size of the power transformers and other components in power supplies, and minimize the heat that amplifiers produce, transformer-coupled output stages should be made as efficient as possible.

Let's see if the situation is any better with transformer-coupled resistive load in Class A. We will assume an ideal transformer (without any losses).

The quiescent point is Q_A, and the load line is AB. $I_{AMAX} = 2I_0$, the anode voltage in the quiescent point is the voltage of the high voltage battery V_{BB}.

η= (AC output power)/(DC input power)*100%

η= $(2V_{BB}-0)(I_{AMAX}-0)/8V_{BB}I_0$*100%

Since $I_{AMAX} = 2I_0$ the efficiency is

η= $2V_{BB}2I_0/8V_{BB}I_0$)*100% = 1/2*100% = 50%

Although this is again a theoretical maximum (for ideal triodes and no transformer losses), that is undoubtedly a big improvement. By eliminating the DC current through the load and by capitalizing on the transformer action, where due to induction, the anode voltage can swing to double the DC voltage of the power supply (V_{BB}), the efficiency is doubled to 50%.

With no signal, all the power is dissipated on the output tube. With a maximum signal, half the power is dissipated as heat in the tube; the other half is converted into the signal power. So, counter-intuitively, power tubes in Class A run cooler with AC signal than when idle (no signal). At the amplifier's maximum output power, the output tubes run coolest.

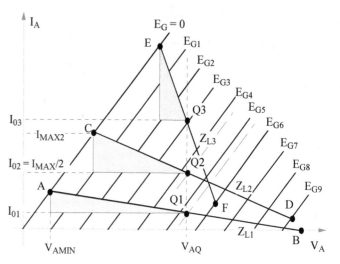

ABOVE: Load line choices for SE triodes: maximizing the area of the power triangle

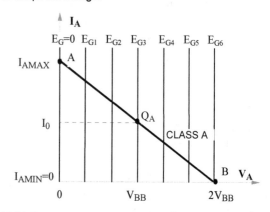

ABOVE: Anode curves and load lines for ideal triode with transformer- coupled resistive load in class A

BELOW: Power tubes in Class A run cooler with AC signal than when idle!

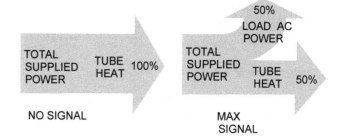

SINGLE-ENDED OUTPUT STAGES WITH PENTODES AND BEAM POWER TUBES

Three most common connections of multigrid output tubes

The existence of the screen grid leads to various ways of connecting it. In a pentode connection of the power tubes the screen is usually tied to the same high voltage point that supplies the output transformer. However, many power pentodes have a much lower maximum screen voltage than anode voltage. In that case a lower source of DC voltage V_S is needed for the screen!

For instance, E130L has the maximum anode voltage of 900V continuous and up to 2,000V in peaks (that is the reason for a top cap anode, so it is physically as far as possible from other pins), but its screen grid is rated at a maximum of only 250V.

Connecting the screen to the anode (usually through a low-value resistor to limit the screen current) turns a pentode or beam tube into a pseudo-triode. Again, if V_{SGMAX} is much smaller than V_{AMAX}, the anode voltage in triode mode will be restricted to the lower value and reduce the available output power. However, we have designed and built many amplifiers with triode-strapped pentodes that happily took screen voltages 50-100V higher than the maximum prescribed in their datasheets! Again, experiment for yourself.

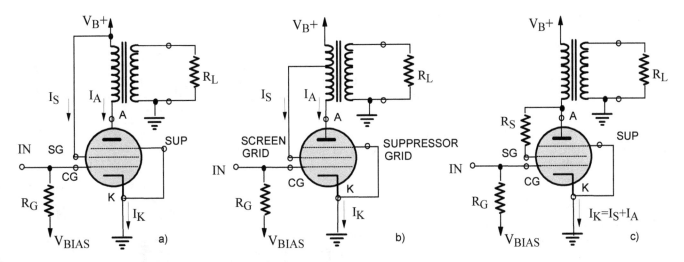

ABOVE: Pentode (a), ultralinear (b), and triode-strapped single-ended output stage (c). All are series-fed and fixed-biased.

Optimal anode load and distortion

For common audio pentodes and beam power tubes (such as EL84, 6V6, 6L6, and EL34), there are numerous graphs and tables available published by tube manufacturers, so guesswork and rules-of-thumb are not necessary. With tubes not initially designed for audio use, there is no data for the optimal load impedance or parameters when triode connected.

One way to calculate the optimal load Z_A for any pentode is to simply divide the DC plate voltage with the idle DC current. For instance, for EL84 with $V_A=250V$ and $I_0=48mA$, the optimal primary load is $Z_{AOPT} \approx V_A/I_A = 250/0.048 = 5,208 \, \Omega$, close to the commonly used value of 5 kΩ.

The second harmonic curve D_2 for pentodes dips to zero at a certain load impedance R_X, at which the output power is at its maximum. That point is thus declared the optimal load (maximum output power). Unfortunately, the discordant 3rd harmonic has no such minimum and keeps rising with increasing load impedance.

Lower load impedance R_Y reduces the harsh-sounding 3rd harmonic, which is now "masked" by the pleasant-sounding 2nd harmonic. The output power is slightly lower, but the sonic benefits are indisputable. Thus, paradoxically, designing pentode output stages for minimum 2nd harmonic distortion may not be a good idea!

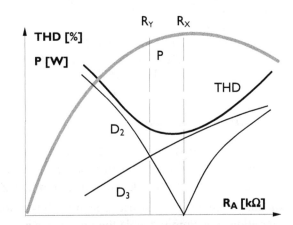

ABOVE: Typical pentode's power and distortion curves as a function of anode load. D_2 and D_3 are the 2nd and 3rd harmonics.

> **OPTIMAL ANODE LOAD & GRID RESISTANCE FOR SE PENTODES**
>
> $\mathbf{R_{AOPT} \approx r_I/5}$
> $\mathbf{R_G \approx 2r_I}$
> r_I = internal resistance of a pentode

SCALING FACTORS

Say you want to design an amplification stage using a different anode voltage from the one specified in the tube's data sheet. Let's use a SE output stage with 6F6 pentode as an example. The data sheet specifies $V_A=285V$, $V_G=-20V$, $I_A=38mA$, $I_S=7mA$, $r_I=78k\Omega$, $R_L= 7k\Omega$, gm=2.55mA/V, $P_{OUT}=4.8W$

We want to use 250V on the anode and the screen. First, we calculate the anode voltage conversion factor K_1: $K_1=250/285 = 0.893$ The control grid bias will be $V_G=-20*K_1 = -20*0.893 = -17.86V$

To estimate anode and screen current we need K_2: $K_2=K_1^{3/2}$ In this case $K_2=0.893^{3/2} = 0.844$, so $I_A= 0.844*38=32mA$ and $I_S=0.844*7 = 5.9mA$

To get the new load resistance we need K_3 which is $K_3=1/\sqrt{K_1} = 1.06$, so the load resistance should be $R_L=7*1.06= 7.4k\Omega$! The internal resistance can also be estimated using K_3 as $r_I=K_3*78k= 82.7k\Omega$

The transconductance in the new operating point can be estimated as gm=$2.55\sqrt{K_1}=0.945*2.55=2.4$ mA/V and output power is estimated by using factor $K_4=K_1K_2= K_1^{5/2}$ so $P_{OUT}= K_1^{5/2}*4.8= 0.754*4.8=3.6W$

DESIGN: EL84 SE PENTODE STAGE

A European answer to the American 6V6 beam tetrode, EL84 (6BQ5), a miniature power pentode, was developed by Philips in the mid-1950s. The tube quickly became popular, not just amongst hi-fi manufacturers, even more so in guitar amps. Vox AC30 combo used four of them in a push-pull arrangement to produce 30 Watts of sweet juicy power.

EL84 also sounds so sweet in SE stages that many guitarists claim that it sounds better than EL34, and I tend to agree. I've never heard an EL84 amplifier I didn't like!

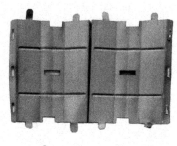

TUBE PROFILE: EL84 (6BQ5)

- Indirectly-heated power pentode
- Heater: 6.3V, 0.75 A
- P_{AMAX}=12W, P_{SMAX}=2W
- TYPICAL SE PENTODE OPERATION:
- V_A= V_S= 250V, V_G= -7.3V
- Signal (grid voltage): 6.1V_P, 4.3 V_{RMS}
- I_{A0} = 48 mA, I_{S0} = 5.5 mA
- Load resistance: 4.5 kΩ
- gm=11.3mA/V, μ=430, r_I=38kΩ
- μ_{G2G1}=19 (triode ampl. factor), gm_{G2}=1.8mA/V
- P_{OUT} = 5.7 W @ THD = 10%

RIGHT: The internal view of the 6BQ5 power pentode. Once the anode is removed (pictured on top in its "unfolded" or flat state), three pairs of grid supports are visible, as is the outer, coarsely spaced suppressor grid. The white-coated cathode is in the middle, and the control grid is the dense spiral right in front of it.

Choosing load impedance

For a single-ended EL84 (6BQ5) pentode, the minimum distortion load is Z_L=4,500–5,000Ω, but the maximum power is at Z_L=6,600 Ω. However, notice the difference in power between the two points is only 0.2 Watt, while the increase in THD is 2%. Our rule-of-thumb says $Z_{AOPT} \approx V_A/I_A$ = 5.3kΩ.

In this example we will use a slightly lower load impedance of 4k5. The same rationale applies to push-pull operation. We simply double the impedance and get the optimal value of Z_{PP}=2*4,500 = 9,000 Ω. Let's choose the operating point given in manufacturers' datasheets, the anode and screen voltage of 250V. We will assume an ideal output transformer with no losses and zero DC primary resistance.

Since the maximum anode dissipation is 12 watts, the maximum anode current in the Q-point is I_A=P_{MAX}/V_0 = 12/250 = 48mA. From the anode curves, we determine that grid bias voltage at that point is -7.3V and that the screen current is 5.5 mA (not shown). Cathode current is the sum of anode and screen current, so I_K=48+5.5=53.5 mA.

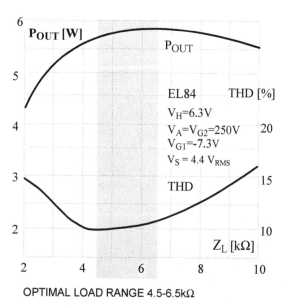

OPTIMAL LOAD RANGE 4.5-6.5kΩ

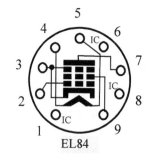

LEFT: Output power and THD curves for EL84 SE pentode stage

RIGHT: The A typical SE output stage using EL84 in pentode mode

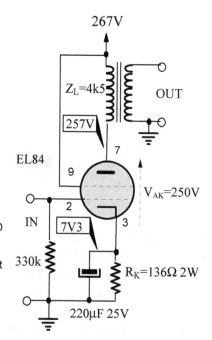

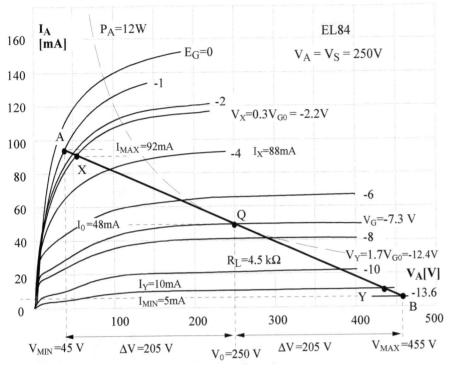

ABOVE: Anode pentode curves of a single-ended stage with EL84 pentode, showing the quiescent operating point Q (with no signal) and the 4k5 impedance load line

To get a 7.3V voltage drop on the cathode resistor, with the cathode current of 53.5 mA flowing through it, its value must be $R_K = V_K/I_K = 7.3/0.0535 = 136\Omega$. Power dissipation on this resistor is $P_{RK} = V_K I_K = 0.4W$, so a 1 Watt resistor is the smallest we should use, a 2 Watt resistor would be better for long-term reliability.

Assuming we decide to limit the grid voltage to -1 Volt, half of the signal swing will be 7.3 - 1 = 6.3 Volts.

We add an identical signal swing in the negative direction to the bias of -7.3 Volts and get -7.3-6.3 = -13.6V.

So, our maximum peak-to-peak grid voltage swing is 12.6V, or $\Delta V_{GEFF} = 4.5\ V_{RMS}$.

We now have endpoints A and B, and can draw vertical lines to get V_{MIN} and V_{MAX} of 45V and 455V respectively. That is our anode peak-to-peak voltage swing, $\Delta V_A = 455-45 = 410V_{PP}$ or $\Delta V_{AEFF} = 410/2.82 = 145.4 V_{RMS}$. The voltage amplification factor of the output stage is $A = -\Delta V_{AEFF}/\Delta V_{GEFF} = -145.4/4.5 = -32.5$! Using peak-to-peak values, the same result must be obtained: $A = -\Delta V_A/\Delta V_G = -410/12.6 = -32.5$!

Estimating output power of SE pentode stages

RULE-OF-THUMB METHOD #1: The anode efficiency of a pentode is approximately 35%. The anode power input is $P_{IN} = V_{A0}I_{A0} = 250*0.048 = 12W$, so $P_{OUT} = 0.35 P_{IN} = 0.35*12 = 4.2W$.

This formula estimates the "clean" power, meaning the total output power levels will be higher.

RULE-OF-THUMB METHOD #2: The output power can also be estimated using the RMS value of the output swing (205/1.41=145.4V), instead of its peak value (205V here).

$P_{OUT} = \Delta V_A^2/R_L = 145.4^2/4,500 = 21,138/4,500 = 4.7W$ This is closer to the output power declared in data sheets (5.7W), which includes significant distortion.

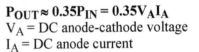

ESTIMATION OF SE PENTODE OUTPUT POWER

$P_{OUT} \approx 0.35 P_{IN} = 0.35 V_A I_A$
V_A = DC anode-cathode voltage
I_A = DC anode current

LOW-POWER OCTAL PENTODES AND BEAM TETRODES: 6V6, 6K6, 6F6, 6Y6

You can reduce the output power levels of your SE guitar amp with the 6V6 tube (such as Fender Champ or VibroChamp) by simply replacing the 6V6 with 6K6. Since 6K6's anode dissipation is 8.5W/12W = 70% of the 6V6 power capability, the output power will be reduced in the same proportion. So, if you get a maximum of 4 Watts of output power with a 6V6, you'll get 0.7*4 = 2.8 Watts with a 6K6.

The 6F6 audio output pentode dates from the mid-1930s. As you can see from the Side-By-Side comparison, it is similar to the ubiquitous 6V6 but is more sensitive; it requires even lower grid driving voltages. However, it needs 55% more heater power than the 6V6, meaning it is less efficient.

Operating in a pentode mode, a pair of 6F6 in class A_1 push-pull can produce 10.5 Watts with only 3% distortion.

6Y6 is a low voltage-high current tube, its max. V_A is almost half that of the other three, but its anode current is doubled! Since it is not interchangeable with the other three, it is included here for comparison purposes only.

All four tube types sound very good, especially when triode connected, and since millions were made in the golden era, they are still plentiful and relatively cheap to buy. NOS are still available as pairs or even quads.

SIDE-BY-SIDE	6V6	6K6	6F6	6Y6
Heater	6.3V/0.45A	6.3V/0.4A	6.3V/ 0.7A	6.3V/ 125A
P_A / P_{G2} max.	12W / 2W	8.5W / 2.8W	11W / 3.75W	12.5W / 1.75W
V_A / V_{G2} max.	315 / 285 V	315 / 285 V	375 / 285 V	200 / 200 V
SE class A_1 pentode P_{OUT}	5.5W	4.5W	4.8W	6.0W
V_A, V_S, V_G	315V, 225V, -13V	315V, 250V, -21V	285V, 285V, -20V	200V, 135V, -14
R_L	8.5kΩ	9kΩ	7kΩ	2.6kΩ
I_{A0}, I_{S0}	34mA, 2.2mA	26mA, 4mA	38mA, 7mA	61mA, 2.2mA
Parameters as a triode	gm= 4mA/V, r_I=2400Ω, μ=9.6	gm= 2.7mA/V, r_I=2,500Ω, μ=6.8	gm= 2.6mA/V, r_I=2600Ω, μ=6.8	gm= 7.3mA/V, r_I=750Ω, μ=5.5
V_A max. triode	300V	315V	350V	250V
P_A + P_{G2} max. total triode	12.5W	7W	10W	14W
V_A, V_G, I_0, R_L	250V, -20V, 39mA, 4.8kΩ	250V, -18V, 38mA, 6kΩ	350V, -38V, 48mA, 6kΩ	250V, -42V, 50mA, 5kΩ
SE class A_1 triode P_{OUT}	1.25W	0.9W	0.9W	2.3W

DESIGNING A 6F6 SE PENTODE OUTPUT STAGE

Since 6F6 was specifically developed for audio applications, there are recommended operating conditions, the most common being V_A =250V and V_S = 250V, with 7kΩ load, so let's stick to that.

Looking at the dynamic transfer curves, the curve for 5k load is the most linear in the low V_G range (from 0 to -10V on the grid). However, when positioned on the anode characteristics, the 5k load line crosses the V_G=0 line way above the knee, meaning the voltage swing will be reduced, and the output power lower. The 7k load seems to be a better compromise, slightly lower current swing but an additional 70V of anode swing (V_{AMIN} around 30V compared to 100V for the 5k load).

L-R: RCA 6F6, Sylvania 6F6, Sylvania 6V6, RCA 6K6 (metal)

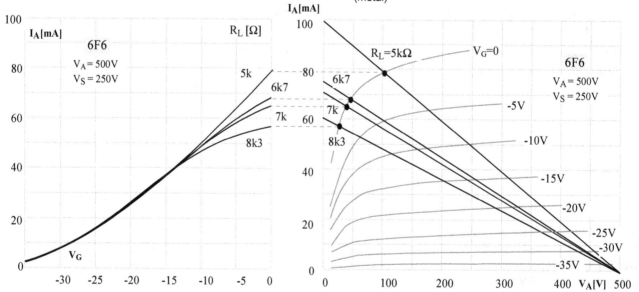

ABOVE: 6F6 pentode's dynamic transfer characteristics and anode characteristics for various load impedances.

With anode current of 35mA and screen current of 6.5mA (specified in data sheets) at the quiescent point Q, the cathode current is 41.5mA. To get the bias voltage of 16.5V the cathode resistor must be R_K=16.5/0.0415= 398Ω.

Estimating the output power

The distortion of pentodes and beam power tubes is usually included in the power output figures. P_{OUT} = $R_L[(I_{AMAX}-I_{AMIN})+1.41(I_X-I_Y)]^2/32 = 7,000*[(0.067-0.005)+1.41(0.061-0.0105)]^2/32 = 3.88$ W

Let's see what we get using the rule-of-thumb approximation, assuming the anode efficiency of a pentode of 35%. For 6F6 output pentode working in a single-ended stage with I_A=35mA and V_A=250V, the input power is $P_{IN}=V_A I_A$ = 250*0.035 = 8.75W so P_{OUT}= 0.35P_{IN} = 3.06 Watts.

SE PENTODE OUTPUT POWER

$$P_{OUT}=R_L[(I_{AMAX}-I_{AMIN})+1.41(I_X-I_Y)]^2/32$$

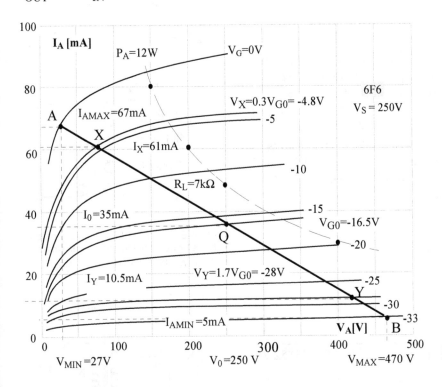

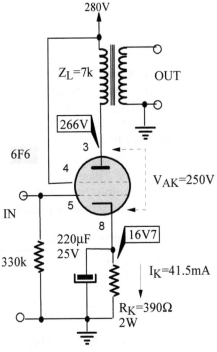

LEFT: 6F6 SE pentode stage (right) and operating conditions (quiescent point, load line, voltage and current swings) on the left

Estimating distortion

Using the "5-point method", illustrated above, the bias voltages of interest are V_X=0.3V_{G0} and V_Y=1.7V_{G0}. In our case they are V_X=-4.8V and V_Y= -28V After marking the intersection of those two curves with the load line, we can read anode currents in those two points as I_X=61mA and I_Y=10.5mA.

As we have learned a few pages back, the formulas for the second and third harmonic distortion coefficients are

D_2= [(I_{AMAX} +I_{AMIN})-2I_0]/[I_{AMAX}-I_{AMIN}+1.41(I_X-I_Y)]*100% and

D_3= [I_{AMAX} -I_{AMIN}-1.41(I_X-I_Y)]/(I_{AMAX}-I_{AMIN}+1.41(I_X-I_Y))*100%

D_2= [67+5- 2*35]/[67-5+1.41(61-10.5)]*100% = 2/131.8*100% = 1.52%

D_3= [67-5 - 1.41(61-10.5)]/(67-5+1.41(61-10.5))*100% = -9.205/133.2*100% = 6.91%, so the distortion from the first two harmonics generated by this power stage is 7.08%. Notice that with harmonics you don't add the percentages directly (you'd get 1.52+6.91= 8.43% distortion), but THD [%] = √(D_2^2+D_3^2+ ...+D_N^2)*100% !

ULTRALINEAR CONNECTION

The name was popularized by David Hafler and Herbert Keroes in the Nov. 1951 issue of "Audio Engineering" magazine, in their article "An Ultralinear Amplifier." U/L circuit requires a tap on the primary winding of the output transformer to which the output tube's screen grid is connected. Instead of the screen grid being held at some stable DC voltage as in the pentode's case, the screen's voltage varies and follows the signal, which means that local negative feedback is applied to the screen. This results in a host of negative feedback benefits:

- The output impedance is reduced to about half of the pentode's circuit
- linearity is improved, resulting in reduced distortion
- the power output is higher than in triode connection, approaching that delivered by a pentode
- the power output is more constant because the output stage behaves somewhere in between the voltage (triode) and current (pentode) source.
- the circuit combines the advantages of both the triode and the pentode stage, without suffering from their respective setbacks.

The illustration on the right tells why this is called an ultralinear circuit. It is a typical example of the "Linearizing Principle": Convex + Concave = Straight! The anode curves of a pentode strapped in U/L circuit are much more linear than those of either pentodes or triodes!

Normally, U/L output stages are used only in hi-fi and bass & PA amps, not guitar amps - it sounds too clean.

U/L stage with EL84

As in our example with EL84 in SE pentode connection, we have the same output transformer with impedance ratio $IR=(N_1/N_2)^2=4,500/8 = 562.5$, or turn ratio $TR=N_1/N_2=23.7$ The only addition is a primary tap at 23% of the primary turns. This means that 23% of the primary or anode AC (signal) voltage is fed back into the screen grid as negative feedback.

The screen tap at 23% of the primary winding turns is the tap that results in minimum distortion. Thus, $N_3=0.23N_1$ and our X-factor is 0.23! We will not go into the tedious mathematics of the ultralinear stage; the final formulas for gain and output impedance are:

$$A = - \frac{(gm+Xgm_{G2})Z_A}{1+(X/\mu_{G2G1}+1/\mu)(gm+Xgm_{G2})Z_A}$$

$$Z_{OUT} = \frac{(N_2/N_1)^2}{(gm+Xgm_{G2})(X/\mu_{G2G1}+1/\mu)}$$

$$A = - \frac{(11.3+0.23*1.8)*4.5}{1+(0.23/19+1/430)*(11.3+0.23*1.8)*4.5} = - \frac{52.713}{1+0.01443*52.713}$$

$$= -52.713/1.76 = -52.713*0.568 = -29.94 \approx -30$$

$$Z_{OUT} = \frac{1/562.5}{(11.3+0.23*1.8)(0.23/19+1/430)}$$

$$= (1,000/562.5)/11.714*0.01443 = 10.5 \ \Omega$$

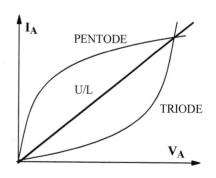

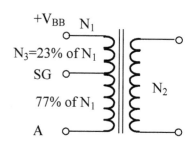

ABOVE: SE EL84 U/L output transformer windings and turns ratios
BELOW: SE EL84 U/L output stage

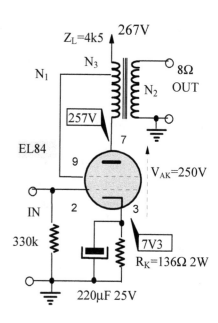

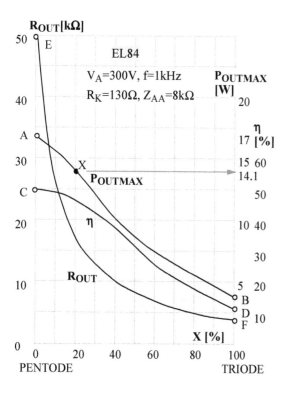

In pentode connection, we had $A=-45$ and $Z_{OUT} = 68 \ \Omega$! With relatively mild screen feedback (X=23%), $A=-30$ and $Z_{OUT} = 10.5 \ \Omega$! The reduction in amplification due to U/L NFB is $A_F/A = 30/45 = 0.667$, or in dB, $20*\log 0.667 = -3.52$ dB! To get a feel for the impact of increased ultralinear feedback, calculate A and Z_{OUT} for X=33%!

Both single-ended and push-pull stages can be operated in the ultralinear mode. The graph shows how the maximum output power of an U/L push-pull stage with EL84 tubes drops from around 17 Watts in point A (pentode) to only 3.5 Watts in triode mode (point B).

The efficiency of the output stage drops from 50% in point C (pentode) to only 12% in point D, triode. A similar situation applies to a single-ended stage.

Pentodes are imperfect current sources (very high internal impedance but not infinite), and triodes are imperfect voltage sources (their internal impedance is low but far from zero). You can see that from the third curve. In point E, the internal or output impedance of the EL84 PP stage is around 50kΩ, and then it drops very rapidly as the stage moves towards triode operation.

Finally, in point F (pure triode connection), its R_{OUT} is only 4 kΩ, more than 12 times lower!

Let's now assume that we have chosen our U/L percentage of 20%, point X, and measured the maximum output power of 14.1 Watts. Notice that the graphs (previous page) are for the anode-to-anode impedance of 8 kΩ. What happens with other values of Z_{AA}? The graphs on the right tell the story.

Again, our point X is marked, and if we go down to the THD curve, we get a distortion of around 1.4%, an excellent result. The THD curve has a minimum at 7k primary impedance (point A), but the maximum output power drops to around 13.2 Watts. A hi-fi designer would most certainly sacrifice a bit of output power to reduce THD to its minimum.

As the anode (load) impedance increases, THD raises in an almost linear fashion, so at 12kΩ, it exceeds 7%.

Although we desire distortion in a guitar amp, keep in mind that this is a push-pull stage in which most of the pleasant and warm sounding even harmonics got canceled.

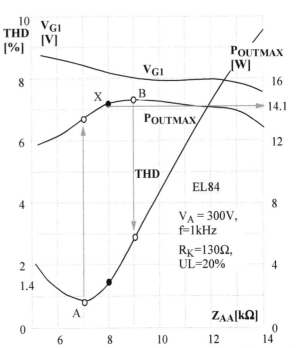

ABOVE: How the maximum output power, THD (total harmonic distortion) and grid drive signal needed for maximum power vary with anode-to-anode impedance for EL84 ultralinear PP output stage (1 kHz test signal, 300V anode supply voltage)

So, this high distortion consists almost exclusively of the shrill and unpleasant odd harmonics, 3rd, 5th, 7th, etc. The other two parameters, the maximum output power and the grid signal required for maximum power, do not vary that much with varying load impedance. P_{OUTMAX} stays above 12 Watts for all loads bar the very low ones, below 5kΩ, and V_{G1} hovers in the 8-9 V range.

6L6 SINGLE-ENDED STAGE: TRIODE, ULTRALINEAR AND PENTODE COMPARISON

6L6 SE pentode stage

To determine the output power and distortion we also need points X and Y: V_{G0}=-13V, V_X=0.3*13=-3.9V, V_Y=1.7*13= -22V, from the graph on the next page we estimate I_X=145mA and I_Y=38mA.

D_2= [(I_{AMAX} +I_{AMIN}) - 2I_0] / [I_{AMAX} -I_{AMIN}+1.41(I_X-I_Y)]*100% =28/303=9.25%

D_3= [I_{AMAX} -I_{AMIN} - 1.41(I_X-I_Y)] / (I_{AMAX} -I_{AMIN}+1.41(I_X-I_Y))*100%= 45/303=14.85%

P_{OUT}=R_L[(I_{AMAX}-I_{AMIN})+1.41(I_X-I_Y)]^2 /32

P_{OUT}= 2,500*[(0.175-0.023)+ 1.41(0.145- 0.038)]^2 /32= 2,500*0.3^2/32= 229/32= 7.17W

The input power is P_{IN}=$I_A V_0$= 0.085*320=27.2W, so efficiency is η = P_{OUT}/P_{IN} = 7.17/27.2 = 26.3%

6L6 beam tetrode is a strange tube. Its sonic merits in guitar amplifiers are unquestionable. In hi-fi, however, the jury is still out after all these decades. THD levels range from 9.7% in a triode connection to 9.25 % second and 14.85 % third harmonic distortion when working as a beam tetrode.

Even in the so-called "ultralinear" mode, the curves predict that THD will still be very high, 8.3%! Calling that ultralinear is delusional.

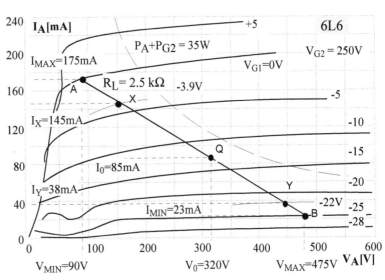

If you can't be bothered with these points X and Y you can get an estimate of the output power by using the triode formula: $P_{OUT} = \Delta V * \Delta I / 8 = (V_{AMAX} - V_{AMIN}) * (I_{AMAX} - I_{AMIN})/8 = (475-90)*(0.175-0.023)/8 = 385*0.152/8 = 58.5/8 = 7.3$ W. As you can see, the difference between the two methods (both approximate) isn't significant, so you may as well use the more straightforward triode formula!

6L6 SE triode stage

$\Delta V = 490-175 = 315V$

$\Delta I = 145-25 = 120$ mA

$P_{IN} = I_A V_0 = 0.8*350 = 28$ W

$P_{OUT} = \Delta V * \Delta I / 8 = 315 * 0.12/8 = 4.7W$

$\eta = P_{OUT}/P_{IN} = 4.7/28 = 16.8\%$

For triode stages, distortion is predominately 2nd harmonic and can be estimated as:

$D_2 = (P_{OUT+} - P_{OUT-})/2P_{OUT} *100 [\%] = (175*0.065/4 - 140*0.055/4)/2*4.7 = (2.84-1.93)/9.4 = 9.7\%$

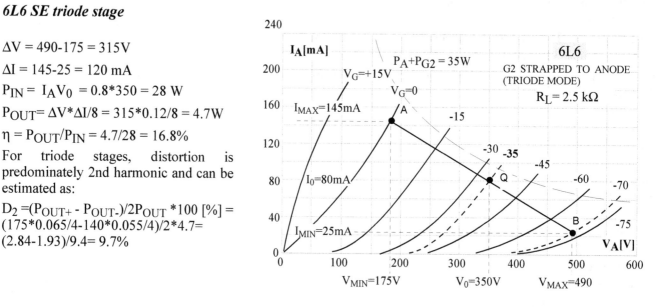

6L6 SE ultralinear stage

Even a superficial glance at the curves tells you that there is nothing linear about 6L6 in this regime. Notice how the spacing between the curves reduces as the grid voltage becomes more and more negative, resulting in the positive anode voltage swing $\Delta V+$ of 245V and the much smaller negative swing $\Delta V-$ of only 175!

$\Delta V=420V$, $\Delta I=155$ mA, $P_{IN} = I_A V_0 = 0.1*300 = 30W$, $P_{OUT} = \Delta V \Delta I / 8 = 420*0.155/8 = 8.1W$

$\eta = P_{OUT}/P_{IN} = 8.1/30 = 27\%$

$D_2 = (\Delta V_{OUT+} - \Delta V_{OUT-})/2\Delta V_{OUT} *100[\%] = (245-175)/2*420 = (5.82-3.06)/8.9 = 8.3 \%$

Notice that 2nd harmonic distortion fell from 9.7% in triode mode to 8.3 %, not a significant reduction.

However, the maximum output power increased from 4.7 Watts to 8.1 Watts, a 72% increase.

The efficiency of the output stage also increased from the very low 16.8% as a triode to 27%, an improvement of almost 61%!

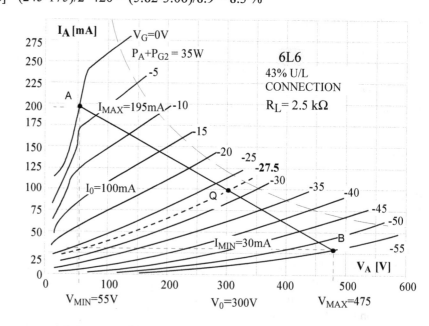

The importance of screen grid voltage on pentode's operation

This graph (next page) from 7868 power pentode's data sheet is unusual. These are not anode current curves for different bias voltages but for the same zero bias voltage ($V_{G1}=0V$) and different screen grid voltages, from 100V to 400V, in 50V steps.

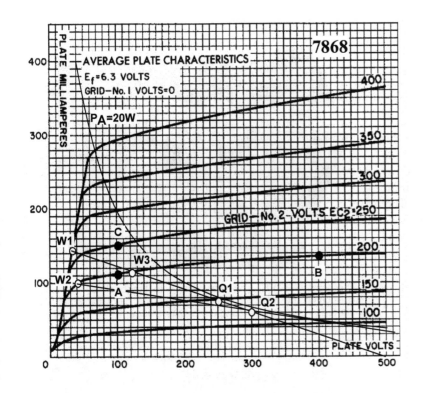

The graph visually illustrates an important point: for pentodes and beam power tubes, the anode current depends much more on the screen grid DC voltage than on the anode DC voltage.

For instance, to move from point A to point B requires an increase in plate (anode) voltage of 400-100=300 Volts. The anode current only raises from 110mA in A to 135mA in B or a 25mA increase for 300V higher anode voltage.

However, keeping the anode voltage steady at 100V and increasing screen voltage by only 50V (from 200 to 250V) represents a jump from A to C. Anode current increases from 110 to 150mA, so a 40mA increase for only 50V higher screen voltage.

Since the screen grid is much closer to the cathode than the anode, this makes physical sense, so its electrostatic field has a much bigger impact on electron flow from cathode to anode.

This type of graph isn't usually published for other power tubes, which is unfortunate since it is very useful in choosing the output tube and designing the output stage of your amp.

With V_A=250V and V_S=300V, we graphically positioned the idle point at Q1. The maximum swing to the left, towards low anode voltages/high anode currents (point W1) is around ΔV_A=250-30=220V.

The anode current increases from 80mA in Q1 to 145mA in W1, so the current swing is ΔI_A=145-80=65mA. We can now conclude that the anode load impedance is $Z_A=\Delta V_A/\Delta I_A$= 220/65=3,384$\Omega$. In real life, a transformer with 3k4 or 3k5 of primary impedance would be used.

Imagine now we choose a lower screen voltage of 200V. The voltage swing would be much shorter, extending only to W3; the anode voltage would not drop below 120V. The current swing is also reduced; the peak is only 115mA. Smaller voltage & current swings mean much lower power, so this would not be wise.

With V_S=200V, we need to reposition our Q-point, move it to Q2 at V_A=300V and I_A=60mA. Of course, in both cases, we must make sure that the product of the idle voltage and current remains below the maximum anode dissipation of the tube, which is 19 Watts (let's assume 20W for simplicity's sake). In Q1, we had 250*0.08= 20 Watts (a borderline case, a 7868 tube would not last very long in this regime), in Q2, we now have 300*60 = 18 Watts, a slightly less stressful situation for this tube.

To optimize our design, we also need to choose a more suitable load impedance. A less steep load line is needed to extend the anode voltage swing to the left, ending in W2. "Less steep" means higher load impedance.

The anode current swing is now ΔI_A=100-60= 40 mA (we had 65mA before), and the anode voltage swing is ΔV_A=300-50= 250V (220V before). This means that the anode load is now much higher $Z_A=\Delta V_A/\Delta I_A$= 250/40= 6,250 Ω.

In terms of maximum output power obtainable, we don't know the other side of the swing, how far the operating point travels in the opposite direction, towards high anode voltages and low anode currents. We can assume an identical swing in both directions (an ideal case). The real swing will be shorter, so the power levels in real life would be lower, but our approximation is pretty close and thus still useful to this discussion, the aim of which is not accuracy but conceptual understanding.

The output power is $P_{OUT}= \Delta V_A*\Delta I_A/8$ = 2*220*0.065/8= 3.58 Watts with 300V on the screen grid, and $P_{OUT}= \Delta V_A*\Delta I_A/8$ = 2*250*0.04/8= 2.5 Watts with 200V on the screen grid.

Despite adjusting the load impedance and an increased anode voltage swing, the output power is much lower with a lower screen voltage. That is due to a much larger drop (in terms of percentages) in the anode current swing, from 65 to only 40mA, when compared to a slight increase in anode voltage 300/250=1.2 or 20% increase!

ANALYZED: BELCAT TUBE-10 AMP

Belcat Co. Ltd has factories in Guangdong in China and Indonesia and Hong Kong offices. They manufacture amplifiers, guitar tuners, effect pedals, preamps, and pickups.

Their products are sold under different names; for instance, the Lorden amplifier also analyzed in this book seems to be made in their factory as well. With all due respect to Belcat, we didn't want another "budget" tube amp, but some positive online comments about this amp and the fact that the eBay seller accepted our law-ball offer of US$80 clinched the deal. Heck, one could not even buy tubes and transformers for that money, not to mention the chassis, cabinet, and speaker.

The amp's online specs mention an 8" Celestion speaker, but our amp had a nondescript cheap-looking driver.

The front (ABOVE) and rear view with the back cover removed (BELOW). The Tolex was not glued properly and not stapled at all so it started to peel off all around the cabinet's edge.

The output transformer

The sticker on the output transformer says "Input: $280V_{AC}$ (50Hz), Output: $10V_{AC}$ 1A". The fact that the secondary current rating is specified means it was designed as a mains transformer.

Its core is rated at only 10 Watts, inadequate for a supposedly 10W amplifier. The primary's DCR was high, 318Ω, while the secondary measured at 0.8Ω. The primary inductance was 8.5H, not a bad result, but leakage inductance was also high, 350mH, so a very low quality factor of Q=170.

Apart from the el-cheapo speaker, the output transformer is the weakest point of this amp. So, the modification path for this amp is obvious, replace both output components with better ones.

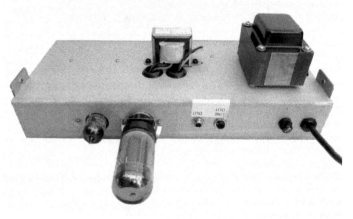

RIGHT: The power transformer (EI76.2 laminations, 40mm stack) is of decent power rating (around 100VA), way too large for such a low power amp. Notice the amateurish paper sticker, sloppily glued, marking "OUT" and "LINE OUT".

BELOW: The Lorden TL-15R and Belcat Tube-10 rear panels. the same design of the metal name plate, same Tolex, same internal electrostatic shield. Conclusion: the same manufacturer.

Belcat Tube-10 OUTPUT TRANSFORMER:

- EI48, a=16 mm, S=20mm
- Center leg cross area $A = aS = 3.2cm^2$
- Power rating: $P = A^2 = 3.2^2 = 10W$
- Z_P (8Ω load)=$6,012\Omega$
- Primary DC resistance: $R_P = 318\ \Omega$
- L_P=10.6H@120Hz, 8.5H@1kHz
- Leakage inductance L_L=50mH
- QF = 8.5/0.05 = 170

The step-down autotransformer

Since the amp's power transformer had only a 115V primary, we installed a step-down transformer. Because the factory DC voltages were on a high side, resulting in the output tube's power dissipation close to its maximum limit, the step-down autotransformer was wound not to produce 115 or 120V on its output tap, but only 105V. This brought the DC voltages down and slightly reduced the amp's output power.

However, the heater voltage was a bit low, too, although the tubes worked well. One noticeable benefit of underheating the input (preamp) tube was a reduction in noise and quieter operation. Subjectively, the sound improved as well.

The output transformer (1) was moved inside the chassis, pretty much in its original position only under the "deck." Since its four leads were of the plug-in spade type, no rewiring or soldering was required. Since it was unshielded and relatively close to the mains transformer when above the chassis, this move also reduced the audible hum. The shielded step-down transformer (2) was positioned right next to the power transformer.

ABOVE LEFT: The output transformer moved inside the chassis
ABOVE RIGHT: A 240V-100V step-down transformer installed next to the power transformer on top of the chassis

Drawing a circuit diagram from the PCB

The component names were screen printed on the PCB, but we didn't have a circuit diagram, so R1, C2, etc., didn't mean anything to us. Thus, so some circuit tracing was in order.

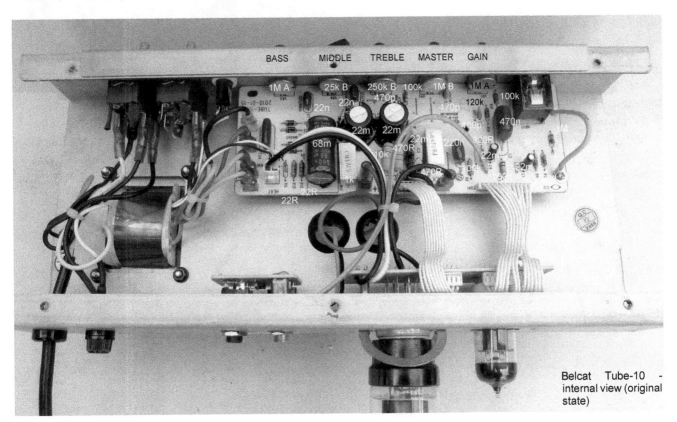

Belcat Tube-10 - internal view (original state)

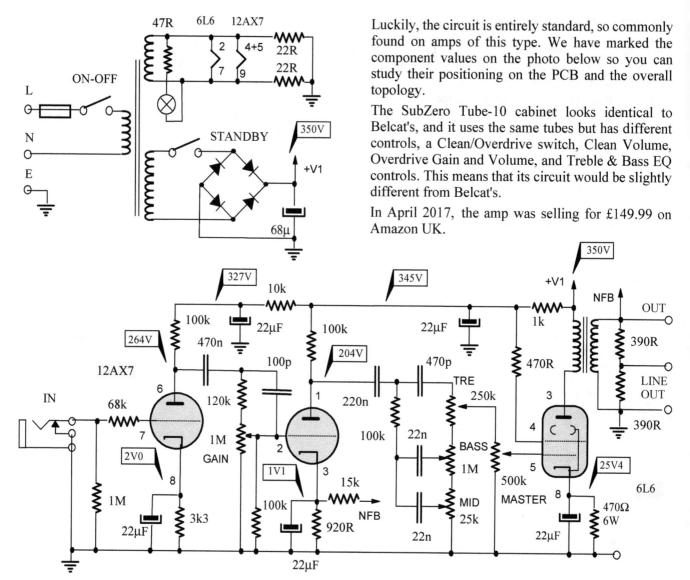

Luckily, the circuit is entirely standard, so commonly found on amps of this type. We have marked the component values on the photo below so you can study their positioning on the PCB and the overall topology.

The SubZero Tube-10 cabinet looks identical to Belcat's, and it uses the same tubes but has different controls, a Clean/Overdrive switch, Clean Volume, Overdrive Gain and Volume, and Treble & Bass EQ controls. This means that its circuit would be slightly different from Belcat's.

In April 2017, the amp was selling for £149.99 on Amazon UK.

Belcat Tube-10 circuit diagram, © Belcat Co. Ltd.

NOTE: The marked voltages are after the installation of the step-down transformer, the original DC voltages were much higher.

The output stage

Since this was supposed to be a quick case study showing a design of a single-ended pentode stage, let's analyze it. 25.4V on the 6L6's cathode means its cathode current is 25.4/470 = 54 mA. The screen current can be deduced through the voltage drop on the 1k series resistor in the decoupling circuit. 350V on one and 345V on the other side means its current is 5V/1k = 5 mA.

Anode (and cathode) current through the 1st stage is 2/3k3= 0.6mA, and that of the 2nd stage is 1.1/0.92 = 1.2mA, so out of 5mA through that resistor 0.6+1.2=1.8mA flow to the preamp stages and 5-1.8 = 3.2mA is the screen current of the output tube. So, its anode current is 54-3.2 = 50.8mA.

The DC resistance of OT's primary winding is 318 Ω and 50.8mA of anode current create a DC voltage drop of 0.0508*318= 16V, so the anode of the output tube is at 350-16 = 334 V.

Finally we can calculate power dissipation on the output tube $P=V_{AK}*I_A = (334-25.4)*0.0508 = 308.6*0.0508 = 15.7$ Watts!

This is very conservative, even for the lowest-rated 6L6 varieties (19 Watts anode dissipation), so they will have a long and stress-free life in this amp.

PHASE SPLITTERS OR INVERTERS

12

- INTERSTAGE TRANSFORMER AS PHASE SPLITTERS
- DC-COUPLED SPLIT-LOAD INVERTER: HOHNER CONTESSA CA300
- PARAPHASE INVERTER: MONTGOMERY WARD (AIRLINE VALCO) GIM-9151A
- CATHODE-COUPLED, "LONG-TAILED PAIR" OR SCHMIDT INVERTER: UNIVOX 1236

Push-pull and parallel PP output stages require a phase splitter or inverter stage to produce two equal but out-of-phase drive signals for their grids. There are many ways to achieve such phase inversion, some simple, others relatively complex. All have their strengths and advantages, but also their drawbacks and weaknesses, so there is no such thing as "the best" inverter or splitter.

In this section, we study the four most common phase splitters, an interstage transformer with two identical secondaries, a split-load inverter (also known as "concertina" or "cathodyne" splitter), a paraphase inverter, and a differential amplifier, also pejoratively known as a "long-tailed pair" or Schmidt inverter.

Before we study the operation of push-pull output stages, we need to understand the preceding stage, which is either a symmetrical (balanced) driver stage driven by a phase inverter stage or the phase inverter or splitter itself. Symmetrical driver stages are relatively rare in guitar amps; even in hi-fi amps, they aren't common, since a whole duo-triode is required, a significant expense according to the penny-pinching attitude is prevalent amongst commercial manufacturers, often run by accountants and with their amps built to a strict price point.

The main goal of a phase splitter is to "split" or invert incoming voltage signal and provide two output signals of the same amplitude and frequency, but "inverted" with respect to one another, of the opposite phase or 180° out of phase.

So, a phase splitter has one input and two outputs. These two voltages should be as well balanced as possible at all audio frequencies, meaning their amplitudes should be equal. Their phase relationship (ideally 180°) should not change with frequency or with the signal's amplitude.

This balance is the main criterion for judging the quality of phase splitters: the amplitudes of signals at the two outputs should remain identical throughout the frequency range of an amplifier, and the phase shift between them should stay constant, exactly 180°.

Unfortunately, one way or another, all real phase splitters fall short of these lofty goals of an ideal splitter! As with all audio circuits, each inverter design has its strengths but also its weaknesses.

The designs can be classified as either phase splitting or phase reversing (inverting), where one signal passes through one tube, while the other signal passes through an additional tube that inverts the signal. Although not (strictly speaking) linguistically correct, for the sake of simplicity we will use "inverter" and "splitter" interchangeably in this book.

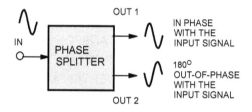

Block diagram of a phase splitter stage

INTERSTAGE TRANSFORMERS AS PHASE SPLITTERS

An interstage transformer with a center-tapped secondary (or two identical secondaries) makes an obvious choice for a phase splitter, and indeed, was favored amongst early push-pull amplifier designers. Since the three ends of the windings marked with dots produce voltages in phase, if we make the CT our referent point, then the voltages at X and Y will be identical in amplitude and frequency but opposite in phase (180° phase shift). CT can be connected to the ground if the output stage is self-biased or connected to a source of bias voltage if fixed bias is chosen.

Most vintage IS transformers by Triad, Stancor or UTC use tiny lamination sizes and thin stacks, so they cannot take any primary DC current. These must be used in a shunt-fed arrangement, with a coupling capacitor in series with the primary winding. However, a couple of models can take up to 8mA of primary DC current, such as UTC A-25 and Stancor WF-35.

Larger-size and better quality vintage IS transformers are those made by two Japanese firms, Tango and Tamura. There are also currently-produced units by Hashimoto, Audio Note, Sowter, Lundahl, Bartolucci, and ElectraPrint, but most are very expensive, costing more than many made-in-China amplifiers!

Those are used exclusively in hi-end (read: extremely expensive) hi-fi amplifiers, their use in guitar amps where distortion is the name of the game cannot be justified, either technically or financially.

Let's see how a transformer phase splitter was used in one of the better-known vintage amp designs.

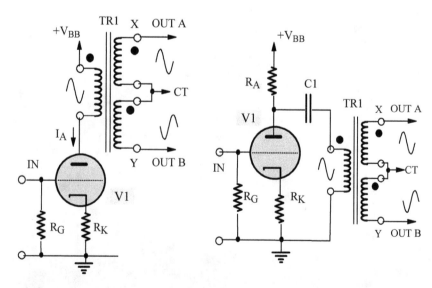

ABOVE: Single-ended driver to push-pull grids, with DC current in transformer's primary (LEFT) and parallel-fed arrangement with no primary DC current (RIGHT)

Fender Musicmaster Bass combo

There are two versions of this amp, one with 6V6 output tubes and 6AQ5 beam power tubes. 6AQ5 is a miniature 7-pin variant of the older octal 6V6, meaning their electrical parameters are identical. The circuit is interesting from an educational perspective. Let's start at the input and follow the signal flow.

Firstly, notice the connection of the two inputs. When a guitar jack is plugged into input #1, the two 68k resistors are effectively in parallel, so the equivalent series resistance is half, or 34kΩ!

However, if input #2 is used, the signal passes through the lower 68k resistor in series with input #2, but the jack shorts the left side of the upper 68k resistor to the ground. A 68k+68k voltage divider is thus formed. At its output (point X), the signal is only 1/2 of the input signal (because the two resistors are of equal value). Therefore, one could call input #1 a "high gain" and input #2 a "low gain" input!

The signal then passes through an RC filter (2), 4.7nF film capacitor, and 470k resistor to ground (as a 1st order high pass filter), which attenuates low frequencies. It is weird that Fender designers would weaken bass frequencies in an amp designed for bass guitar!

The -3dB frequency of this filter is $f_L = 1/(2\pi RC) = 72$ Hz! This means that a signal of 72 Hz would be attenuated by 3dB, so its amplitude would be down to 0.71 or 71% of the midrange value or the value without such a filter.

Calling this feeble amp of only around 10 Watts output power a bass amp was either a marketing gimmick or a serious lack of thought by Fender!

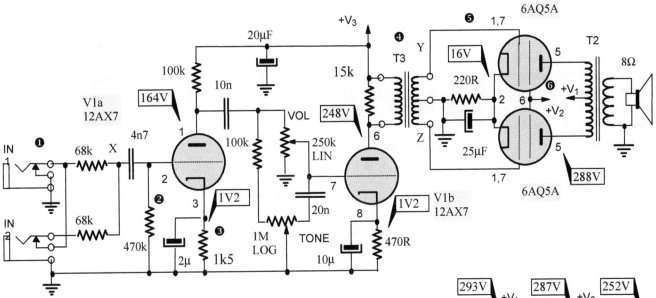

Musicmaster Bass circuit diagram, © Fender Musical instruments

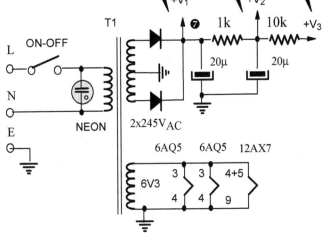

Luckily, this amp makes quite a decent small guitar amp. Moving on to the first gain stage with 12 AX7 triode, notice a very low value of the cathode bypass capacitor, only 2μF.

The lower -3dB frequency of the RC filter in the cathode is $f_L = 1/(2\pi R_K C_K) = 1/(2\pi*1,500*2*10^{-6}) = 53$ Hz!

Again, that is quite a high cutoff frequency, OK for a lead guitar but not for bass. This amp would not have much bass below 50 or so Hertz even without the input RC filter.

The preamp stages are fairly standard, as are the controls, a single tone control and a volume control. The most interesting design feature of this amp is the use of an interstage transformer T3 as a phase splitter (4).

The secondary winding has two equal halves, and, since its CT (center tap) is grounded, the signal voltages in points Y and Z are of the opposite phase, just what the output push pull stage needs!

The primary winding is paralleled with a 15k resistor, but there is still some DC current flowing through it. We have two DC voltages across the primary, 252V at one and 248V at the other end.

We can calculate the anode current of the 2nd stage by the voltage drop across its 470R cathode resistor, so $I_2 = 1.2/470 = 2.6$mA. Since the voltage drop across the 15k anode resistor is 252-248 = 4V, its current is 4V/15 = 0.27mA. The rest of the current (2.6-0.27 = 2.33mA) flows through the primary winding of the IS transformer.

As a matter of interest and to get some feel for the magnitudes and parameters involved, we can calculate its DCR (DC resistance) as DCR= 4V/2.33mA = 1.72kΩ! That is quite a high DC resistance, meaning that the primary has many turns, the winding wire of a very small diameter (high resistance) was used, or both.

AC signal levels were not given on the original Fender drawing, and we don't have the winding data for this transformer, so we cannot ascertain its impedance and voltage ratios.

The output stage is self- or cathode-biased. 16V across the common 220Ω cathode resistor means the cathode current is 72mA or 36mA per tube.

DC-COUPLED SPLIT-LOAD INVERTER: HOHNER CONTESSA (Model CA300)

Concertina, split-load or cathodyne phase splitter

The simplest tube phase splitter to comprehend and to build is the split load inverter, also known as "cathodyne" splitter and "concertina" (probably because of an analogy with an accordion expanding and contracting). Its simplicity and the fact that only a single triode is needed per channel gives it an edge over other topologies.

It is simply a common-cathode stage with an un-bypassed cathode resistor. There is local negative feedback since that un-bypassed resistor is in both input (grid circuit) and output or anode circuit. Thus, the first disadvantage of this phase splitter is that it provides no gain. Depending on the tube used and the values of anode and cathode resistors, it attenuates the signal to around 80% of its input value.

With the same DC power supply, say 280V, two-tube inverters can provide outputs of up to a maximum of +/-140V (neglecting distortion and unbalances for a moment). Since each output in the split load inverter only has half of the V_{BB} voltage "available", or 140V, it can only provide outputs up to a maximum of +/-70V!

On the positive side, since the circuit uses only one tube, any changes in tube parameters will affect both outputs equally, so the outputs will stay balanced!

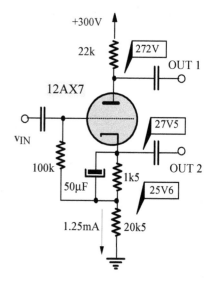

ABOVE: AC-coupled split-load inverter with ECC83 triode

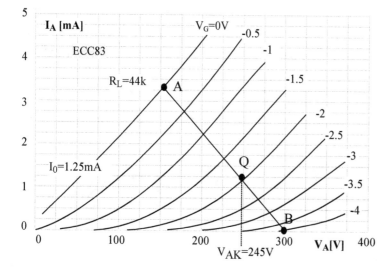

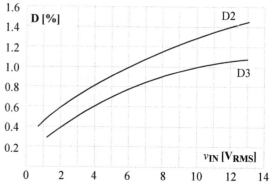

ABOVE: The 2nd and 3rd harmonic distortion coefficients for the ECC83 split-load inverter

LEFT: The quiescent point and load line for the ECC83 split-load inverter

Hohner is a reputable European maker of accordions and keyboards, but not known for their amplifiers, except a few models of keyboard amps. So, when this beautiful amp was listed on eBay USA, we just had to have it.

The amp's large case fell apart in transport, not helped by the fact that the seller wrapped it in cellophane. No bubble wrap, no Styrofoam, no cardboard box, nothing, except a few microns thin cellophane! Luckily, the chassis and the tubes were OK, as was the huge 15" Alnico speaker. We got all of our money back from eBay, US$203 purchase price, and US$108 transport cost.

The four preamp tubes (12AX7) were made by Amperex in the USA, while the two output tubes (7591) were labeled Westinghouse USA. All components were original, and all looked distinctly American; no European parts were used. After some Google-ing, our diagnosis was confirmed. Hohner did not make this amp, after all. It was manufactured in Irvington, New Jersey, by a company called Sano.

Model CA200 (with two EL84 in the output stage) was released with the Sano badge shortly after.

There are two channels with two inputs each. Channel one features a tremolo and reverb (Accutronics unit mounted on the sidewall) plus 2-way tone controls (bass & treble). Tremolo's intensity and speed are both adjustable, as is reverb time. Channel two is without any effects, only volume and bass & treble tone controls.

Both channels share the power supply, CT secondary, rectifier with two SS diodes (1) and two dual section elcos (2). There is no filtering choke.

The amp uses a pair of 7591 pentodes in push-pull, producing around 30 Watts of output power.

The heater wiring

The heater wiring was done in the worst possible fashion. One end (pins 2 of 7591 and pin 4 of 6CS7) was run on one side of all tubes, the other wire on the opposite side, all at ninety degrees. Follow the green wire (photo on the next page) from pin 2 of one out tube to the other, from (4) to (5), and then to the 6SC7 heater (6).

ABOVE: The front baffle, the speaker and the chassis were not damaged.

BELOW: The power supply is very simple, and Channel 2 controls (3) were all wired in a true point-to-point fashion.

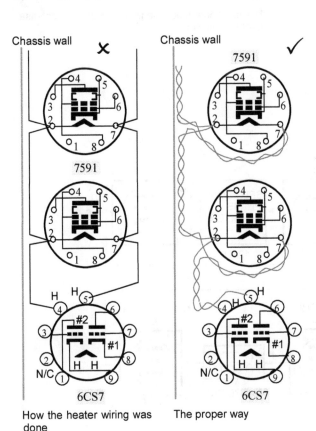

How the heater wiring was done   The proper way

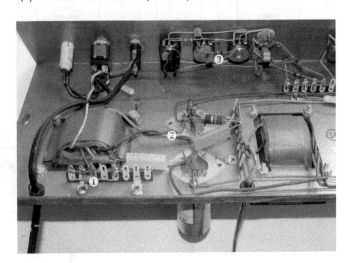

The same "square" loops were on the other side but cannot be seen due to the angle at which the photo was taken and the fact that these wires are down at the bottom of the chassis. The same wiring style was used for the heaters of the four preamp tubes.

This places each tube socket and all hum-sensitive pins, esp. the grids, inside a huge inductive loop. The proper way is to twist the wires, so their magnetic fields cancel and run them on one side only, so there are no loops around any socket.

Topology

The 1st gain stage, tone control, and 2nd stage are practically identical in both channels. Both their outputs come together at the "mixer stage" grid, the 1st half of the 12AU7 triode, which is DC-coupled to its other half, wired as a split-load phase inverter.

The tremolo oscillator feeds a simple J-FET switch which switches a cathode bypass capacitor of channel 2's first stage in and out and thus changes its gain by roughly 3dB. Therefore, this tremolo design provides a very "mild" amplitude modulation.

The signal after Ch1 volume control feeds two amplification stages of 6CS7 duo-triode, which drive the reverb pan. Its delayed output ("wet" signal) is amplified in a single 12AX7 recovery stage and fed back into the same point, just after the series 220k resistor. The design is very clever; even though the cathodyne phase splitter has no gain (actually attenuates the signal 10-20%), there is plenty of overall voltage gain, primarily due to the addition of the 12AU7 mixer stage.

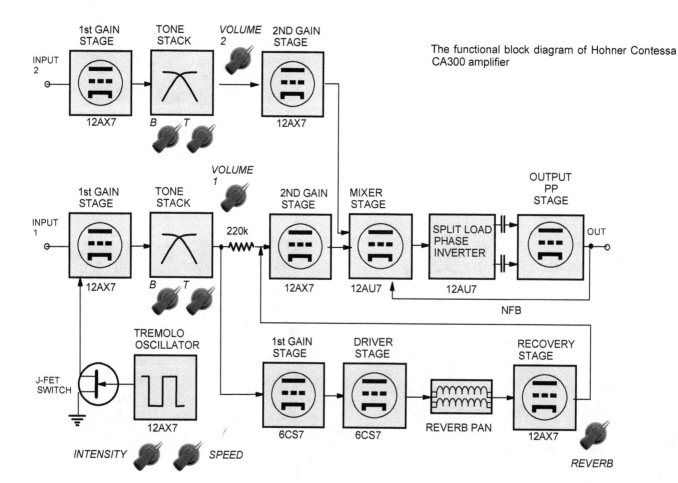

The functional block diagram of Hohner Contessa CA300 amplifier

The 6CS7 dissimilar duo-triode

This duo-triode was only made in the USA and Japan. Although a few hi-fi and guitar manufacturers have used it over the years (Leben, for instance, in their CS600 stereo amp), cheap new old stocks (NOS) are still available (about $5).

Both triodes are in the low μ category, under 20. Triode #1 is a low anode dissipation design, but triode #2 has a healthy 6W+ power rating and low internal impedance of 3-4kΩ, which makes it a great driver tube.

TUBE PROFILE: 6CS7

- Indirectly-heated duo-triode, Noval socket
- Heater: 6.3V/600 mA
- V_{AMAX}=500V, V_{HKMAX}= 200V$_{DC}$
- Tr. #1: P_{AMAX}=1.25 W, gm=2.2 mA/V, r_I=7.7kΩ, μ=17
- Tr. #2: P_{AMAX}=6.25 W, gm=4.5 mA/V, r_I=3.45kΩ, μ=15.5

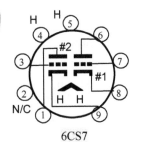

6CS7

7591 beam power tube

7591 was initially designed to operate as an audio tube with high efficiency and high power sensitivity, meaning very low grid signals were needed for maximum power output. While it uses an octal socket, it is not pin-compatible with the 6V6-6L6 family of octal beam power tubes.

Physically, the vintage 7591s are relatively small, so good ventilation is needed for efficient cooling. The modern replacements produced in Slovakia (JJ) and Russia (Electro-Harmonix) are taller.

7591 was never as widely used as 6L6, especially not in guitar amps, but a few vintage hi-fi brands such as Eico, Harman Kardon, Fischer, and H.H. Scott amps used them successfully.

Circuit analysis

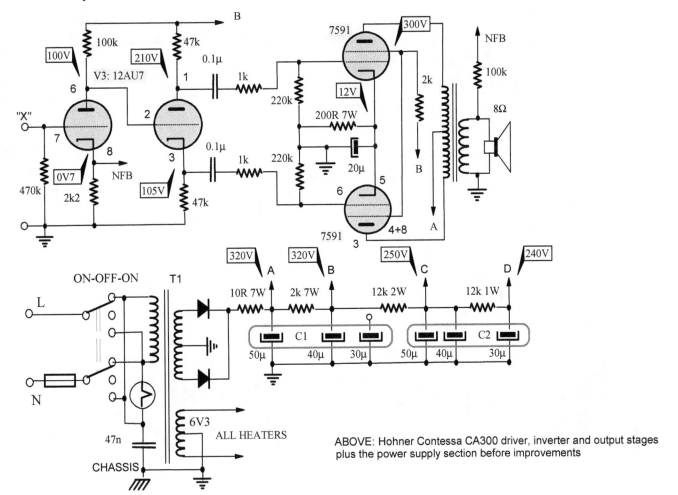

ABOVE: Hohner Contessa CA300 driver, inverter and output stages plus the power supply section before improvements

Replace the old power cord with a modern 3-pin one, ditch the dangerous mains capacitor, and earth the chassis; we no longer need the polarity reversal feature. The unused half of the mains switch can now become a standby switch. The middle position is "off."

Another possible location for the standby switch is the CT (center tap) connection to the ground. If the switch opens that connection, there is no high voltage in point "Z"!

There was plenty of room on the chassis, so we added a filtering choke. The inductance is not critical; 3-10 Henry would do, whatever you can get, but its current rating has to be at least 200mA. The first two elcos were "moved" or shifted towards the choke; now, we are also utilizing the previously unused 30μF section of C1!

The lower -3dB frequency of the output stage is determined by the $R_K C_K$ time constant of the cathode circuit (1), so $f_L = 1/(2\pi R_K C_K) = 40$Hz, which is fine, since a guitar does not produce frequencies below 80Hz.

NFB from the output is taken via a 100k resistor to the 2k2 cathode resistor of the mixer stage. Thus, the NFB is very mild; only 2.15% of the output signal is fed back (2.2/102.2 = 0.0215 or 2.15%).

Since the cathode of a split-load phase splitter is at an elevated DC voltage, usually in the 50-150V range (105V here), that makes it easy to directly couple its grid to the anode of the previous stage. Of course, the grid must remain negative vis-a-vis the cathode, so the anode voltage of the previous stage must be lower than the cathode voltage of the phase splitter tube (100V here, meaning the grid of the phase inverter is biased at -5V)!

Adding "Presence" control

With 220nF capacitor, the lower -3dB of this filter would be $f_L = 1/(2\pi*2,200*220*10-9) = 329$ Hz so that bass frequencies would be mildly attenuated by NFB, but midrange and high frequencies would not. A 470n cap would lower that frequency to around 154 Hz. This makes it very easy to add presence control to this and similar amps. As pictured, replace the 2k2 fixed cathode resistor with a 2k potentiometer and wire a 220, 330, or 470nF cap between its wiper and ground. If you find the stock amp too bright for your taste, the presence will make it even brighter, so it may not be a good idea. If you want a cleaner sound, remove the 25μF cap completely. That will make NFB active across all frequencies!

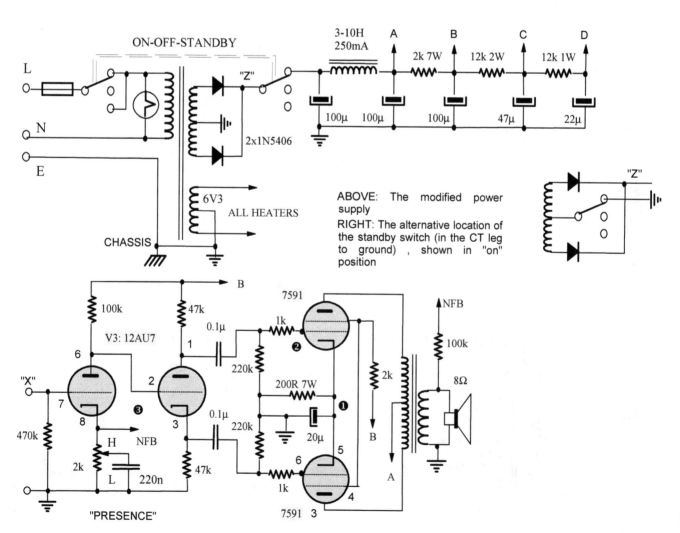

ABOVE: The modified power supply

RIGHT: The alternative location of the standby switch (in the CT leg to ground) , shown in "on" position

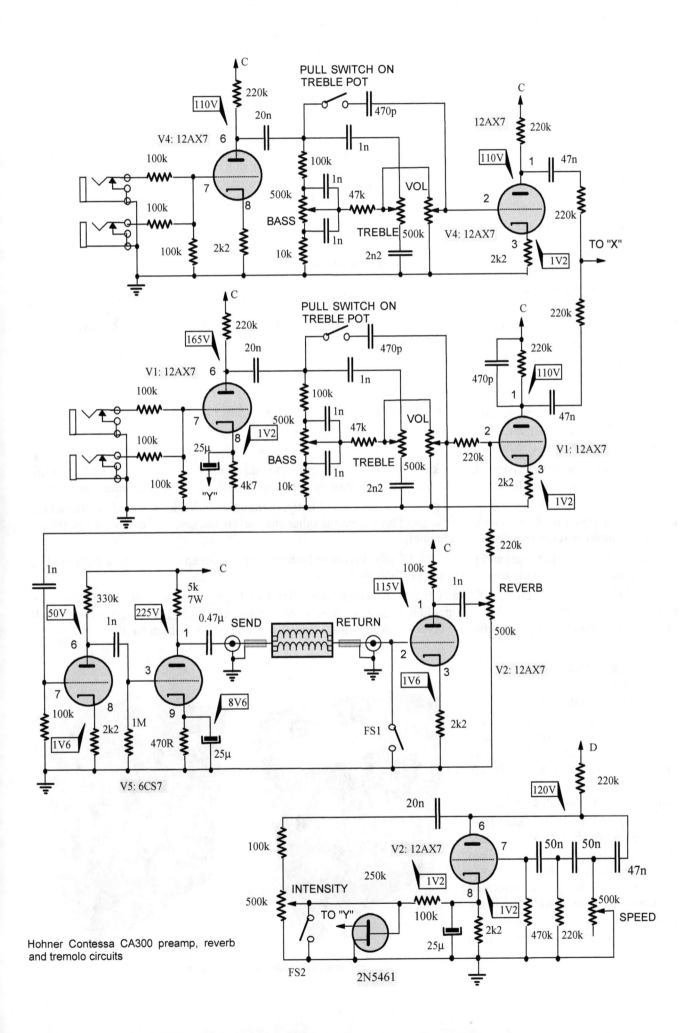

Hohner Contessa CA300 preamp, reverb and tremolo circuits

The input, reverb and tremolo circuitry

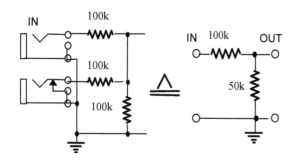

Since both guitar inputs are shorted to ground, if only one input is used, the other jack adds its 100k resistor in parallel with the 100k resistor to ground, a total of 50k. This means that a voltage divider is formed, 100k in series and 50k to GND, so only 1/3 of the signal is passed onto the first grid. That is quite a strong attenuation (66%)!

A better arrangement is keeping one input as a low gain and making the other a higher gain input. The low gain has the same topology as before, but this time its attenuation is 50% since the two series resistors are equal (68k is typically used on Fender amps). The high gain signal is not attenuated.

Both channel's treble pots have a pull-activated switch that bypasses the tone control stack with a 47nF capacitor, so the signal goes straight to the wiper of the relevant volume control potentiometer.

The 1n capacitor and 100k resistor to ground form a high pass filter at the input of the reverb circuit, with lower -3dB frequency $f_L = 1/(2\pi RC) = 1,592$ Hz.

Normally only bass frequencies below 100-200Hz are attenuated so as not to swamp the reverb circuitry.

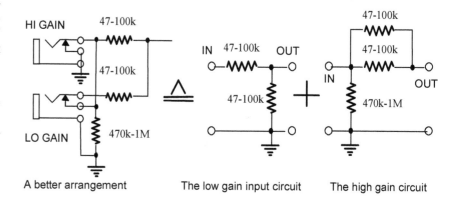

A better arrangement   The low gain input circuit   The high gain circuit

You can replace the 100k resistor with 220k or 330k, but it's easier just to add another capacitor in parallel, 10-47nF. Since that will change the character of the reverb signal, experiment to determine the value you prefer sonically.

The arrangement of the tubes was made in a user-friendly way. Firstly, if either V1 or V4 fails, the other channel isn't affected, and, if you wish, you can swap the good tube from the other channel (from, say, the "dry" channel, the one without tremolo or reverb) to the "wet" channel.

Secondly, you can experiment and plug a 12AT7 tube in one or both channels (instead of V1, V4, or both) and see how you like the sound. This amp makes tube rolling fun!

Tube V4 should always stay 12AX7 since it is in the tremolo oscillator. A high gain is needed for oscillations (a theoretical minimum of 29), so low and medium m tubes such as 12AU7 and even 12AT7 are out of the question. Its other triode is the recovery amp after the reverb tank, and that stage also needs 12AX7's high gain!

Rebuilding the cabinet

Providing its timber is structurally sound, a broken in transport cabinet could be rejoined. However, this amp suffered water damage at its bottom - the particleboard was all puffed up, the bottom joints simply crumbled apart, making repair impossible.

We bought 1st grade (knot-free) 19mm pine boards for $50 and made the "frame" ourselves. Since we had neither the skills nor the special tools required, no complicated "professional" joints were used, the four sides were joined with thick internal aluminum L-profiles.

ABOVE: The Hohner Contessa amplifier in a rebuilt cabinet

PARAPHASE INVERTER: MONTGOMERY WARD (AIRLINE VALCO) GIM-9151A

Both tubes in the paraphase inverter work in the common cathode mode. Output 1 is direct from V1, and part of the output voltage (voltage divider R_A and R_B) is fed into the grid of V2. The output voltage OUT1 passes through one tube and is 180° out of phase with the input signal; the output voltage OUT2 passes through both tubes, and since both invert the phase, it is in phase with the input signal.

With a 1V input, assuming the amplification of the V1 stage is 30, the voltage across R_1+R_2 is 30V. The input into grid V2 must also be 1V (since its A=30 as well), meaning we need to cancel the gain of the V1 stage before we feed the signal into the V2 stage. We must attenuate V1's output 30 times; therefore, R_2 must be 1/30 of the total resistance R_1+R_2 of the voltage divider.

This inverter provides voltage gain, but it requires two tubes instead of one for the cathodyne option. Since it is common to both outputs, the balance is not affected by the changes in V1, but it is susceptible to any changes in tube V2 and its gain.

The circuit is not self-balancing. Due to the effect of two coupling capacitors in the V2 circuit and only one in the V1 circuit, low frequencies are unbalanced. Plus, unmatched tubes or unequal aging of tubes can cause further serious unbalance in the outputs.

The floating paraphase inverter

This inverter derives the grid voltage for V2 from a resistor common to the outputs of both tubes, R_2. Assuming positive-going input signal, the output of V1 will be of the opposite phase; the anode current I_{A1} will flow through R_1 and R_2 to the ground (solid arrow), making point X negative. This negative-going or reducing voltage is fed into the grid of V2, so its anode current I_{A2} will decrease, and this negative current will flow from the ground through R_2 and R_3 into the anode, shown with a dashed arrow.

Both anode currents flow through R_2 but are in opposition, so the total current through R_2 decreases. I_{A1} is slightly larger than I_{A2}. If I_{A2} tries to increase for some reason, point X and the voltage drop across R_2 will become more negative, offsetting this increase in I_{A2}. Ditto, if I_{A2} decreases, the voltage drop across R_2 would become more positive (less negative), increasing I_{A2}.

Thus, the floating paraphase inverter is a self-balancing circuit. Paradoxically, that does not mean that the circuit is balanced; it is always slightly off-balance; otherwise, the self-balancing mechanism would not work since the grid drive voltage for V2 is proportional to the difference or unbalance between two AC anode currents. Thus, R3 should be slightly larger than R_1.

This imbalance can be reduced by using high gain tubes and making R_1 slightly smaller than R_2.

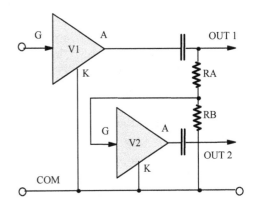

ABOVE: Block diagram of the paraphase inverter
BELOW: Circuit diagram of the paraphase inverter

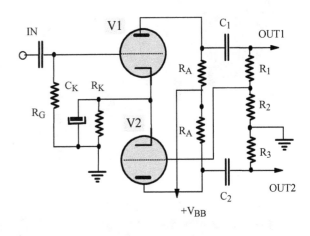

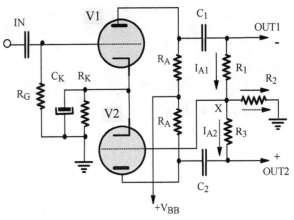

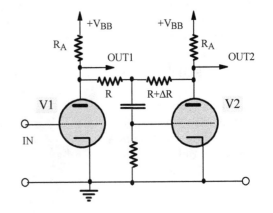

RIGHT: The basic circuit of the floating paraphase inverter showing NFB from the anode to the grid of V2 (anode follower)

Then V_{OUT2} can be equal to V_{OUT1} even though I_{A1} is still slightly smaller than I_{A2}.

Our older American readers will remember Montgomery Ward, one of the largest mail order and department store retailers in the USA in its heyday. Just as they did for Sears, Roebuck & Co., various amplifier manufacturers made guitar amplifiers for Montgomery Ward. Model GIM-9151A is one of the whole series of heads and combos made by Valco. The other models were GIM 9111A (pair of 6L6, no tremolo or reverb), GIM-9131 (9111A with tremolo and reverb), and GIM-9171A, a 9151A with tremolo and reverb.

Valco started operations in the 1930s and went into the tube guitar amp manufacturing business in the 1940s, making tube amps under Supro, Airline, Oahu, and National brands. Harmony, Gretsch, and Kay contracted out the manufacture of some of their amps to Valco. Kay Musical Instrument Company eventually merged with Valco in 1967, but that did not improve their financial position since it went bust only a year later!

The textured silver finish of the fascia cannot be seen in the photo but is something to be truly appreciated. Large black plastic knobs have brighter silver inserts, which stand out, creating a nice contrast with the darker background. Overall, this must be one of the most elegant and sleek-looking heads.

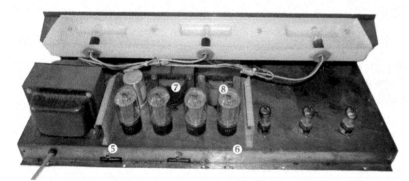

The three large top-mounted lights are a great idea; fumbling around black knobs and hard-to-see jacks on a dark stage is every guitarist's nightmare.

There are two channels with two inputs each and three switches, on-off (1), line reverse (2), and panel lights on-off (3). The amp uses a resettable circuit breaker instead of a fuse, another great idea (4)!

The chassis is large and spacious, with shields between the power transformer and the power supply/output section (5) and the output and preamp sections (6).

Judging by the surface corrosion and patina, the two output transformers, (7) and (8), are original, although of different construction and slightly different sizes. Electrically, however, they had almost identical parameters.

The wiring is very easy to follow. Four discrete diodes form an HV rectifying bridge (1), followed by a 3-section elco (2). The first two sections are used in a simple two-stage RC filter; the third is used as a cathode capacitor of one of the output stages (3).

Notice the poorly wired heater supply (4). The white wires are simply looped on one side, the green wires looped in the same fashion on the other side of the sockets, thus encompassing all sockets in a huge inductive loop (photo next page).

Those wires should be unsoldered and a twisted pair of wires used on one side of the sockets only, routed along the chassis edge.

The preamp section is well laid out on six large terminal strips. While resistors are easy to access (should you need to change them), the capacitors are at the bottom and not so easy to get to without disturbing the resistors (5).

Notice that 3 pins of one duo-triode (6) are not used, so only five of the six 12AX7 triodes are used.

Although this amp has been referred to as a stereo amp and despite the fact that it has two identical output push-pull stages and even two output transformers, unfortunately that does not make it a stereo amp. The clue is in the preamp topology.

The input stages, volume controls and treble and bass controls are separate, but the two signals are then combined, before passing through the common 2nd gain stage and the common paraphase splitter. Thus, both output stages pass the same signal.

The "Main speaker" and "Extension speaker" stereo jacks are also wired together, but there is a difference. The main jack is shorting while the "Ext" jack is not, and for some reason the "Ext" jack is wired in reverse.

RIGHT: The functional block diagram of Montgomery Ward GIM-9151A amp

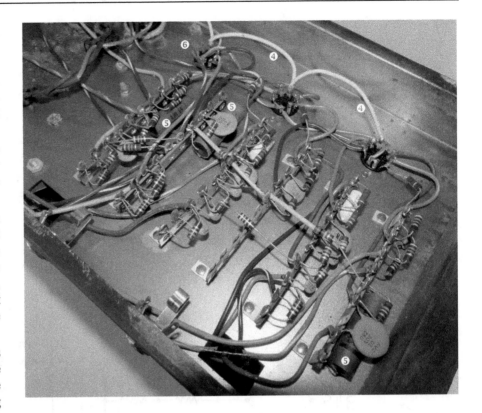

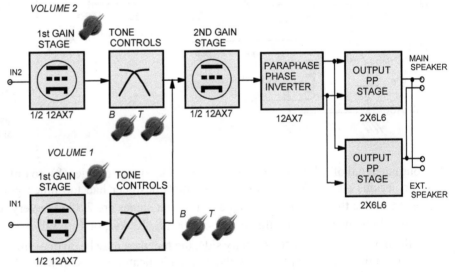

The circuit analysis

This amp uses solid-state rectification but has no standby switch, so the line reverse switch (1) should be converted into a standby switch in position (2).

The three panel lights are on the primary side and are not shown here for clarity. The single filtering elco (3) has four sections, three are used in the HV section, and the fourth (also rated at $450V_{DC}$) is used as the cathode bypass cap for one of the output stages (4). Notice the total capacitance of the HV filtering is only 85µF, which is extremely low for an amplifier of this size and power level.

Thus, connect the first two sections in parallel (40+25µF) and use the fourth section (4) instead of the 25µF section. Then simply add a 20-47µF low voltage cathode bypass cap under the chassis, as was done for the other output stage (5). This will improve the energy reserves a bit, but you should also parallel a few elcos in points A and B (add 47µF cap in each under the chassis) to really stiffen the power supply.

There are no screen resistors; add 470R resistor to each screen grid (6), four in total.

In the paraphase inverter, part of the output signal from the top triode is taken from the 270k + 12k voltage divider (7) and fed to the lower triode. The upper phase signal passes through one triode, but the lower phase signal passes through two triodes. This type of phase inverter is never used in hi-fi since it is invariably unbalanced and distorts significantly, but it is sometimes used in guitar amps. However, there are much better inverter designs around.

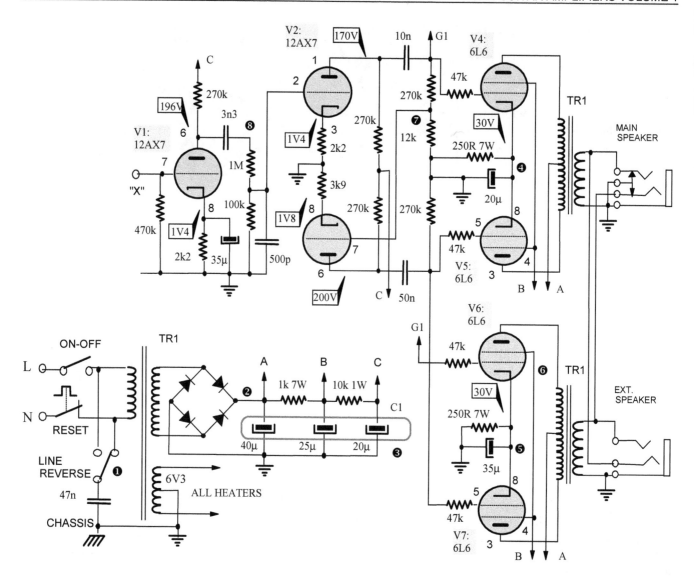

The output signal of V1 passes through a 1M+100k voltage divider (8), so only 0.1/1.1 = 0.091 or 9.1% is used. This is some serious attenuation; obviously, the amp's designer had too much gain and decided to reduce it this way. However, notice that the anode resistors of the first stages are 270kΩ, a very high value. The designer should have reduced gain by using lower value resistors, 100kΩ, 120kΩ, or 150kΩ!

Since the panel lights are on the primary side, the fact they may be on does not indicate the health of the power supply chain. So, in the improved power supply, a green indicator comes on when the standby switch is in the S/B position and goes off when the "operate" position is selected. Then the red neon comes on, indicating the health of the whole HV chain since it is placed at its end.

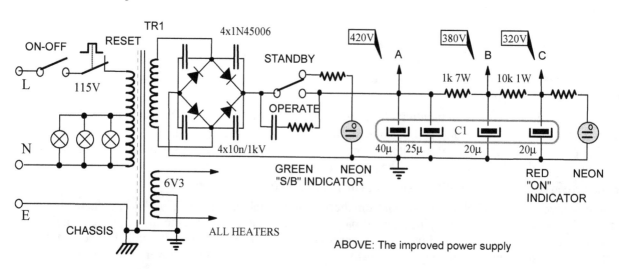

ABOVE: The improved power supply

The two preamp stages are identical for both channels, except that the cathode resistor in channel one is bypassed by a 35mF capacitor, while channel two's isn't. Thus, channel one would be approx. 3dB louder! However, notice that the output of channel one passes through a 470k series resistor, which, together with the 470k grid-to-ground resistor of the next stage (previous page), forms a voltage divider, attenuating the signal by 50% (cutting the gain in half). There is no such attenuation at channel two's output, thus equalizing their voltage gains.

The 2-gang tone controls are unusual. The output of volume control feeds both circuits. Treble control is achieved first via a fixed CR high pass filter (1n+270k) with a turnaround (-3dB) frequency of 590 Hz, which attenuates bass frequencies regardless of the position of tone controls. This filter is followed by a simple RC bleeder to the ground. With the wiper in the uppermost position, the 5n cap shunts the 500k pot, so treble attenuation is at its maximum.

With the wiper at the bottom end of the resistor, the resistor is in series with the shunt capacitor, and there isn't much treble attenuation.

The output from the Bass pot feeds two low-pass RC filters in cascade, the first with an upper -3dB frequency of 321 Hz, the second with f_U= 234 Hz.

With the slider in the uppermost position, the treble is severely attenuated (a.k.a. "bass boost"!); in the lowest position, the filters are out of the circuit since their input is grounded and the 500k bass pot is simply a fixed resistor in parallel with the lower portion of the volume control pot.

With both being passive circuits in parallel with each other, the tone controls are interactive.

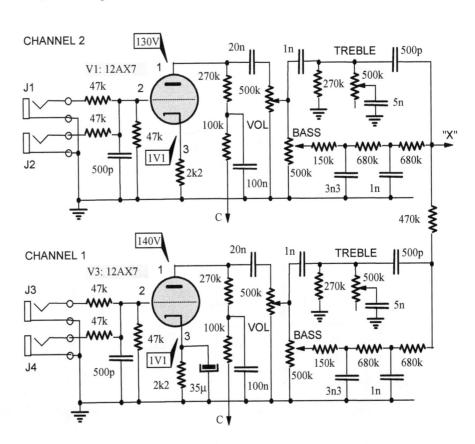

The modified preamp stages

Notice the 500pF capacitor to ground at the input. It forms a low pass filter with the 47k series resistor; its -3dB turnaround frequency is only 6,773 Hz. This means that all harmonics above 7kHz will be seriously attenuated, and the amp will lack "sparkle"!

To not drill any holes and make the mods reversible, one input jack to each channel should be removed, and the hole used to mount a potentiometer or a switch, depending on which mod you choose to implement. One possibility is to install two master volume pots, one for each channel, or one master volume and a balance. Both would make the outputs of the two channels blendable in any ratio.

The input shunting capacitor and 47k resistors are removed, a 1M grid leak resistor and 10k grid stopper resistor are installed instead. A 120k resistor replaces the 270k anode resistor. If there is too much bass for your liking, reduce the value of the cathode bypass capacitor from 35µF down to 10µF or even 2.2µF!

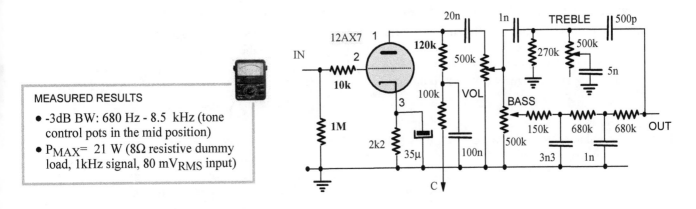

MEASURED RESULTS
- -3dB BW: 680 Hz - 8.5 kHz (tone control pots in the mid position)
- P_{MAX}= 21 W (8Ω resistive dummy load, 1kHz signal, 80 mV$_{RMS}$ input)

Conversion into a true stereo amp

You may have noticed an odd number of preamp triodes in 9151A's topology. One duo triode serves as Ch. 2 preamp tube, the other as a phase splitter, but only one of Ch.1 12AX7 duo-triode is used, the other one isn't. That is the good news - we have one spare 12AX7 triode, which we can use as a separate 2nd gain stage for channel 1!

However, we still need two triodes for paraphase or long-tail pair phase splitter, or at least one triode (one half of 12AX7) for a split-load phase inverter stage. Adding a tube is a major exercise in drilling & punching the chassis, extending heater wiring, adding terminal strips, etc. This assumes that the power transformer can supply the additional heater load of 0.6A, which usually isn't the case!

There is a way out, and that is using an interstage transformer with its single-ended primary winding as the 2nd stage's anode load. Two identical but out-of-phase secondaries make a simple but very effective phase splitter!

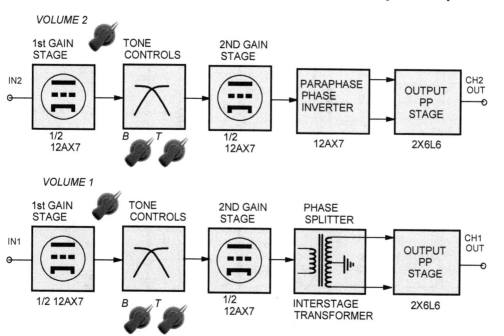

LEFT: One way of converting GIM-9151A to a true "stereo" (or more accurately, a "2-channel") amplifier.

CATHODE-COUPLED, "LONG-TAILED PAIR" OR SCHMIDT INVERTER: UNIVOX 1236

The cathode-coupled inverter or " long-tailed pair"

Triode V1 in the "Long-Tailed Pair" works as a common-cathode stage, while V2 is a common-grid amplifier. Since the two cathodes are coupled via the common cathode resistor, any signal variation at the cathode of the first triode will be fed into the cathode (or input) of the second stage.

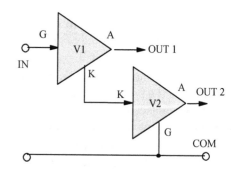

The common cathode resistor provides a negative feedback mechanism. For a balanced output, R_K must be selected so that the feedback voltage is equal to half of the input signal amplitude so that the input voltage to the grid of V1 is also half of the input signal.

Due to NFB, the gain of the V1 stage is reduced to half of its gain as a common cathode stage without feedback.

Both anode currents flow through R_K and their difference determines the feedback voltage by its voltage drop across R_K. As in the paraphase inverter, these two currents are never the same, so the circuit is never in perfect balance, but it is self-balancing. The grid voltages are $V_{G1}=V_S-(I_1-I_2)R_K$ and $V_{G2}= -(I_1-I_2)R_K$. With equal anode resistors, this difference in the grid voltages will result in unequal outputs. Their ratio is

$$V_1/V_2 = \frac{[r_I+R_{A2}+(\mu+1)R_K]R_{A1}}{(\mu+1)R_K R_{A2}}$$

Ideally, the value of R_K should be infinite, which would result in the perfect coupling between the two tubes. However, since R_K is of finite value, the gains of the common cathode and common-grid stage are not the same. To bring the outputs closer to balance, the gain of the CC stage needs to be reduced by lowering the value of its anode resistor, for instance, from 100k down to 82k or 91k (compared to R_A=100k for the second stage).

The advantage of this phase splitter over the cathodyne or split-load inverter is that it provides voltage gain, while the cathodyne phase inverter reduces the gain of the previous stage 10-20%.

In practice, as in this inverter from Legacy Blues Twin (Epiphone Blues Custom) amplifier, R_K is usually split into two resistors. Only the upper resistor (1k2) determines the bias voltage. Sometimes called the Schmidt inverter, this variant must be AC- or capacitively coupled to the previous stage. R_{A2} is selected first based on the desired gain and the chosen operating point, and then R_{A1} can be calculated from

$$R_{A1} = \frac{(\mu+1)R_{A2}R_K}{(\mu+1)R_K+R_{A2}+r_I}$$

Substituting $R_K=11,200$, $R_{A2}=100k$, $\mu=100$ and $r_I=66k\Omega$, we get $R_{A1}=87.1k\Omega$. Instead of using a slightly higher standard value of 89k, Gibson designers used a lower standard value of $82k\Omega$.

Two-tube paraphase and cathode-coupled inverters provide much larger output signals than the split-load inverter, a distinct advantage when high grid driving voltages are needed.

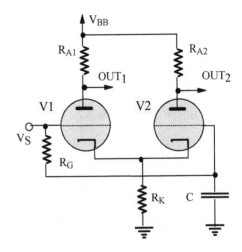

ABOVE: DC-coupled   long-tailed   phase inverter

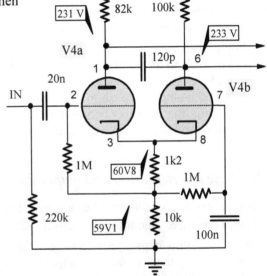

ABOVE: Legacy Blues Twin & Epiphone Blues Custom: Phase inverter and output stage, © Gibson

Judging by the very few online photos of this relatively obscure amp, the output transformer (3) is not original. Neither is the large power resistor (in the HV filtering circuit) bolted onto the chassis (4).

Despite having two tone-control switches, marked Deep and Sharp (1), somebody added a third at the very input (2). It switches a bypass cap in and out of the circuit in order to boost treble frequencies.

A few non-original components are noticeable, the last elco in the filtering chain (5), two larger filtering elcos ahead of it (6), and one of the rectifying solid-state diodes (7). Japanese-made paper in oil coupling capacitors are used extensively (8).

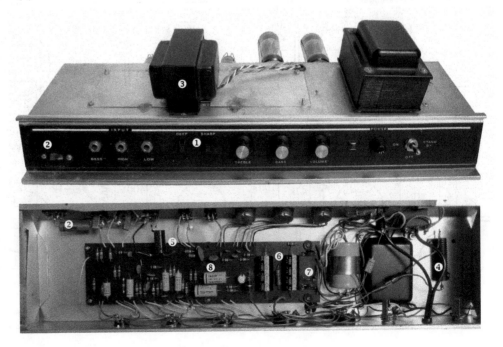

There is a sticker proclaiming this to be a "100 Watt" bass amplifier. A snowball in hell has more chance of survival than this being a 100 Watt amp. Two 5881 (a variety of 6L6) in push-pull cannot produce more than 50 Watts.

The types of resistors, capacitors, and diodes used, plus the amp's overall look & feel, indicate that it was made in Japan. Its circuit diagram (available online) was drawn using Japanese manufacturers' symbols and overall style.

The original power supply

The solid-state voltage doubler (2) provides 660V for the anodes and half that value from its midpoint for screen grids. Notice that the upper elco is 47µF while the bottom one is 100µF. These elcos must be of the same value; why the designer/manufacturer chose such values is beyond comprehension.

The large 10 Watt rated 330Ω resistor (3) was added by someone in the past since it is not on the original Univox diagram, and the wires that used to connect the plus of the upper elco in the voltage doubler with the output transformer CT were modified and extended.

This was probably done to reduce the high voltage (660V) on the anodes of output tubes. Perhaps one of the amp's previous owners noticed a shortened tube life and increased failure rate and attributed it to such high anode voltage.

A simple half-wave rectifier (3) supplies a fixed bias voltage of -32V to the control grids of output tubes. A 10R resistor (5) enables the measurement of total cathode current of both output tubes, however, there is no provision for adjusting the balance between slightly unmatched power tubes, and there is no way of adjusting bias at all.

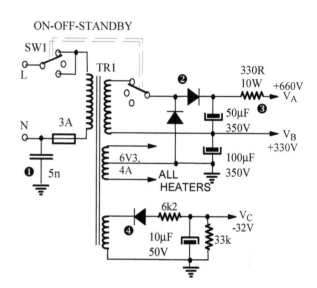

The phase inverter, driver and output stage

The grid resistors of output tubes are only 100k each, so a higher capacitance coupling capacitor had to be used, 0.1µF in this case. That means the lower -3dB frequency of this CR high pass filter is f = 1/(2πRC) = 16 Hz.

Notice two 10nF caps bypassing the grid resistors of the driver stage (6). This will severely decrease the upper-frequency extension of the amp and attenuate the treble. If you want more chime and sparkle from this amp', reduce the value of these caps to 1n or even less, or remove them altogether. Some experimentation is needed here. This low gain 12AU7 driver stage is unusual in guitar amps but is very common in hi-fi designs.

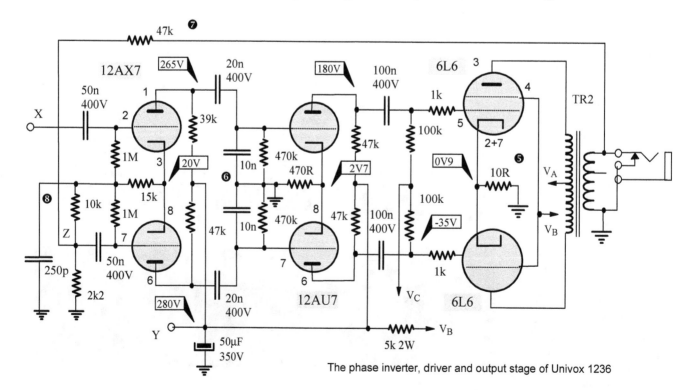

The phase inverter, driver and output stage of Univox 1236

The phase inverter is our old friend, the long-tail pair of 12AX7 triodes. To equalize gains, notice that the upper anode resistor is 39kΩ while the lower one is 47kΩ! These are pretty low values for this circuit, half of what is usually used (89k and 100k).

The negative feedback loop encompassing these three stages is closed by the 47k feedback resistor (7), back to the 2k2 resistor in the phase inverter. The feedback ratio is $\beta = R_2/(R_1+R_2) \times 100\% = 2.2/(47+2.2) \times 100\% = 4.5\%$, which is a moderate amount of feedback, meaning 4.5% of the output signal fed back.

Finally, notice another HF shunting resistor to ground, 250pF from the midpoint in the phase inverter (8). Whoever designed this amp tried hard to kill the highs, as if designing a bass amp. Perhaps the clue lies in one of the three inputs, which is indeed marked "Bass."

The preamp stages

The single-channel has two 12AX7 stages, three inputs, and Treble & Bass tone control. Compared to the usual 100k value, both anode resistors are of high 220k resistance, which would result in an increased voltage gain, but notice that the cathode resistors are also of high resistance, 3k3, instead of the more common 1k5 or 2k2. Both were left un-bypassed, providing local negative feedback and reducing gain.

The bias voltage for both stages is also high, 2V and 2V2, respectively. All of that indicates the designer's desire for lots of clean headroom, like a hi-fi, PA, or bass amp. However, notice the lowish anode supply voltage of 160V and the very low actual DC voltage on the anodes, 40V and 45V, which is usually chosen when a designer wants early distortion. So, some mixed messages here.

The Sharp switch adds a 100n cathode bypass capacitor in parallel with the 1st stage's cathode resistor, increasing gain for higher frequencies. In contrast, the Deep switch changes the configuration of the negative feedback loop from the 2nd stage's output back to the point already mentioned, namely the cathode of the 1st stage.

In one position it is a resistive divider, 470k and 3k3, provifing a mild NFB, $\beta = R_2/(R_1+R_2) \times 100\% = 3.3/(470+3.3) \times 100\% = 0.7\%$, so only 0.7% of the signal is fed back. In the "Deep" position, the NFB signal first passes a low pass RC filter with $f_H = 1/(2\pi RC) = 6,773$ Hz (upper -3dB frequency), thus attenuating HF content.

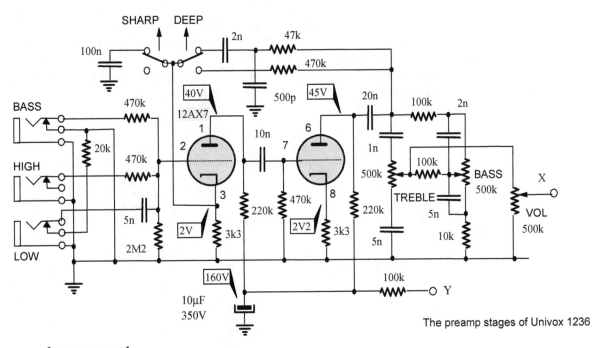

The preamp stages of Univox 1236

The improved power supply

The fuse was moved to the "Live" side of the primary (where it must be by law) after the "On-Off-Standby" switch. A 240V/120V autotransformer was installed, the chassis earthed (grounded), and a 150kΩ/3W bleeder resistor was added after the series voltage dropping 330R resistor.

The bleeder acts as a simple voltage regulator and also discharges the dangerously high DC voltage from the power supply elcos once the amp is turned off. Without that bleeder resistor, the high voltage would remain on those elcos for up to a few hours! Getting zapped by more than 700 Volts isn't just scary and unpleasant; it is often fatal!

The unequal voltage doubling caps were replaced by equal caps of higher capacitance for increased energy storage and better filtration.

The simple biasing circuit was retained. Various ways of providing either two individual bias adjustments or a common bias adjustment plus a fine bias balance were described elsewhere in this book.

With an added autotransformer, the DC voltages were a bit higher, 720V instead of 660V. Still, the bias was -36V instead of -32V, which compensated for things in the power stage and reduced the idle cathode current.

The anode supply voltage of the first two stages increased from 150V to 180V, and in the PI and driver stage, the anode supply increased from 260V to 280V.

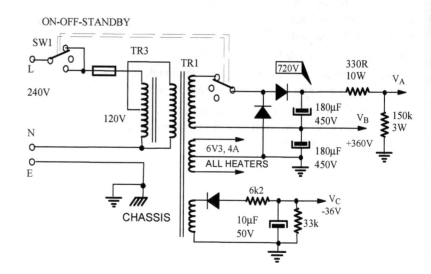

The modified audio circuit

Adding another 10Ω cathode resistor (1) is easy. It still does not make it possible to balance the bias between the two output tubes, but at least now you will know how (un)balanced the existing and any replacement tubes are. Who needs a tube tester when a 2 cent resistor will do?

If you are happy with the amp's tone, the circuit was designed correctly and works fine.

Following the advice of our "consultant" guitarist, the treble-sapping 10n caps were replaced by an order-of-magnitude smaller capacitance (2) for added top sparkle. Along the same lines, adding a Presence control isn't difficult (3), just one pot and one 100n cap parallel with the existing 2k2 resistor. The 250p treble bypass cap from the top of the long-tail resistor to the ground is also gone (4).

Another easy mod would be adding an NFB ON-OFF switch (5). A choice of three differently voiced inputs, two voicing switches (4 combinations), and an added bright switch at the input gives one 3x4x2= 24 different voicings. The addition of the ON-OFF switch for NFB would double that to 48 tone combinations. If that is not enough for you, you should switch from playing guitar to a synthesizer!

While the amp sounded like a crossbreed between a Marshall and Fender, now it is also possible to get a hint of that VOX sparkle and mix it into this subtle blend in various proportions.

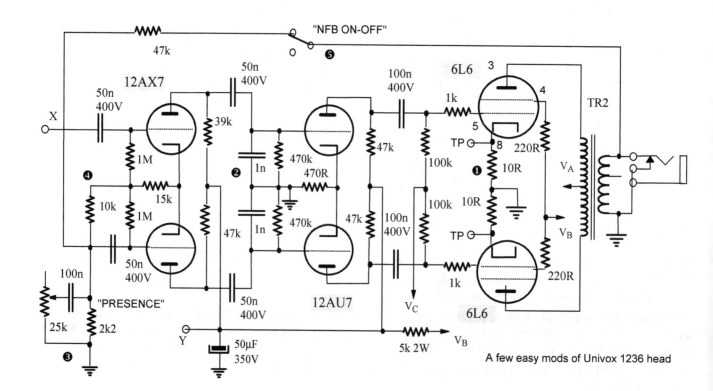

A few easy mods of Univox 1236 head

PUSH-PULL OUTPUT STAGES

13

- CLASSES OF OPERATION AND BIASING OF THE OUTPUT STAGE
- CLASS A PUSH-PULL OUTPUT STAGE
- BIASING AND AC-DC BALANCING METHODS
- THE WORST BIAS CIRCUIT EVER: EPIPHONE GALAXIE 25
- SELF-INVERTING PUSH-PULL OUTPUT STAGE: KUSTOM DEFENDER V15
- PARALLEL PUSH-PULL STAGES

Single-ended output stages are inefficient, require large output transformers with air gaps (due to high DC currents flowing through their primary windings), and are limited to low power levels, generally 5-15 Watts. Push-pull and parallel PP output stages are more efficient, use smaller and cheaper output transformers, and easily achieve 50-200 Watts power levels.

However, they bring to fore their own problems and requirements. Apart from the need for a phase splitter or inverter (covered in the previous chapter), there are issues such as signal balance, crossover distortion, DC and AC balancing and output tube matching & unequal aging.

CLASSES OF OPERATION AND BIASING OF THE OUTPUT STAGE

For a given anode and screen DC supply voltage, the operation of a push-pull output stage is determined by the negative bias on tubes' control grids. Let's use a 6L6 PP stage as an example.

The bias needs to be selected for Class A operation to achieve a maximum symmetrical grid voltage swing. In this example, that bias is -16V. The graph shows the dynamic transfer curve and signal waveforms for one tube only, so it is evident that both tubes conduct at all times in Class A.

By changing the bias from -16 to -22V, the operation moves from Class A to Class AB. In Class AB_1, no grid current flows, the signal's amplitude is smaller than the bias voltage. Once the negative peak of the input (grid) voltage exceeds the tube's cutoff point (line C-C), the tube stops conducting. The other tube still conducts, and the stage operates in Class B, defined as a regime where one tube conducts while the other one does not.

If the signal's amplitude exceeds the bias voltage for class A (-22V), the operation changes into Class AB_2. The signal has crossed into the positive grid voltage territory and grid current flows. The quiescent point Q_{AB} is close to Q_B. This raises a question about the real meaning of Class AB.

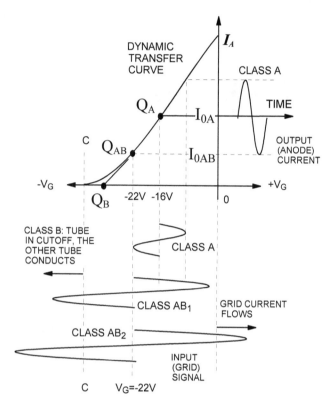

ABOVE: Three biasing options for 6L6 PP stage, Class A, AB and B. Class AB_2 signal is also show crossing into positive grid voltages and causing grid current to flow!

What does Class AB really mean?

Strictly speaking, there is no such thing as Class AB. For smaller signals, tubes in the Class AB stage operate in Class A (both conducting simultaneously), while for larger signals, they operate in Class B; one tube is not conducting (in "cutoff" state), while the other one conducts on its own.

So what is meant by Class AB can be "mostly Class A with some excursions into Class B" if the stage is biased closer to Class A, or "Class A for smaller signals (lower power levels) and Class B for higher power levels" if the stage is biased closer to Class B.

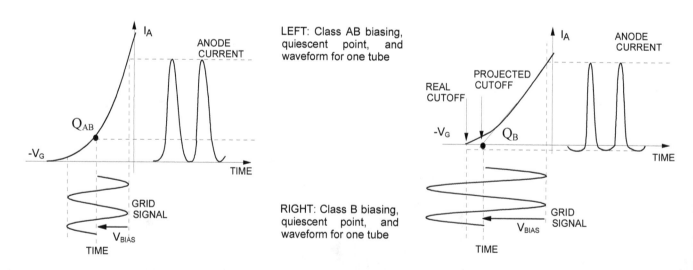

LEFT: Class AB biasing, quiescent point, and waveform for one tube

RIGHT: Class B biasing, quiescent point, and waveform for one tube

Practical example: 6L6 beam tetrode in class A, AB_1 and AB_2

The illustration on the next page summarizes three possibilities of a pair of 6L6 beam tubes in push-pull operation, all except the Class B operation. Please note that the anode currents indicated are for both tubes, meaning the total draw on the power supply for one channel.

With a relatively low 250V on anodes and screens, almost 15 Watts is available in Class A. In Class AB, to get more power, the anode voltage is raised to 360V, with 270V on the screens. With the same bias level of -22V, a pair of 6L6 can produce 27 Watts in Class AB_1 and 47 Watts in AB_2.

Increased bias reduces the quiescent current I_0 from higher-level I_{0A} to I_{0AB}, which allows higher screen and anode voltages, thus increasing the anode efficiency of the output stage. As a result, much higher output power levels can be obtained in class AB than class A designs, but slightly increased distortion levels.

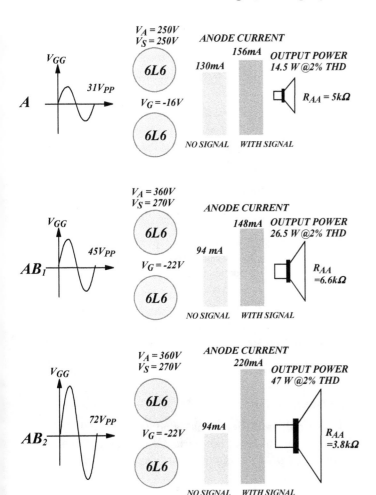

ABOVE: Fixed-bias push-pull amplifier using a pair of 6L6 tubes in various classes of operation

Of course, a larger grid-to-grid drive signal is required in Class AB_1; in our 6L6 push-pull example here, $32V_{PP}$ for Class AB compared to $22V_{PP}$ for Class A.

In Class AB_1 stages, no grid current flows, so the same transformer designs as for class A are used, although their impedance ratios may be slightly different, 5k primary impedance for class A and 6k6 for class AB_1.

The peak-to-peak drive signal must be much larger for AB_2 operation, $72V_{PP}$ on the grids compared to only $45V_{PP}$ for Class AB_1. Of course, there is a need for an additional low impedance driver stage for Class AB_2, to supply the grid current without any additional distortion. Notice that in both AB_1 and AB_2 classes, the quoted figures are at only 2% distortion!

The same applies to Class AB_2 and Class B. In class AB_2, both tubes are biased closer to cutoff but not as far as in the Class B stages. Grid currents flow, and, for a part of each period, tubes act independently as in Class B; therefore, class B transformer designs should apply. Even larger grid-to-grid drive signal is required, in our 6L6 push-pull example, $52V_{PP}$ for Class AB_2 compared to $32V_{PP}$ for Class AB_1.

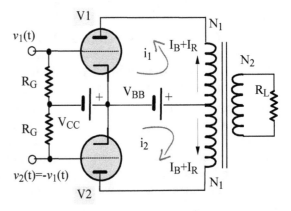

SE VERSUS PP OUTPUT POWER
CLASS A PUSH-PULL: $P_{PP} \approx 2.5 P_{SE}$
CLASS AB_1 PUSH-PULL: $P_{PP} \approx 4 P_{SE} - 5 P_{SE}$

CLASS A PUSH-PULL OUTPUT STAGE

The sinusoidal signal voltages v_1 and $v_2 = -v_1$ at the two grids are of equal frequency ω and amplitude V_M: $v_1 = V_M \cos\omega t$ and $v_2 = V_M \cos(\omega t + \pi)$. Remember that a phase shift of 180 degrees is equivalent to π radians!

The anode current of V1 is not purely sinusoidal. Due to its nonlinear transfer characteristics, harmonics are generated in this output stage. By using Fourier analysis, which says that a signal of any periodic waveform can be expressed as an infinite series of sine waves (harmonics) of the fundamental, V1's anode current is $i_1 = I_B + I_R + B_1\cos\omega t + B_2\cos2\omega t + B_3\cos3\omega t + ...$

B_1 to B_N are the amplitudes of the harmonics. B_1 is the first harmonic amplitude, also called the fundamental, since it is of the same frequency as the input signal.

ABOVE: Push-pull output stage with fixed bias V_{CC}

B_2 is the amplitude of the 2nd harmonics whose frequency 2ω is twice the original signal frequency ω, and so on. B_0 (which we call I_R here) is the DC component of the waveform or the *rectified* component of the amplified signal! The other tube's (V2) anode current is $i_2 = I_B + I_R + B_1\cos(\omega t+\pi) + B_2\cos 2(\omega t+\pi) + B_3\cos 3(\omega t+\pi) + ...$

From basic trigonometry we know that $\cos(\omega t+\pi)=\cos 3(\omega t+\pi)=\cos 5(\omega t+\pi) = -\cos\omega t$ (for all *odd* harmonics) and that $\cos 2(\omega t+\pi)=\cos 4(\omega t+\pi)=\cos 6(\omega t+\pi) = \cos\omega t$ (for all *even* harmonics), so we can write $i_2 = I_B + I_R - B_1\cos\omega t + B_2\cos 2\omega t - B_3\cos 3\omega t + B_4\cos 4\omega t - ...$

Cancellation of DC components and even-order harmonics generated in the output stage

Now comes the crucial aspect of the push-pull operation. The two anode currents flow through the output transformer's primary winding in opposite directions, so all components with the same sign (positive) will be canceled. These are the DC components I_B and I_R and all even harmonics, $B_2\cos 2\omega t$, $B_4\cos 4\omega t$, etc. The total signal current in the anode circuit is thus $i=i_1-i_2= 2B_1\cos\omega t + 2B_3\cos 3\omega t + 2B_5\cos 3\omega t + ...$

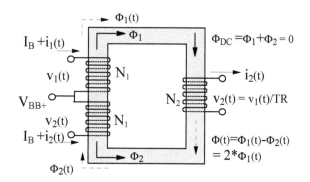

However, the amplitudes of odd harmonics are now doubled, and since these are discordant (unpleasant) sounding, this is not good news from the sonic perspective!

This analysis assumes that the two power tubes are perfectly balanced (identical), which is *never* the case in practice, so some even harmonics will be present, although with reduced amplitudes compared to their odd counterparts!

DC fluxes Φ_1 and Φ_2, produced by the two primary winding halves, cancel each other, while AC fluxes $\Phi_1(t)$ and $\Phi_2(t)$ add to one another in a PP output transformer.

Since the DC components of the output tubes' currents flow through the primary in opposite directions, their magnetomotive forces and fluxes in the transformer's core are canceled. This allows PP output transformers to be stacked without any air gap (which increases permeability). Also, smaller size EI laminations or C-cores can be used compared to their SE counterparts, making PP transformers cheaper and easier to make.

THE MYTH OF COMPLETE EVEN HARMONICS CANCELLATION IN PP AMPLIFIERS

It is often claimed that *all* even harmonics are canceled in push-pull amplifiers. That is not true. Only the even harmonics generated in the output stage are canceled *if* the output tubes are perfectly matched. *The even harmonic distortion generated in previous (preamplifier and driver) stages is not canceled even if output tubes are perfectly matched!*

The implications of cathode biasing

If cathode- or self-biasing is used, the situation in the cathode resistor is the opposite of that in the output transformer. The fundamental and the odd harmonics are canceled so that the cathode AC current i_K contains only even harmonics. If the tubes are operating in the linear portion of their characteristics, as in Class A_1 operation discussed here, and for small signals and low output power levels, the even harmonics are zero (no distortion) and the DC current through the bias resistor is $I_K=2I_A = 2I_B R_K$.

However, for larger input signals, the tubes will be pushed into nonlinear portions of their curves, and distortion will increase. The rectified component of the signal will increase and will be added to the DC bias, which is now $V_G = -V_K = -2R_K(I_B+I_R)$!

The greater the grid input voltage, the more negative the bias of the output tubes, and more negative bias means lower anode current. So, the rectified component of the signal and the increased anode DC current conspire together, lower the anode current and reduce the output power through increased bias. When we say increased bias, we mean a more negative value!

This is the exact mechanism as in single-ended amplifiers, a kind of negative feedback at higher power levels or dynamic limiting if you wish.

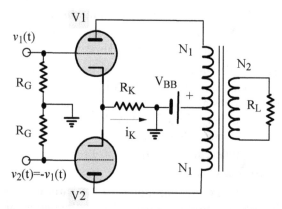

Push-pull output stage using cathode- or self-bias through the voltage drop on R_K

Secondly, the amplitudes of even harmonics will increase, but their phase is such that an un-bypassed cathode resistor introduces negative current feedback, reducing even harmonic distortion. It may seem a good idea to leave RK un-bypassed, but if tubes are not perfectly balanced, odd harmonics will also flow through the cathode resistor. Due to their phase relation, positive feedback will be created, accentuating the imbalance and increasing the odd harmonic distortion!

Since unmatched tubes are likely, and since odd harmonics are clearly detrimental to sound quality, a cathode capacitor in parallel with R_K should *always* be used to bypass any harmonics to ground.

Do push-pull amps really need less filtering in the anode supply than single-ended ones?

The conventional "wisdom" is that push-pull amplifiers don't need much filtration and can tolerate higher levels of ripple and hum on the high voltage supply lines. The argument is that hum would be canceled in a well-balanced PP output stage, and that is true, but only under quiescent (idle or zero-signal) conditions.

As soon as a music signal comes through, the output stage becomes unbalanced. The hum frequencies (100 or 120Hz) will interact with the signal frequencies and produce significant IM (intermodulation) distortion.

The same effect happens when the HV power supply voltage drops under heavy load. Secondly, input stages require superior filtration in any case, so *always* aim for the best filtering and the stiffest power supply possible.

Why self-bias cannot be used in Class B amplifiers and should not be used even in Class AB designs?

In Class B power stages, fixed bias must be used. Since the quiescent (without a signal) current is very low (often even zero) and suddenly jumps to very high levels, the rectification component of the signal is so significant that a cathode bias would change from zero to a very high level. For that reason, its use in Class B amps is not possible.

Class AB operation using fixed bias achieves higher output power levels and lower distortion than cathode bias.

Fixed- or cathode-bias?

To elaborate on the technical, sonic, and reliability differences between fixed and cathode-biased output stages would take at least a whole page of text, so I've summarized it all in the table below.

In a nutshell, cathode bias is easier and cheaper to implement, so most commercial amps use it. The fixed bias is often adjustable by the user, leading to an oxymoron term "adjustable fixed bias."

However, there is a danger that the user (guitar player) will under-bias the output tube(s), resulting in anode currents and anode power dissipation levels that are too high. This may have sonic benefits (higher current stages usually sound better) but significantly shortens the life of power tubes and increases warranty claims for manufacturers, another reason for them to prefer the cathode bias.

FIXED BIAS	SELF- OR CATHODE-BIAS
Additional bias supply needed, more complex power supply (-)	Simple design, additional bias power supply not required (+)
Clips in an abrupt manner, causing sudden distortion (-)	Softer clipping, overloads in a milder manner (+)
Unblocking (recovery from overload) takes a significant time (-)	No blocking or recovery issues (+)
As tubes age, periodic adjustments by the user are required (-)	Maintenance- and adjustment free design (+)
Reliability issues: if the bias is lost, tubes would be fully open (excessive anode current would flow) and would get destroyed unless the amplifier is promptly switched off (-)	Self-protection provided by the negative feedback (dynamic limiting) mechanism: increased cathode current produces a larger voltage drop on the cathode resistor, increasing negative grid bias, resulting in automatic reduction in cathode current! (+) or (-)
Unless prevented by the design, maladjustment by the user can result in underbiasing of output tubes, resulting in excessive anode currents and drastically reduced tube life. In other words, the concept is not foolproof! (-)	Foolproof design, no possibility of users causing any damage (+)
Fixed bias tube amplifiers can sound sterile and even harsh (-)	Generally softer and more "musical" sounding than fixed-bias designs (+)
Lower anode supply voltages are needed since there is no voltage drop on the cathode resistor (+)	Higher +V_{BB} voltage needed to compensate for the voltage drop on the cathode resistor (-)
Higher efficiency (no power loss on the cathode resistor) (+)	Lower efficiency, in some cases a significant power loss on R_K and more heat generated by the amplifier (-)
No electrolytic capacitor in the cathode circuit, resulting in more open and transparent sound (+)	Cathode bypass elco and cathode resistor significantly impacting the sound, can sound "muddy" and "constipated" (-)
Small variations in output tubes can be compensated for, matched pairs aren't absolutely necessary in push-pull stages (+)	No compensation is possible by the user, matched tubes must be used in push-pull stages (-)

INVESTIGATION: VOX AC15 and AC30 - are they truly "Class A" amplifiers?

Some modern books on tube guitar amps talk about these as the most famous examples of Class A amps. However, VOX does not mention Class A at all. How would we confirm or debunk that claim?

VOX AC15 circuit diagram shows $310V_{DC}$ on the anodes of EL84 output tubes and $315V_{DC}$ on the screens. The cathode voltage is not specified, but VOX AC30 shows 10V idle cathode voltage and 12.5V at the maximum power of 30 Watts, so we could assume a similar situation with AC15, whose output stage is like half of AC30's!

We need anode characteristics for the screen-to-cathode voltage of 300V, the closest published curves to 315-10 = 305V in AC15. The difference is small and has no impact on the conclusion.

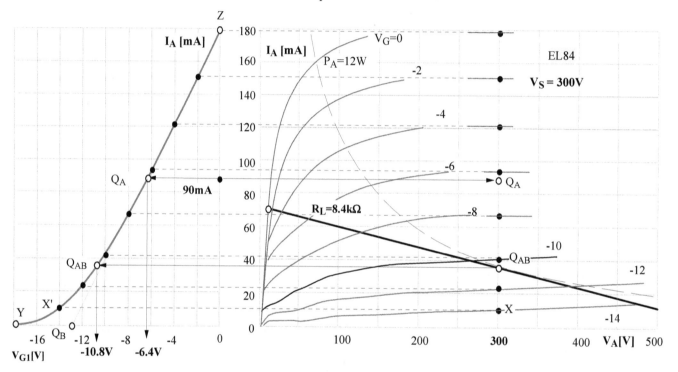

The load impedance is 8.4kΩ, as confirmed by the VOX AC15 output transformer. We will position the idle operating point Q_{AB} at the maximum allowed dissipation (right on the P_A=12W curve); the required bias is around -10.8V_{DC}. This is the closest to Class A operation we can get, as you will see. Any higher bias, for instance -11.5V, would take the operation away from class A and into class B. How do I know that?

It pays to draw a transfer curve for the anode voltage of 300V. This curve is not published in data sheets, since operating EL84 at its maximum allowed anode voltage is not recommended and is not a good design practice. Guitar amp designers are generally oblivious to those "minor" details.

However, we can construct the transfer curve from anode curves. Simply draw a vertical line through the anode voltage of 300V, mark the intersection with anode curves with dots, and then draw horizontal lines to the left and mark corresponding points at the intersections of each horizontal line with its vertical grid bias voltage.

For example, moving up from 300V on the V_A axis, we hit the curve for V_G= -14V, marked X. Draw a horizontal dotted line until you hit the vertical line from V_G= -14V and mark that dot X'. Repeat the process for intersections with the -12V curve, -10V curve, etc., join all the dots, and voila, you will get the transfer curve. The intersection for biases between -8V and 0V is not very precise since the published curves don't reach 300V on the anode since that is deep inside the tube overload territory. Again, don't worry; an error of 5-15% will not change anything.

Now we draw the R_L=8.4kΩ load line through Q-point and draw a horizontal line to the left to find Q-point's position on the transfer curve, marked as Q_{AB}. The Class B operating point QB is positioned in the "projected" cutoff point, with zero idle anode current! The real cutoff point is point Y, where the transfer curve ends.

Had VOX AC15 been a class A design, its operating point would sit roughly halfway between the maximum anode current (around 180mA, point Z) and zero, at around I_A=90mA. That would allow for maximum anode current swing, and neither of the two tubes in the push-pull stage would stop conducting, which is, remember, the definition of Class A operation. That is marked as quiescent point Q_A!

However, because its Q point is much closer to QB than to QA, we can conclude that VOX AC15 is biased very close to class B operation and is not even close to Class A design. So, as most push-pull amps do, it will only operate in Class A at very low power levels and will cross into Class B operation soon afterward. That is what is meant by the "Class AB" operation. The more accurate name would be "a little of class A at very low output power levels and then Class B all the way to the maximum power output"!

INVESTIGATION: Converting VOX AC15 & AC30 into "Class A" amplifiers

So, I guess your next question would be "Is it possible to convert VOX AC15 or AC30 output stage to operate in Class A *all the time*?"

To answer that question, we have to move the other way (to the right) from QA on the TX curve and mark its position on the anode graph. So, you can see that with 300V on both anode and screen, it is impossible to operate this output stage in class A. The anode power dissipation would be 300V * 90mA = 27 Watts, or 225% of tube's power rating (27/12 = 2.25)! That tube would only last a couple of minutes and would literally melt away!

So, the operating voltages are way too high for class A operation. Both anode and screen DC voltages would need to be reduced towards the 220-250V_{DC} region, and the bias would need to be suitably chosen, meaning in the middle of the transfer curve, to allow for a maximum grid voltage swing.

BIASING AND AC-DC BALANCING METHODS

We have already discussed various biasing methods used in single-ended (preamplifier) circuits. The same methods apply to push-pull output stages, with an additional design choice of independent biasing or common biasing for the output pair(s) of tubes.

For cathode biasing one common resistor can be used for both tubes, necessitating the use of matched pairs of tubes, or each tube can have its own cathode resistor.

Matched pairs are still recommended but precise matching isn't as critical since each tube will "find" its own operating point. Separate cathode resistors also enable the inclusion of a balancing trimmer potentiometer, as illustrated above right.

Rotating the wiper in one direction lowers the portion of R_{BAL} in that tube's cathode and reduces its total cathode resistance while at the same time increasing the portion of R_{BAL} in the other tube's cathode, whose total cathode resistance is thus increased. This changes the magnitudes of idle DC currents through the two tubes.

The first fixed bias option (below left) shows one trimmer potentiometer per tube, meaning the biasing is independent. The other approach, below right, uses a single trimmer for both tubes. With this balance-type arrangement, as the negative voltage on one grid increases, the bias voltage on the other tube's grid decreases, so the adjustments are not independent.

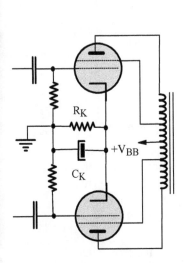
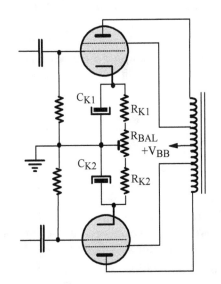
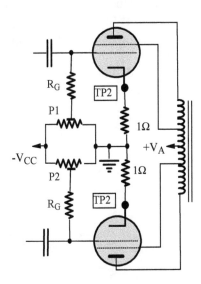

LEFT: Self- or cathode-biasing using a common cathode resistor
MIDDLE: Self- or cathode-biasing using separate cathode resistors and a DC balance trimmer potentiometer R_{BAL}
RIGHT: Independent fixed biasing and DC balancing for each output tube

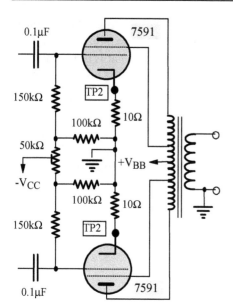

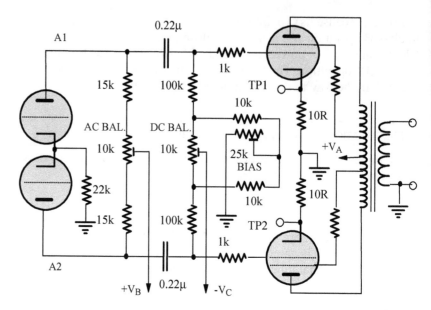

ABOVE: Common fixed biasing and DC balancing

ABOVE: A sophisticated circuit with AC balancing, bias adjustment and DC balancing between the output tubes. Most guitar amps designers don't go to such trouble.

During biasing, cathode currents must be measured, and for that purpose, small value non-inductive resistors are inserted between the cathodes and the ground. Any value can be used, but 1Ω or 10Ω resistors enable an easy mental conversion from the DC voltages measured in test points TP to cathode currents - each mA is equivalent to either 1mV or 10mV.

In the circuit illustrated on the right, the AC balancing is accomplished by adjusting the anode resistances of the phase splitter (10k trimmer), making them slightly unequal until the AC signals on the anodes A1 and A2 are equal in amplitude. Obviously, an AC voltmeter or an oscilloscope is needed for these adjustments.

The bias adjustment changes the negative voltage to the grids, while the DC balance of the two output tube currents is done by adjusting the balance of the negative bias voltages on the two grids until the voltages in test points TP1 and TP2 are equal. This is done without any signal, of course.

DC balance adjustment changes the amount of the negative bias for each of the power tubes slightly. The more negative the bias voltage (higher bias), the lower the tube's DC current. The less negative the bias voltage (lower bias), the higher the tube's idle current.

AC balancing changes the amplitude of the two out-of-phase signals at the output tube's grids. The method of AC balancing depends on the type of phase splitter used. For a split-load (cathodyne) inverter, it's only a matter of adjusting a single trimmer in the cathode circuit (right).

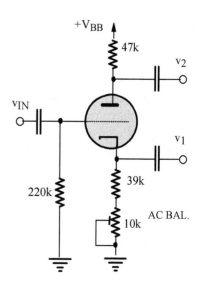

ABOVE: AC balancing for the split-load phase splitter is simple, a trimpot in series with the cathode resistor.

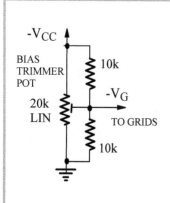

PROTECTION AGAINST BIAS TRIMPOT FAILURE   TRADE TRICKS

One of the most common primary faults, especially in vintage amplifiers, is the failure of a bias potentiometer. The predominant failure mode is the open circuit due to poor contact between the wiper and the track. This leaves the grids of the output tubes without bias voltage, turns the tubes "fully on," resulting in excessive anode currents, overheating, and eventual tube destruction (secondary faults).

This simple modification ensures that power tubes are protected from such a fault. In case of contact failure, control grids would be biased at exactly half of the maximum bias voltage V_G due to the voltage divider action of two identical 10k resistors.

DC imbalance as a cause of distortion

In a typical push-pull stage, if the tubes are unbalanced in DC terms, their plate currents at the same negative bias voltage will be different, and there will be a resultant DC primary current through the output transformer. This current causes two problems.

First, push-pull output transformers are designed on the premise of no resultant DC primary current, so any DC imbalance will reduce the permeability of the magnetic core and may even saturate it if the core is of a small volume or made of the material with low B_{MAX} (maximum magnetic flux density). That would have a detrimental impact on the output power which would drop significantly, especially in the bass or low-frequency region, and on harmonic and intermodulation distortion, which would dramatically increase.

Secondly, the imbalance of anode currents at the same bias voltage indicates the mismatch of transfer characteristics of the two power tubes, and that means that the overall transfer curve of the output stage is not a straight line, as it ideally should be, but a curve; a curved or distorted transfer function means added distortion!

The graph below shows two individual dynamic characteristics, or transfer curves (for V1 and V2), mirror images of one another. The grid bias point is in the middle line X-X. Imagine an incoming sine wave centered around line X-X as a zero line. The resultant (or composite) dynamic characteristic of the whole push-pull stage is the sum of both and is a straight line. This is the ideal case when the two power tubes are perfectly matched.

The unmatched situation shows both transfer curves shifted upwards. Consequently, the resultant transfer characteristic is not a straight line anymore but has a pronounced double bend or kink, and that kink is the cause of crossover distortion. Paradoxically, this crossover distortion is more significant at low volumes (lover signal amplitudes).

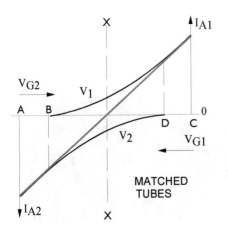
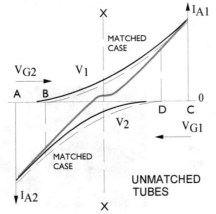
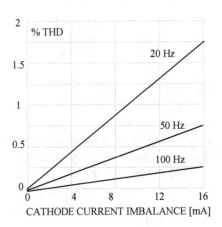

Matched tubes result in a linear composite characteristic and no distortion in a PP stage. The unbalanced case shows stronger V1 (higher anode current) and weaker V2 (lower I_A), resulting in a composite characteristics with a significant "kink" and offset.

Harmonic distortion in a U/L push-pull output stage using 6L6 tubes with cathode bias, as a result of imbalance in DC cathode currents.

THE WORST BIAS CIRCUIT EVER: EPIPHONE GALAXIE 25

After its release in 2001, Gibson produced this amp for only three years. This tells you something, or, as any psychologist worth their salt would know, everything counts!

While the amp was too dull to some guitar players, other buyers found the amp too bright, even shrill. What most agreed on was the amp's extremely annoying hum.

The amp's circuit is reminiscent of Fender Princeton Reverb, a 12AX7 gain stage with a 100k anode resistor, biased at 1.8V, followed by a tone stack and another gain stage biased at 1.7V.

Neither the power (1) nor the output transformer (2) is very large, so there is plenty of room on top of the chassis (next page). The output tubes are unnecessarily closely cramped against both transformers, so installing a choke would require moving the OT to position (3) and bolting a choke in its current place.

> **SPECS**
>
> Epiphone Galaxie 25
>
> - 25W PP combo amplifier
> - Construction: PCB
> - Rectifier: solid state diodes
> - Tube-driven spring reverb
> - 3-band EQ (Bass-Middle-Treble)
> - Controls: Gain, Volume, Reverb level
> - 10" 8Ω Celestion 70/80 speaker
> - Tubes: 2 x EL84, 3 x 12AX7, 1x12AT7
> - Dimensions: 18"W x 16"H x 11"D

The reverb transformer (4) is at an angle, but it is unclear if that is a coincidence or the position of the minimum hum, found experimentally. Although shielded cables were used for grid signals (7), all the interconnecting wires between the PCB and tube sockets were strapped together using cable ties. That may make the amp's innards look a bit tidier, but it is a poor practice from the hum & hiss point-of-view.

The PCB is neat, and components are laid out logically, with power flowing from left to the right, starting with the rectifier and the first two filtering elcos (5), followed by four more filtering capacitors, neatly tucked away at the back (6). Strangely, the component numbers on the drawing (circuit diagram) did not match the component numbers printed on the circuit board.

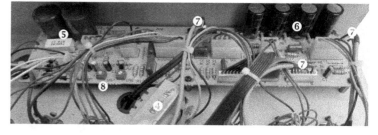

The power supply and audio section

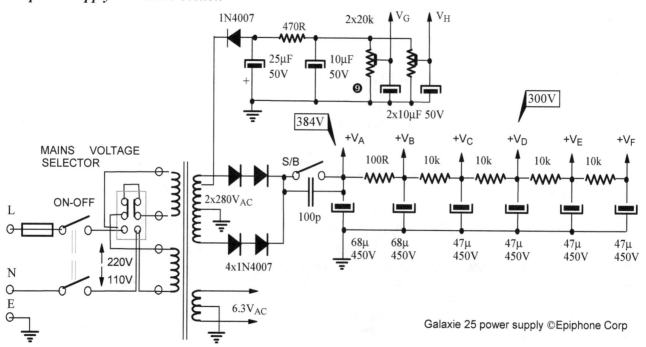

Galaxie 25 power supply ©Epiphone Corp

The amp had a 110-220V mains switch. However, our mains voltage of 245-247V caused all voltages to be proportionally higher. The heater voltage was $7.1V_{AC}$, so we installed a 0.47Ω power resistor in series.

The two preamp stages with 12AX7 are pretty standard and are not shown. The two paralleled triodes inside V2 (12AT7) drive the reverb transformer.

The recovery stage is one-half of 12AX7 (V3a), followed by its 2nd triode as a driver stage and a long-tail phase inverter. Not counting the output stage, there are four gain stages in this amp, the most likely reason for a very strong negative feedback!

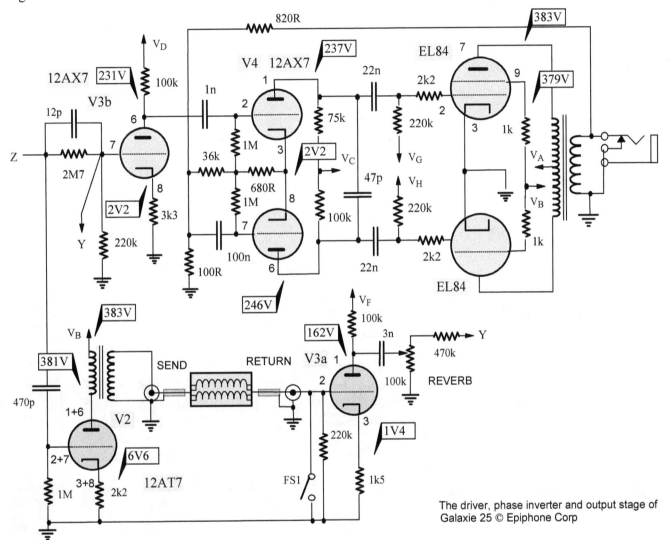

The driver, phase inverter and output stage of Galaxie 25 © Epiphone Corp

The shocker of a bias circuit

The two bias trimmer pots are easily accessible (8). While the factory circuit diagram specifies 10-turn pots, our amp had single-turn pots. Moving them a tiny bit to one side increases the cathode current a lot, making biasing way too sensitive and frustrating.

Also, while each output tube has its own bias adjustment pot, the pots can be turned all the way down for zero bias (9), meaning the tubes would be fully open, overheat and self-destruct!

There is no provision for measuring each tube's anode or cathode current while biasing them. Remember, you don't bias tubes by making their negative grid bias voltage equal; you adjust those voltages to the value needed to make their idle currents equal!

Furthermore, for one tube turning its trimmer clockwise increases the cathode current, while with the other, it decreases it, an unforgivable blunder!

The amp we got was biased, so one tube's cathode current was way too high, 42mA, while the other tube was biased at 11.5 mA! With $383V_{DC}$ on the anode and 40 mA of anode current, the first tube's anode dissipation was $383*0.04$ = 15.3 Watts, way above the allowed 12 Watts. Whoever designed this biasing circuit either knew nothing about tube amps or did not care about the amp they were designing! Have they ever biased this amp as a trial, before its final design approval, to see things from the user's perspective? I doubt.

Circuit diagram after improvements

By adding a 22k fixed resistor in series with each bias trim pot (1), the lowest bias that can be set (slider in the lowest position) is roughly half of the maximum bias value (slider in the topmost position), so tubes can never be underbiased and damaged.

10-ohm resistors and test points were added in each output tube's cathode circuit (2), so their idle currents could be precisely adjusted. Each 10mV between TP and GND is equivalent to 1mA of current. Biased at 28.5 mA, the idle power dissipation was 383*0.027 = 10.3 Watts!

A master volume is a necessity on more powerful amps, 10 Watts or over. Since we don't really want the long tail pair inverter to distort and didn't want to mess around with the PCB and dual pots too much, the easiest way to add it was before the phase inverter, as shown (3).

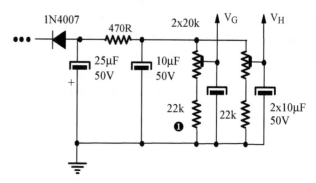

ABOVE: The improved biasing circuit
BELOW: Master Volume control and switchable NFB were implemented and bias test points added

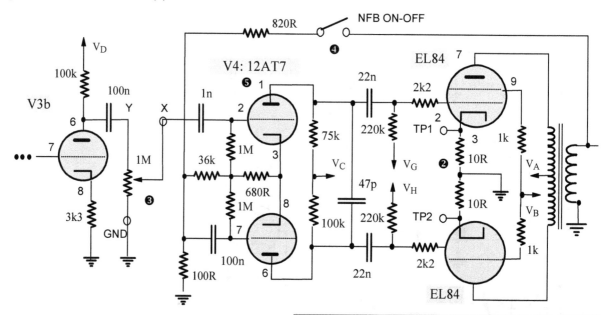

Due to its very strong negative feedback ($\beta=100/920 = 10.9\%$), the amp sounded too clean and controlled, so to make it rowdier and raunchier, the negative feedback was made switchable (4). The tone improved, but the hum increased somewhat as well. However, there was now way too much overall gain, so 12AX7 in the phase inverter was replaced with 12AT7. Even 12AU7 can be used.

Reducing hum

There are many factors affecting hum in any amp. In this case, the heater wiring was a contributing factor. Rewire it using tightly and closely twisted wires, right against the chassis and with no tube sockets inside its loops! Remove cable ties and unbundle the interconnecting wires between the tube sockets and PCB.

Use matched power tubes and then bias them for identical idle currents. Replace stock preamp tubes with the best NOS you can afford, especially in preamp stages. Use low noise 12AD7 instead of standard 12AX7 and 12AT7 in the phase inverter.

ABOVE: This amp is a textbook example of how not to do the heater wiring! BELOW: Heater wiring redone

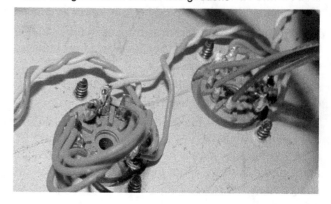

Measurements

With the three tone controls set at "5" (middle) and Gain at maximum, a 1kHz 100 mV input signal produced 13.5 V_{RMS} on a resistive 8Ω load, so the maximum output power was $P=V^2/R=13.52/8 = 22.8$ Watts.

At 1 Watt output power level, the -3dB frequency limits were 720 Hz and 21 kHz. This means the bass frequencies of the first three octaves a guitar can produce (80Hz- 160Hz- 320Hz- 640Hz) were severely attenuated—no wonder the amp had quickly acquired a reputation for a shrill tone.

Galaxie 25 OUTPUT TRANSFORMER:

- EI60 laminations
- a=20 mm, stack thickness S=30mm
- Power rating: $P=A^2= 6^2= 36W$
- Primary impedance: 7,500 Ω
- Primary DCR: 128Ω
- Prim. inductance @120Hz L_P= 7.9H
- Prim. inductance @1kHz L_P= 9.6H
- Leakage inductance L_L=7.5mH
- QF = 1,280

SELF-INVERTING PUSH-PULL OUTPUT STAGE: KUSTOM DEFENDER V15

This cute but heavy combo seems well made and has some interesting features, most notable a self-inverting output stage, which means there is no separate phase inverter at all! The four-position low-frequency response switch allows "control for American to British low end response and blended settings in-between." There are 4, 8, and 16-ohm outputs and a speaker-emulated, XLR balanced line output.

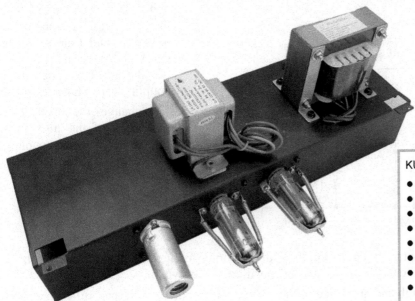

KUSTOM DEFENDER V15

- Construction: PCB
- Rectifier: solid state diodes
- Controls: Volume, Tone, Response (bass)
- 15W-4W switch
- 2x 6BQ5, 1 x 12AX7
- 10 "Celestion speaker
- Dimensions: 402 x 418 x 215 mm
- Weight 11.8 kg

Output transformer

The output transformer is a smallish affair, rated at only 23 Watts, but its primary impedance is very high, at least at 120Hz (38H), and then at high frequencies, it drops rapidly, so at 1kHz is precisely half that value (19H). This indicates that instead of GOSS (Grain Oriented Silicon Steel) laminations, a cheaper ordinary 3-4% silicon steel magnetic core was used, as in power transformers.

The leakage inductance is relatively low at 22mH, resulting in a fair quality factor of 914, one of the better results amongst similar amps.

KUSTOM DEFENDER V15 OUTPUT TRANSFORMER:

- EI54 laminations, a=18 mm, S=30mm
- Center leg cross section $A=aS = 4.8cm^2$
- Power rating: $P=A^2= 4.8^2= 23W$
- Primary impedance with 8Ω speaker: 10.6kΩ
- Primary DCR: $R_P= 396$ (193+203) Ω
- Primary inductance @ 120Hz/1kHz $L_P= 38.6H/19.2H$
- Leakage inductance @1kHz $L_L=21mH$
- Quality factor: QF = 19.2/0.021 = 914

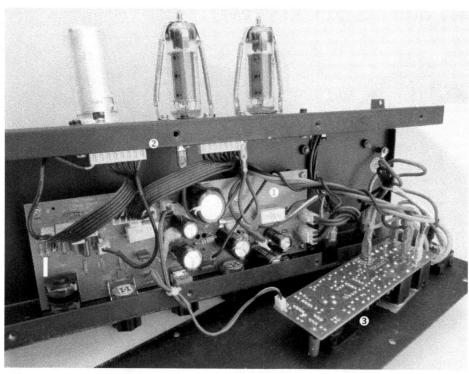

ABOVE: There are three PCBs, the main one (1) with the power supply and preamp circuitry, the one on which tube sockets are mounted (2) and the output board with speaker emulation circuitry, XLR and speaker output jacks (3)

BELOW: V15 "Bass response" circuit © Kustom Musical Amplification Inc.

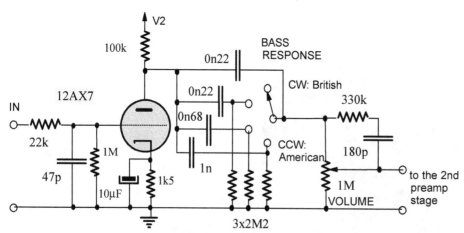

Circuit details

We won't reproduce the whole circuit diagram here, only a few interesting details (sections).

All tubes are 12 V_{DC} heated, a bridge rectifier, followed by a 4,700µF elco.

The 4-position "Bass response" switch is marked "American" at the CCW end (counterclockwise) and British in the CW position. It switches different coupling caps in parallel with the permanently connected 220p capacitor. In the CW position (British), no other cap is switched in (as shown).

Moving in the CCW direction, another 22p cap is added in parallel to the permanent cap first, for a total of 440pF.

When 680p is added to 220p, the total coupling capacitance is 900pF, and finally, in the CCW or "American" position of the switch, the total coupling capacitance is 1n22.

This changes the lower -3dB frequency of the high pass filter formed by the coupling cap, 2M2 resistor (if any), and the volume control pot.

For simplicity's sake, assuming Volume control at maximum, the -3dB or corner frequencies range from 723 Hz in the CW or "British" position to 231 Hz in the CCW or "American" position of the switch.

So, more bass frequencies are passed through and amplified in the "American" mode for a deeper, fuller tone, and less bass in the "British" mode for a thinner, brighter tone.

Cathode-coupled self-inverting push-pull stage

Just as the two tubes in preamp stages, such as a differential amp or a long-tail phase inverter, are coupled through their cathodes, the two tubes in a push-pull output stage can also be cathode-coupled (next page). This saves the manufacturer one preamp tube (or interstage transformer), its socket, and associated components.

A single-ended signal drives V1 in the usual way through its control grid. The two cathodes share a common biasing resistor (91R). Since the driving signal for the other output tube (V2) develops across this resistor, it must NOT be bypassed by a capacitor. The control grid (pin 2) of the cathode-driven tube is usually grounded directly (since it works in the common grid mode), here Kustom grounded it through a 47Ω resistor.

This circuit is never fully balanced; the two output tubes behave differently. However, it works reasonably well with high mutual conductance tubes such as EL84.

With 9.3V idle DC voltage on the cathodes, the total cathode current is $I_K=9.3/91 = 102mA$. With 4 mA screen current per tube, each tube's anode current is $(102-8)/2 = 47$ mA. The anode voltage is 314V, the anode-to-cathode voltage is 314-9.3 = 304.7V, resulting in power dissipation of $P_A=I_A*V_{AK} = 0.047*304.7 = 14.3$ Watts on each tube. This is way too high for a 12 Watt-rated pentode! These poor tubes will have a short and stressful life.

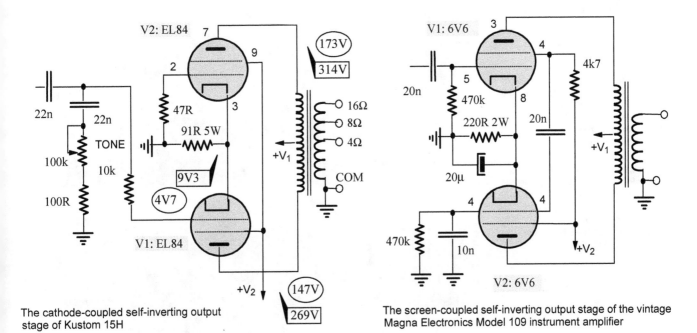

The cathode-coupled self-inverting output stage of Kustom 15H

The screen-coupled self-inverting output stage of the vintage Magna Electronics Model 109 instrument amplifier

Screen-coupled self-inverting push-pull stage

While triodes can only be coupled through their cathodes, pentode output tubes afford us another possible way of coupling the self-inverting output stage, and that is through their screen grids.

Referring to the Magna Electronics' Model 109 instrument amp output stage illustrated above, V1 is driven by a single-ended signal in the usual way through its control grid. The two cathodes share a common biasing resistor (220R) and its 20µF bypass capacitor. The screens share the same DC supply $+V_2$. So far, nothing out of the ordinary. However, the 20nF film capacitor feeds the AC signal from V1's screen grid to the screen of V2 (pin 4).

Just as with control grid circuits, its time constant is also determined by the value of the screen grid resistor to ground, in this case 470kΩ, so the lower -3dB frequency is $f_L=1/(2\pi CR) = 16.9Hz$.

PARALLEL PUSH-PULL STAGES

Two or more tubes can be paralleled in each of the two push-pull arms to achieve higher output power levels. Usually, all tubes are identical and operate in unison. However, a few creative designers devised push-pull stages where different tube types are used in parallel, in the same class of operation, or even in different classes. We will study some examples of these unusual topologies, but first, let's look at a more conventional example.

Fender PA100 This was a smaller of the two Public Address (PA) amps from Fender's "silverface" era, the CBS days of the late 1960s. The other model, PA135, achieved a slightly higher output power of 135 Watts using the same tube complement. There are two versions, with separate reverb switches and without. Our amp did not have separate reverb switches for each channel.

Fender PA100 - a close relative of the famous Twin Reverb amp

Fender PA100 has four identical channels, each with their Gain, Bass, and Treble controls. Master volume and reverb control are common to all channels.

There are two speaker output jacks at the back (1), a jack for the footswitch to turn reverb on and off (2), and access to the hum balance pot (3).

Power-related features and controls are on the other side, the on-off and standby switches (4), fuse holder, and mains voltage selector switch (5).

RIGHT: Top view of the chassis. Mains transformer (6), filtering choke (7), output transformer (8), and reverb driver transformer (9), with the filtering capacitor compartment right above it, under the steel cover.

This cover prevents accidental contact with and shock from exposed axial electrolytic capacitor leads. Should a capacitor overheat and explode, the cover also prevents the liquid electrolyte ("gunk") from splashing and spluttering all over the amplifier, a great idea.

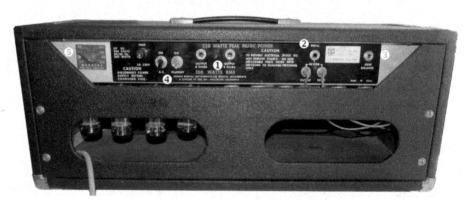

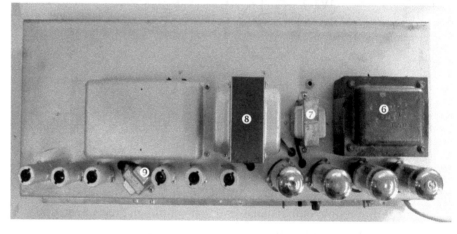

Circuit diagram - power section

The two elcos in the bias circuit (next page) were not original (1). The bias balance trimmer potentiometer (2) is accessible through the hole on top of the chassis, next to the choke (photo on the previous page).

Notice a very careless attitude and a significant departure from the best practice. Firstly, the chassis is used as a ground bus; multiple components are soldered onto it at various points. This was a common practice in the 50s and 60s but is the most common cause of hum and should not be practiced.

Speaking of hum, to keep it a minimum, the heater wiring should be done first - those wires must be laid at the bottom against the chassis and in corners. Here the heater wires are on top of all other wiring (3)!

Not a single shielded cable in sight; notice long run from the reverb and master volume controls back to the board (4), three bundled wires from each pot. These should be replaced by low capacitance shielded cable and grounded at one end. That will significantly improve the reverb sound!

Fender did not care about wire color-coding since most are white. This makes wire tracing and visual troubleshooting more difficult.

ABOVE: Under the chassis view of the components.

It is hard to see why Fender amps became a de facto standard that many other brands (such as Marshall) copied or used as a starting point for their designs. Their circuits are of standard "garden variety," straight from textbooks and tube datasheets, and their messy wiring leaves a lot to be desired.

Circuit diagram - power section

The circuit blocks in PA100 and its topology are almost identical to one of the most famous and coveted Fender guitar amps, the Twin Reverb (TR). The values of components are identical or very similar to TR AA769 and CC568 versions but very different from the AB763 variety. However, this is where the similarities end, and numerous differences in component values and even in some circuits start.

BELOW: Fender PA100 phase inverter and output stages, with Twin reverb component values also marked for comparison © FMIC

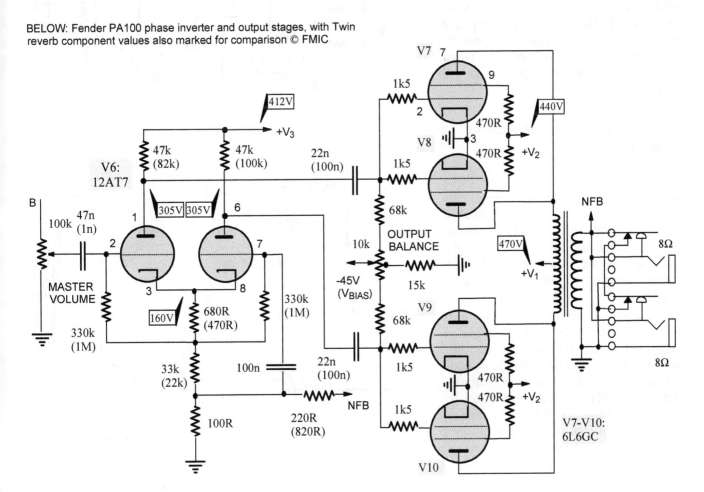

For instance, the biasing arrangements of the two amps are totally different. Twin Reverb uses an ordinary trimmer pot for bias adjustment. There is no balancing provision, so all four tubes get the same bias, $-52\,V_{DC}$. PA100 has no bias adjustment; the bias voltage is fixed at around $-45\,V_{DC}$.

The "output balance" pot gives the user the capability of decreasing the bias for two tubes on one side (V7 and V8) and increasing the bias for the two tubes (V9 and V10) on the other side. Presumably, this feature was added to ensure as little distortion in the output stage as possible. After all, PA100 is a public address amp and is meant to amplify clean, microphone-derived signals.

The circuit diagram lists both PA100 component values and Twin Reverb (TR) values (assuming the stages are identical). That is why there are no TR values given for the biasing components since the two circuits are very different, as mentioned.

Going back to the issue of clean headroom, notice that both amps use a 100Ω resistor in the tail of the "long tail" phase inverter, but the feedback resistors differ, 820Ω for TR and only 22Ω for PA100. In both cases, the negative feedback is very strong but is even stronger in the case of PA100, where $\beta=100/(100+220) = 0.3125$ or 31.25%!

The power transformer uses EI114 laminations with 5cm stack thickness, so $A=3.8*5=19\,cm^2$, and $P=19^2= 360$ Watts. The output transformer uses EI96 laminations with 4cm stack thickness, so $A=3.2*4=12.8\,cm^2$, and $P=12.8^2 = 164$ Watts. The quality factor is high for an instrument amplifier, around 2,800, but, after all, this is meant to be a clean PA amp with a wide frequency range and low distortion, so a great result here.

Extended triode operation

The output stage of the extended triode amplifier consists of two pairs of pentodes or beam power tubes, in this case, 6AR6 lovelies. One pair works as beam power tubes ("pentode" mode) with 300V on the screen and 450V on the anodes; the other pair is triode-strapped.

The -50V fixed bias is chosen so that the triodes start conducting first. At lower power levels, the output stage operates in the triode mode, while at higher power, the pentodes provide most of the output power. The maximum output power with 807 or 5881 tubes is 45-47 watts.

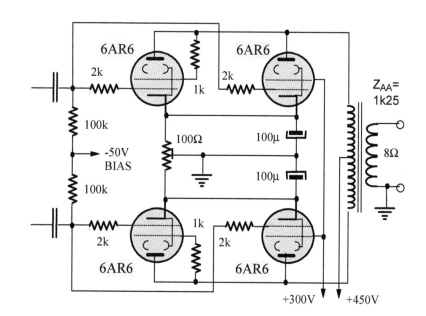

Switchable pairs: Matchless DC30 power level selector

Since DC30 uses two pairs of EL84 pentodes in its power stage, its full-power/half-power selection (marked Hi/Lo) is made by disconnecting cathodes of one output pair (V3 & V4) from the cathodes of the other pair (V1&V2), which remains in operation in both modes.

Since the 62R cathode bias resistor is correct for two pairs of EL84 tubes, it needs to be doubled in the Lo mode when only one pair is operating, and the cathode current is halved, so the bias (cathode voltage) remains constant. That is done by the 2nd contact of the Hi-Lo switch, which switches the 68R resistor in series with 62R, for a total of 130R in Lo mode.

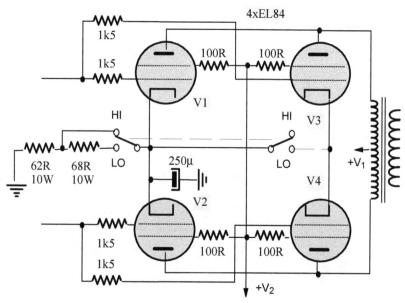

DC30 output stage, © Matchless

Mesa Boogie's Simul-Class® Circuit

Mesa Boogie is one of the more creative and adventurous tube guitar manufacturers. In 1985 they patented their "Power amplifier capable of simultaneous operation in two classes," patent number US4532476.

According to the patent, the outer pair of tubes, V5 and V6 (usually but not always EL34, sometimes 6L6) operate as triodes in class A, while the inner pair of pentodes, V7 and V8 (usually 6L6), operates in class AB_1. Since triodes have a lower internal impedance than pentodes, the plates of the outer pair are connected closer to the center tap on the output transformer's primary (lower load impedance). In comparison, plates of the class AB pentodes are connected to a higher plate-to-plate impedance, the outer terminals (full primary winding).

There seems to be some confusion in online discussions about Mesa's Simul-Class® operation, primarily due to Mesa's inconsistency and differences between their patent and their published circuits.

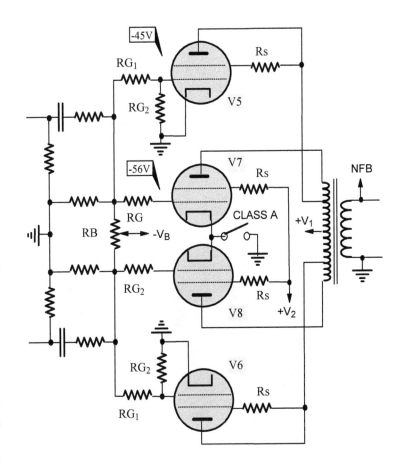

Some amplifiers use triode connected pair of EL34 and pentode-connected 6L6, for instance, the Mark IIC+ model. However, Mesa's hand-drawn diagram shows the opposite connection. The triode-strapped tubes (V8 & V9) should be connected to the inner taps (lower load impedance), while the pentodes (V6 and V7) should be tied to the ends (higher anode impedance).

Also, the center tap on the output transformer's primary ($+V_1$) is missing entirely on Mesa's drawing!

The clue is in the DC voltages on the plates. Class AB tubes have 473 Volts while Class A tubes have more, 478 Volts on anodes, meaning they have to be closer to the center tap (lower voltage drop on that section of the primary), the source of plate voltage!

Voltage V2, to which screen grids of pentode tubes are connected, is not 480 V, but 473 V, voltage V1 is 480 V!

These amps have a "Class A - Simul-Class®"switch that in "Class A" position (as shown here) simply disconnects the cathodes of the pentode pair and leaves only the triode pair operating.

EL34 screens should never be directly strapped onto their anodes but via a 1kΩ resistor, as shown in our drawing here, but such resistors are missing on the Mesa'a Mark IIC+ diagram.

There are questionable output power claims, too. In its promotional literature, Mesa states: "The 30 watts of vintage Class A tone combines with 65 watts of efficient Class AB power to produce a total of 95 watts per channel ..."

As we have seen on page 40, only up to 18 watts is possible with the 6L6 push-pull stage in Class A_1, not 30 Watts!

The data for 6L6 shows that with 450 V on plates up to 55 Watts in Class AB_1 is possible, so with 480 V that Mesa uses, 65W could be possible, but the tubes are pushed very hard, so they may not last long.

ABOVE: The output stage of Mesa Simul-Class® amplifiers with separate primary taps for the triode and the pentode pair of output tubes

BELOW: Operating conditions for a pair of 6L6 tubes in Class AB_1

PUSH-PULL CLASS AB_1 AMPLIFIER

Plate Voltage	360	450	Volts
Screen Voltage	270	400	Volts
Grid-Number 1 Voltage	−22.5	−37	Volts
Peak AF Grid-to-Grid Voltage	45	70	Volts
Zero-Signal Plate Current	88	116	mA
Maximum-Signal Plate Current	132	210	mA
Zero-Signal Screen Current	5.0	5.6	mA
Maximum-Signal Screen Current	15	22	mA
Effective Load Resistance, Plate-to-Plate	6600	5600	Ohms
Total Harmonic Distortion	2	1.8	Percent
Maximum-Signal Power Output	26.5	55	Watts

When triode strapped, EL34 can produce only about 5-6W in single-ended amps, and for class A push-pull, the power cannot be more than twice that, or 10-12W, not 30W as claimed. Even in class AB_1 push-pull, the power specified in the EL34 datasheet is only 16.5W when connected as triodes, almost half of the claimed 30 Watts!

The clue is in the bias voltage. With -45V on the triode's grids, these tubes (EL34) are biased deep in class AB, not in class A, which would require a much lower bias of around -27V.

Another source of confusion is that other amp models don't use a pair of triode-strapped EL34s but all four pentode-connected 6L6s, but with a Simul-Class® transformer (model 290 for instance). The outer 6L6 pair is biased at -47V and the inner pair at -52V, from the common bias supply of -59V. Interestingly, there is no "Class A-Simul-Class® " switch on that amp.

Finally, some owners of Simul-Class® amps have tested and opened their output transformers and concluded that their primary only has no primary taps; there are only three terminals - a CT and two anodes.

The bottom line is that different primary impedances are not required for Simul-Class® operation. One could connect both the triode and pentode pair's anodes to the same primary points, to a single primary winding. Unique Simul-Class® output transformers are only necessary when the amp's designer wants the triode and pentode pairs to work with different primary impedances (loads).

Without different primary taps, the two pairs of output tubes would be loaded by the same load impedance, but they can still operate in different regimes, so again, two pairs of primary taps are not a crucial aspect of Simul-Class® operation; they simply provide different primary impedances for the two pair of tubes.

NEGATIVE FEEDBACK

14

- HOW NEGATIVE FEEDBACK WORKS
- THE BENEFITS AND DRAWBACKS OF NEGATIVE FEEDBACK
- A CASE STUDY IN NEGATIVE FEEDBACK: PEAVEY VALVEKING 100
- REDUCING UNWANTED POSITIVE & NEGATIVE FEEDBACK
- LOCAL NEGATIVE FEEDBACK IN THE OUTPUT STAGE: ULTRALINEAR CONNECTION

Negative feedback is simply a design feature or method where a portion of the output signal is brought back to the input of a gain stage (local feedback) or a whole amplifier (global feedback). While this reduces gain, it (technically at least, if not sonically) improves quite a few other amplification aspects.

As with most other audio concepts, both hi-fi buffs and guitar players are divided over its merits. While some consider the negative feedback to be mostly beneficial to the overall sound, others claim that it makes things worse. As always in such cases, you need to trust your own judgment and make up your own mind.

HOW NEGATIVE FEEDBACK WORKS

Feedback ratio β

We have already learned a bit about NFB, so let's broaden our analysis and draw some general conclusions. Global NFB is usually taken from the secondary of the output transformer to the cathode of the first stage, as illustrated.

The feedback voltage is determined by the ratio of the feedback resistor R_1 and the cathode resistor R_2, which must be left un-bypassed. Only the un-bypassed cathode resistance counts, in this case, R_K is bypassed for audio signal by C_K, so it does not enter the NFB calculations.

The circuit is a simple voltage divider. The lower that ratio, the weaker the feedback. This feedback ratio is $\beta = R_2/(R_1+R_2) \times 100\%$ Usually $R_2 \ll R_1$, so $R_2/R_1 \times 100\%$.

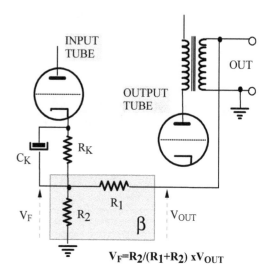

ABOVE: The most common way of applying global negative voltage feedback in a tube amplifier, in series with the input voltage (across R_2).

Feedback factor FF

If we take an amplifier with amplification factor A as a starting point, to get amplitude A at its output, we need a signal amplitude of 1 at its input. The NFB network has its own amplification (or rather attenuation) factor β; it takes the output voltage A and multiplies it by b, so the signal amplitude at its output is βA.

This signal is added in series to the original input (1), so now the new input voltage into the amplifier with feedback, 1+βA is needed to get the output voltage A! 1+βA is called Feedback Factor (FF).

The new amplification factor (with feedback) is $A_F = V_{OUT}/V_{IN} = A/1+\beta A = A/FF$, so $FF = A/A_F$.

The factor βA is called *open-loop gain*, where the "loop" is the path from the input point X to point Y, which includes the two blocks (A and β), but is *not* closed. Negative feedback does NOT change the amplification factor A of the amplifier; it reduces its input voltage instead, thus reducing the output voltage.

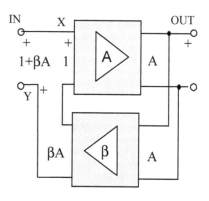

ABOVE: The general block diagram of a feedback amplifier

Negative feedback in dB

When you read that an amplifier has a certain number of dB of negative feedback, what does that really mean? Let's use an easy and convenient example, say 20dB of feedback is used in a particular design. We have $20\log(A/A_F) = 20\log(1+\beta A) = 20$dB, so $\log(1+\beta A)=1$ which means the $FF=(1+\beta A)=10$ (because $\log 10=1$). Finally, we have the open-loop gain $\beta A=10-1=9$!

Assuming the global NFB arrangement illustrated above, back to the cathode of the input stage, the feedback ratio is $\beta = R_1/(R_1+R_2)$. Let's say that the amplification factor of our amplifier without feedback is $A=18$ (18V out for 1V in, about 40.5 Watts into an 8Ω load). Applying 20dB of NFB would reduce the gain to $A_F=A/10 = 18/10 = 1.8$, which would produce an output of 1.8V for a 1V input signal.

This is obviously far too much, so let's see what a mere 3dB of NFB would do: $20\log(A/A_F) = 3$ so $A/A_F = 10^{0.15} = 1.41$, or $A_F=A/1.41 = 18/1.41 = 12.74$. On an 8Ω load, this is equivalent to $P=V^2/R = 20.25$ Watts, so a mere 3dB of NFB reduced the voltage to 71% of its previous value and halved the output power!

TYPES OF NEGATIVE FEEDBACK

Negative feedback can be local (un-bypassed cathode resistor in one stage) or global (two or more stages inside the feedback loop). Global feedback can include the preamp stages (without the output stage), or, if taken from these secondary of the output transformer), it will include ALL stages.

There are four types of negative feedback, classified as per how the feedback signal is derived (is it proportional to the output voltage or output current) and how it is applied in the input circuit (in series with the signal or in parallel with it).

The four arrangements are illustrated here in a block diagram form.

Since vacuum tubes are voltage-controlled devices, even if current feedback is used, what is fed back is still a voltage, a voltage drop on a resistor in series with the load, so the feedback voltage is proportional to the output current.

The voltage feedback lowers the output impedance, while the current feedback increases it. The series-applied feedback increases the input impedance, while the parallel-applied feedback reduces the input impedance.

Simple NFB circuits

We have already analyzed simple examples of NFB, such as the common cathode stage with an un-bypassed cathode resistor and its special case (without the anode resistor), called cathode follower.

Likewise, with resistive feedback from the anode to the grid of the common cathode stage, the circuit is sometimes called "anode follower."

RIGHT: Series-applied voltage feedback: Cathode follower. The whole output voltage signal is fed back into the cathode circuit and added in series with the input voltage.

FAR RIGHT: Parallel-applied voltage feedback in the so called "anode follower". The whole output voltage signal is fed back into the grid circuit.

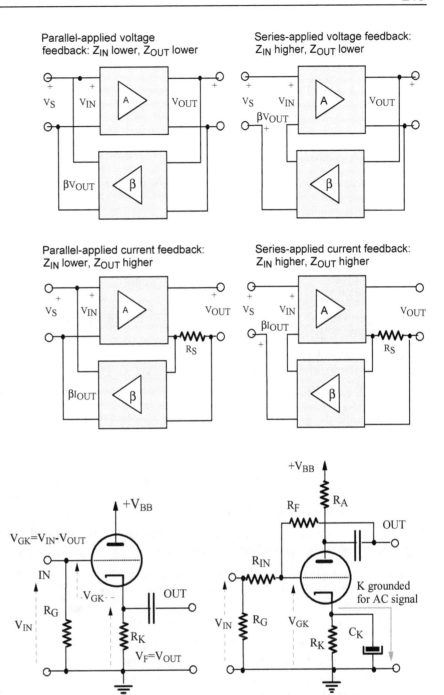

THE BENEFITS AND DRAWBACKS OF NEGATIVE FEEDBACK

NFB has many advantages:

- widens the frequency range (lowers the lower -3dB frequency and increases the upper -3dB frequency)
- reduces distortion and noise
- increases input impedance
- reduces output impedance, improves damping factor
- provides phase correction and reduces phase distortion
- stabilizes the amplification factor and reduces changes due to drift, component aging, and mismatching

We will not elaborate on most of the points just made; for deeper coverage of negative feedback, please refer to my book "Audiophile Vacuum Tube Amplifiers," Volume 1.

Let's just repeat here the famous maxim "There is no such thing as a free lunch," meaning there is always a price to be paid in engineering and life. Thus, for each of the advantages listed above, there are disadvantages or problems that negative feedback causes. The most important of these in guitar amps is that NFB reduces gain.

Negative feedback widens the frequency range but reduces gain

As an example, let's take a simple case of an amplifier with a 1st-order amplitude characteristics (low pass filter). $A = V_{OUT}/V_{IN} = A_0(1+jf/f_U)$ Assuming resistive feedback network resulting in real feedback ratio β (independent of frequency), the transfer function after the application of NFB becomes:

$$A_F = A/(1+\beta A) = \cfrac{\cfrac{A_0}{1+jf/f_U}}{1+\cfrac{A_0}{1+jf/f_U}} = \frac{A_{F0}}{1+jf/f_{UF}}$$

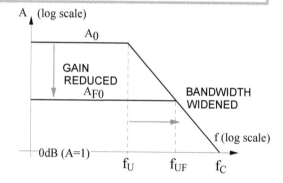

NFB REDUCES GAIN BUT WIDENS THE FREQUENCY RANGE

$$f_{UF} = f_U(1 + \beta A_0) \qquad\qquad A_{F0} = \frac{A_0}{1 + \beta A_0}$$

Three important conclusions can be made from this analysis:

1. When applied to 1st-order systems, NFB does not change the shape or the character of the system or its transfer function.

2. The midrange gain A_0 is reduced to A_{F0}, in proportion to the feedback factor $1+\beta A_0$.

3. The upper -3dB (half-power) frequency is increased by the same feedback factor.

Notice that f_C, the frequency at which the amplification drops to 1 or 0dB (log1=0), has not changed. $f_C = A_0 f_U = A_{0F} f_{UF}$. The gain has decreased, but the upper corner frequency has increased by the same factor, so we can conclude that the application of NFB does not change the gain-bandwidth product of an amplifier!

NFB reduces the midrange gain but widens the bandwidth, keeping the GBW product constant!

Voltage feedback reduces output impedance

To determine the output resistance of an amp, with input test signal V_S and *without NFB*, we measure the output voltage without any load connected and then with the load R_L. In the second step, we repeat the same two measurements but this time *with NFB*. We get: $V_{UL} = A_F V_S R_L/(R_{OUTF}+R_L)$ and $V_L = A V_S R_L/(R_{OUT}+R_L)$.

$A_F V_S$ and $A V_S$ are output voltages in unloaded cases (with and without NFB). Likewise, R_{OUTF} and R_{OUT} are output resistances with and without feedback.

The output voltage in the unloaded case is higher than when R_L is connected. With NFB, the connection of the load reduces the output voltage, but it also reduces the voltage that is fed back (βV_L), thus increasing the input voltage $V_{IN} = V_S - \beta V_L$.

This means that with NFB, the output voltage is less dependent on the load resistance. The voltage drop on the output resistance of the amplifier is lower, and lower voltage drop means that the output resistance with NFB is lower.

With NFB the unloaded output voltage is

$V_{UL} = -AV_{IN}R_L/(R_{OUT}+R_L)$, so if we substitute $V_S - \beta V_L$ for V_{IN} and rearrange, we get

$$V_{UL} = \frac{-A}{1 + \beta A} \;\; \frac{R_L}{R_L + \cfrac{R_{OUT}}{1 + \beta A}}$$

The factor in bold is the new, lower output resistance with NFB, $R_{OUTF} = R_{OUT}/(1 + \beta A)$!

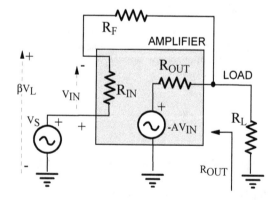

Determining output resistance with NFB applied R_{OUTF}

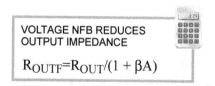

VOLTAGE NFB REDUCES OUTPUT IMPEDANCE

$$R_{OUTF} = R_{OUT}/(1 + \beta A)$$

With NFB, the amplifier's output voltage is less dependent on the load (speaker's) resistance or impedance. This is because this type of NFB reduces the output impedance of an amplifier and thus the voltage divider effect between the amp and the speaker. In other words, with NFB, the amplifier moves closer to being an ideal voltage signal source (which would have zero internal or output impedance).

After all this theory, let's look at a real amplifier and see how its designer implemented negative feedback!

A CASE STUDY IN NEGATIVE FEEDBACK: PEAVEY VALVEKING 212

We got this large and heavy beast cheaply from a local seller, pretty much the cost of the speaker drivers alone. Indeed, the twin 12" speakers were the main attraction of this model.

Initially, we thought about selling off the printed circuit boards as spares to ValveKing owners who needed a replacement and building a stereo amp with dissimilar output tubes (along Kendrick lines). The logic was that PCBs would recoup the cost of the whole amp, so the rest (cabinet, chassis, transformers, etc.) would be free.

However, the amp would be too large and heavy (31kg!) for a boutique amp aimed chiefly for recording purposes, so we decided to retain it and use it solely as a case study in this book.

The chassis is quite spacious, both above and below deck, so adding a power supply choke or anything else would be very easy, should you wish to improve high voltage filtration and regulation.

PEAVEY ValveKing 212 | **SPECS**
- 100W push-pull combo amp
- Construction: PCB
- Rectifier: solid state diodes
- Spring reverb: Yes, Tremolo: No
- 3-band EQ (Bass-Middle-Treble)
- Presence, Resonance and Texture™ controls
- Two foot-switchable channels, switchable gain/boost on the Lead channel
- Buffered effects loop
- Two 12" 16Ω speakers (in parallel)
- Power tubes: 4 x 6L6
- Preamp tubes: 3 x 12AX7
- Weight 68.5lb/31kg

ABOVE: The amp is at the limit of one person handling (carrying) capability, and castors certainly help when rolling it around flat paths.

BELOW: The internal (under-the-chassis) view of Peavey ValveKing 212 shows a carefully thought-out layout resulting in minimized lengths of wire looms between the three boards. The main (preamp and power supply) PCB layout follows the signal flow from left to right and the power flow in the opposite direction (right to left).

A few obvious sections and components. 1) Bias supply section 2) Heater supply fuse and rectifiers 3) Heater filtering capacitors and 1Ω resistor 4) HV supply fuse and rectifiers 5) HV filtering capacitors 6) 400Ω resistor in the HV filter (instead of a choke) 7) input 12AX7 tube's socket 8) Second preamp tube's socket 9) Driver and phase inverter tube's pins

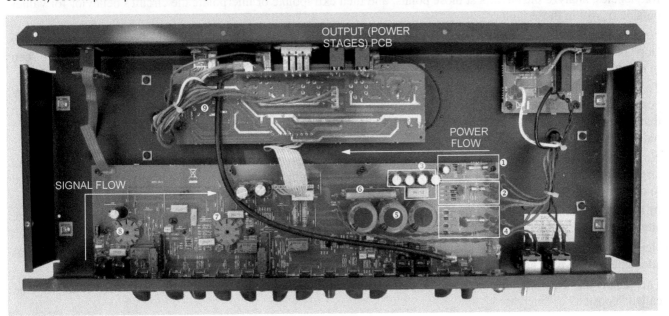

RIGHT: Tube-side view of the chassis. The four power tubes are too close to each other, impede air circulation and increase temperature levels.

The output transformer

EI96 laminations make up its magnetic stack rated at 125 Watts, which doesn't leave that much spare capacity considering that it has to transmit 100 Watts of music power to the speakers.

The primary inductance is only 6.9 Henry; however, the leakage inductance is very low, around 4 mH, so the transformer's quality factor is high, around 1,600; its high-frequency response should extend way up into ultrasonic frequencies.

The primary PTP (plate-to-plate) impedance is around 1.8 kΩ or 3.6 kΩ for each pair of 6L6 tubes.

The two primary halves have slightly different DCR figures, 25 and 32 ohms. Such very low levels of primary DCR and low primary inductance indicate that the primary winding was wound with a large diameter wire and a low number of turns.

PEAVEY VALVEKING 212 OUTPUT TRANSFORMER:

- EI96 laminations
- a=32mm, stack thickness S=35mm
- Center leg cross section A=aS = 11.2cm^2
- Power rating: P=A^2= 125 W
- Primary resistance RED-BRN: 31.5Ω
- Primary resistance RED-BLU: 24.6Ω
- Primary impedance with: around 1.8kΩ
- Primary inductance L$_P$= 6.9H
- Leakage inductance L$_L$=4.3mH
- Quality factor: QF = 1,604

Presence, Resonance and Texture™ controls

ValveKing features Peavey's triple tone-shaping controls. According to Peavey's promotional literature, while *Presence* and *Resonance* controls "adjust speaker damping from tight to loose in the high and low-frequency ranges, respectively," *Texture*™ varies the operation of the output stage "from modern Class A/B push-pull to vintage Class A sounds." Since the rest of the design is pretty standard, we will focus on these three features in our analysis.

The first time I learned this trick was in mathematical analysis, but it can also be applied to the conceptual analysis of electronic circuits that involve variable resistors (potentiometers) or capacitors, or fixed component circuits analyzed across a range of frequencies.

The trick is to look at each case separately, consider the extreme values of variable components or the extreme of frequencies, analyze the circuits at those points, and then extrapolate or interpolate the circuit's behavior for values outside or within those limits.

For example, look at the 10k "*Presence*" potentiometer. The output signal from the 8W speaker tap is fed back to the "top" of that pot via the "Resonance" pot and its 10n bypass capacitor. This is a typical negative feedback loop seen in many hi-fi and guitar amps.

To understand how "*Presence*" works in this case, imagine the slider or wiper of the "*Presence*" pot in the L or LOW position. At that end, the capacitor wired between the wiper and the ground is short-circuited, as if it never existed. In that position, the signals of all frequencies are equally attenuated (or weakened) because that is what NFB does - it reduces the gain of the stages it encompasses.

Now imagine we are turning the pot towards the H or "High" position. The bypass capacitor shunts a larger and larger percentage of the 10k potentiometer, meaning it shunts higher frequencies to ground more and more. It is vital to distinguish between the operation of the tone control circuit from this circuit.

In tone controls, when we say that more and more high frequencies (HFs) are shunted to the ground, HFs are attenuated more and more; in other words, that "treble cut" is being achieved.

Here the situation is opposite; the shunting means that HFs are shunted away from the NFB circuit, so they are not as attenuated as the bass and middle range frequencies are by negative feedback.

This means that HFs are boosted when the pot moves from L to H! Relatively speaking, "less attenuated" may be called "boosted" or "increased."

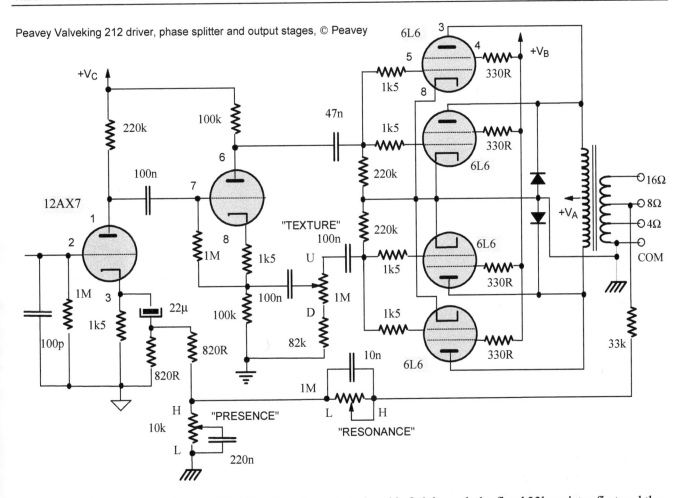

Peavey Valveking 212 driver, phase splitter and output stages, © Peavey

What about the *"Resonance"* control? Notice that the output signal is fed through the fixed 33k resistor first and then through the 1M potentiometer and the shunting resistance of 10k to GND. The circuit is a simple voltage divider.

The lower the ratio of two resistances, shunt versus total, the weaker the feedback. With the wiper of the *"Resonance"* pot in L position, the 1M pot is short-circuited and the NFB ratio is $\beta = R_2/(R_1+R_2) \times 100\% = 10/(10+33) \times 100\% = 23.26\%$. This is very strong feedback, meaning the output signal will be strongly attenuated.

As the *"Resonance"* control is turned towards the H end, more and more of the 1M resistance is added in series to the fixed 33k series resistor. In other words, the resistance of our R_1 from the simple circuit on the right is increasing fast. At the H end R_1 is 1M+33k = 1,033k and now $\beta = R_2/(R_1+R_2) \times 100\% = 10/(1,033) \times 100\% = 0.97\%$, which is a very weak feedback, with less than 1% of the output signal fed back!

That discussion is valid for low (LFs) and midrange frequencies (MFs). Since the capacitor impedance is inversely proportional to the capacitance and the frequency, the 10nF bypass capacitor bypasses the high-frequency HF signals around the 1M feedback resistor.

At high frequencies, the capacitor's impedance is much smaller than the 1M resistor, so it forms a lower resistance path, which means that our feedback ratio remains much higher than for low frequencies.

If you've been paying close attention, you should have noticed an apparent contradiction here. Your question to me should now be, "Wait a minute, Igor, are you saying that Peavey designers included the *"Presence"* control, which boosts HFs when cranked up, but they've also included the *"Resonance"* control, which attenuates HFs when cranked up? Isn't that insane?" The answers are YES and NO, respectively.

Yes, the two controls seem to cancel each other out. However, the clue is the "Resonance" name they've given the other control and in the adjustable NFB just explained. Just what resonance is being controlled here?

NFB causes an undesirable gain increase at frequency extremes

Can Peavey's control knob really control the speaker resonance? No. A speaker is just as it is, as our Buddhist friends would rightfully point out; an amplifier cannot change its resonant frequency or even its resonant impedance peak.

However, varying negative feedback's strength affects the output impedance of an amplifier and changes its damping factor, thus affecting the way the amp "interacts" with a speaker. If that is what "resonance" means in this context, then the answer is qualified yes.

There is another issue at play here. This illuminating comparison of amplifiers with various amounts of NFB applied. The trend is unmistakable - the stronger the NFB, the higher the peaks at frequency extremes.

These peaks in amplification factor mean that negative feedback at midrange is turning into positive feedback at high and low frequencies due to the phase shift in those regions. Positive feedback increases gain and potential instability, turning a stable amplifier into an unstable oscillator.

As the NFB gets stronger, the peaks at frequency extremes also get bigger and bigger, especially in cases D) and E). This is why Peavey included that 10n bypass capacitor across their Resonance control.

The NFB is much weaker for high frequencies, and the peaks in the amp's A-f characteristics (such as the one illustrated on the right) cannot happen!

In the bass region, the amp's gain drops way before the sub-10Hz frequency where peaks would develop, so there is no danger of bass instability, as illustrated on the curves here. This issue in the bass region applies mainly to hi-fi amps that go deep down in their frequency response, not to guitar amps.

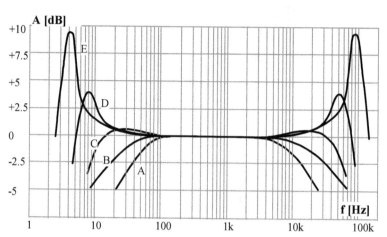

ABOVE: Relative gain (referenced to midrange gain) versus frequency of an amplifier A) without any feedback, B) with mild negative feedback, C) with moderate NFB, D) with strong NFB, E) with excessive NFB

NFB widens the linear region but makes clipping harsher and more abrupt

When a global NFB is brought back to the input stage of a tube amplifier, the linear region of input (grid) voltages (signals) with NFB is wider than without it. The amp will be able to accept larger input signals without clipping.

Notice how the transfer characteristics are "straightened" and how the use of NFB widens the linear region.

However, once clipping occurs, as indicated by the waveform of the anode current, it is more abrupt with NFB, and the output signal will resemble a square wave.

This results in much higher distortion levels and introduces the harsher sounding odd harmonics, so the high gain sound of the amp will sound different with and without feedback.

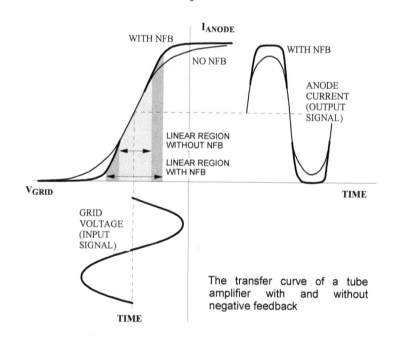

The transfer curve of a tube amplifier with and without negative feedback

REDUCING UNWANTED POSITIVE & NEGATIVE FEEDBACK

Many amplifier designers deliberately use NFB to achieve specific design goals. Still, there is also unwanted feedback, the most common example being signal coupling and propagation between gain stages through the common power supply.

Decoupling between amplification stages

The audio signal from all stages passes through the power supply, so the DC anode supply voltage for all stages is "modulated" by the signal AC voltage, which is superimposed on top of it. This unwanted signal appears on the anode resistors or directly on the anodes of cathode followers. Depending on the amplifier design, this unwanted feedback signal can be in-phase (positive feedback) or out-of-phase with the amplified ("wanted") signal (negative feedback), as indicated by the pulse waveforms indicated on the block diagram in A, B, and C.

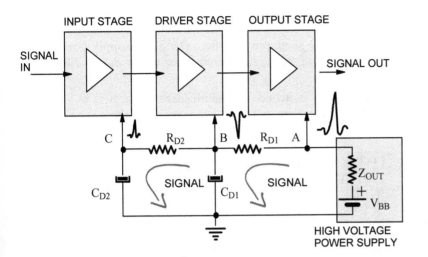

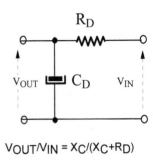

$$V_{OUT}/V_{IN} = X_C/(X_C+R_D)$$

ABOVE: A typical RC power supply decoupling circuit

LEFT: Signal coupling through the common power supply and the operating principle of decoupling filters

The block diagram illustrates a three-stage amplifier and its power supply with a finite internal impedance Z_{OUT}. The highest anode voltage is usually in point A, the output stage supply, and then it gets progressively smaller for the driver stage (point B) and the preamp stage (point A). This is achieved by voltage dropping resistors or chokes.

If out of phase, the power-supply-propagated signal acts as negative feedback in that particular stage, which may be undesirable but is not detrimental to the operation of the amplifier. However, positive feedback occurs if the two signals are in phase, and the amplifier may start oscillating.

"Motorboating" is a popular name for low-frequency oscillations (a few Hertz) since the sound the amplifier makes is similar to the sound of the outboard boat engine (put-put-put).

High-frequency oscillations above 20 kHz are even more troublesome. Since they cannot be heard, they may go undetected by the inexperienced amplifier builder. Only careful examination with an oscilloscope will reveal the presence of an HF signal at the output.

In one amplifier, we noticed that the output transformers were getting warm. This was not normal unless a very thin wire was used for the windings or the amp played at very high volume levels for hours. After further investigation, high frequency (ultrasonic) oscillations were discovered, increasing thermal losses in transformers. After the HF instability was fixed, the output transformers were cool as cucumbers.

A decoupling circuit is an RC filter that acts as a voltage divider and reduces the feedback due to the common power supply impedance in the ratio of V_{OUT}/V_{IN}. It also reduces the ripple voltage (AC component superimposed on the DC voltage bus).

This voltage divider is only effective if the capacitive reactance $X_C=1/(\omega C_D)$ is small compared to the decoupling resistance R_D at low frequencies. A typical frequency at which such effectiveness is evaluated is the lower -3dB frequency f_L.

Of course, the DC anode current(s) of the previous stage(s) must pass through R_D so the resistance of R_D cannot be too high; otherwise too much of the supply voltage V+ would be lost on such a series resistor.

Circuits are usually drawn from left (input) to the right (output), but because decoupling filters are drawn in the opposite direction, their inputs are on the right and outputs on the left side (the power flow is right-to-left).

HOW EFFECTIVE ARE DECOUPLING CIRCUITS AT REDUCING RIPPLE?   CALCULATION

Determine ripple voltage V_{ROUT} at the output of the typical decoupling circuit, if the ripple voltage at its input is $V_{RIN}=1V_{PP}$. Assume 50Hz mains frequency.

Since $X_C=1/(\omega C)$, at 50Hz mains frequency the ripple frequency is double that or 100 Hz, so we have $X_C=1/(2*\pi*100*22*10^{-6}) = 72\Omega$

The attenuation of ripple is $A=V_{OUT}/V_{IN} = X_C/(X_C+R_D) = 72/(72+10,000)$ $= 0.00715$, or in dB: $A= 20\log0.00715 = 20*-2.146 = -42.9$ dB

Finally, $V_{ROUT}= V_{RIN}A = 1*0.00715 = 7.15$ mV$_{PP}$

The peak-to-peak value of the ripple will be reduced from 1 Volt or 1,000 mV to 7.15 mV!

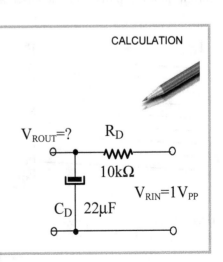

Typical decoupling circuit and the bleeder resistor

This partial diagram shows typical decoupling circuits from a 2-stage preamplifier. The 340V supply goes to the output stage and then supplies the second voltage amplification stage, illustrated here, which draws 2.8mA of current, and the input stage, which draws 1mA.

Apart from discharging the filtering capacitors once the amp is switched off, the bleeder resistor acts as a simple voltage regulator. Some designers use a high voltage Zener diode there, but from our experience, there is no need to go to such extremes.

A typical dual-decoupling circuit with a 120k bleeder resistor

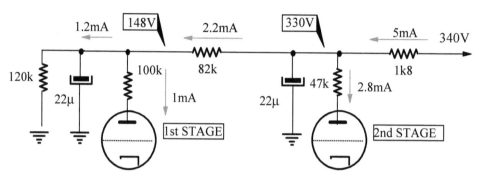

LOCAL NEGATIVE FEEDBACK IN THE OUTPUT STAGE: ULTRALINEAR CONNECTION

In two common situations, the screen grid in pentodes and beam power tubes can be connected to its power supply, making the tube operate as a pentode or strapped to its anode, which turns a pentode or a beam power tube into a pseudo-triode. However, providing the primary winding of the output transformer has a tap between $+V_{BB}$ and the tube's anode, connecting the screen grid to such an "in-between" tap results in the so-called "ultralinear" connection.

The name was popularized by Hafler and Keroes in Nov. 1951 issue of "Audio Engineering" magazine, in their article "An Ultralinear Amplifier." Instead of the screen grid being held at some stable DC voltage as in the pentode's case, the screen's voltage varies and follows the signal, which means that local negative feedback is applied to the screen. This results in a host of negative feedback benefits, as discussed in earlier chapters:

- the output impedance is reduced to about half of the triode's circuit
- linearity is improved, resulting in reduced distortion
- the power output is higher than in triode connection, approaching that delivered by a pentode
- the power output is more constant because the output stage behaves somewhere in between the voltage (triode) and current (pentode) source.
- the circuit combines the advantages of both the triode and the pentode stage without suffering from their respective setbacks.

A visual comparison of the three major operational modes

The almost horizontal curves of a pentode indicate very little anode current change with an increased anode voltage, thus a very high internal resistance, typical of current sources. On the other end of the spectrum, a triode's anode current jumps rapidly with increased anode voltage, due to its low internal resistance.

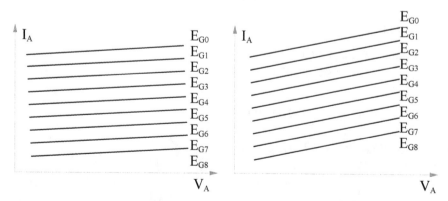

PENTODE: μ=gm*r$_l$, high r$_l$, high μ, low grid drive voltages (high sensitivity)

ULTRALINEAR CONNECTION : μ=gm*r$_l$, medium r$_l$, medium μ, medium grid drive voltages (medium sensitivity)

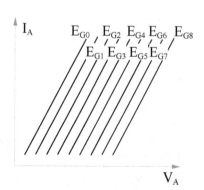

TRIODE: μ=gm*r$_l$, low r$_l$, low μ, high grid drive voltages (low sensitivity)

Since transconductance does not differ much between triodes and pentodes, the amplification factor and sensitivity track the internal resistance, so pentodes with high m need a minimal signal to drive them (high sensitivity). In contrast, low μ triodes are "lazy" and need large grid voltages! The same applies to grid bias voltages, which are very small for pentodes (typ. around -10V) and much higher for triodes (typ. -50 to -70V).

The distributed load is an "in-between" arrangement. Tubes so connected exhibit much lower internal resistance than pentodes but higher than triodes. Thus they are easier to drive than when triode-strapped but require higher negative bias voltages and larger grid signals for full power than when used as pentodes.

The ultralinear screen grid tap can be positioned anywhere between 0% (pentode connection) and 100%, where the pentode is connected as a triode. The higher the %, the stronger the local NFB is, the less output power we get at lower distortion.

The reduction in tap % (moving towards the pentode operation) brings two positives - the power stage is easier to drive, places fewer demands on the driver stage, and produces higher output power. It also brings two negatives, higher THD and higher output impedance Z_{OUT} (lower damping factor).

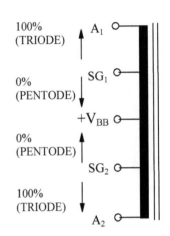

There is confusion about what this percentage is, a percent of the primary winding turns or percentage of the primary impedance. It is the former. The tap percentages published refer to % of primary turns counted from the power supply connection ($+V_{BB}$), which is the center tap for push-pull stages and the end connected to $+V_{BB}$ power supply for single-ended stages.

There is, however, a direct relationship between these numbers. For instance, the optimal percentage for 6L6 tubes, 43% of the winding, is equivalent to 18.5% of the primary impedance. For 6V6 tubes, 23% of the winding is 5.3% of the primary impedance.

The optimal U/L tap percentage varies for different output tubes and depends on what is to be achieved, minimum distortion or maximum power. It also depends on the load, reflected to the primary as the primary impedance Z_P.

For EL34 with primary anode-to-anode impedance Z_{AA}=4.3kΩ, the U/L tap should be at 33%, but with Z_{AA}=6kΩ, the optimum is 40%.

Notice a significantly lower THD (total harmonic distortion) levels of the U/L stage compared to the beam tetrode stage (for KT66 push-pull amplifier below), under 0.5% until $V_{PEAK} = V_{BIAS}$, when distortion rises due to operation in class AB_2 (grid current flowing). The transfer curve is more linear (very close to ideal) in the aptly-named ultralinear case.

TUBE	Screen tap % for minimum distortion
6V6	23-24
EL84	23
6L6	40-43
EL34	33

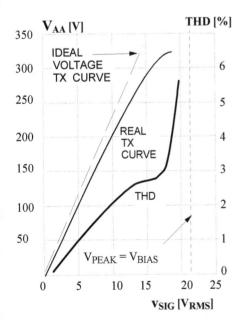

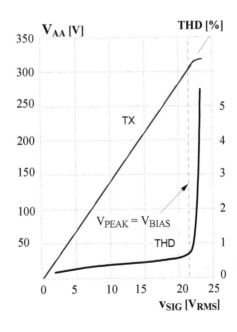

FAR LEFT: Voltage transfer curve (TX) and THD levels of KT66 push-pull amplifier (beam tetrode connection). The signal voltage is per tube.

LEFT: Voltage transfer curve (TX) and THD of the same KT66 push-pull amplifier in U/L connection (20% tap).

THIS PAGE WAS DELIBERATELY LEFT BLANK

TRANSISTOR AND HYBRID
GUITAR AMPLIFIERS

15

- KALAMAZOO MODEL 4
- JORDAN PERFORMER J100
- CRATE PALOMINO V8
- FENDER VAPORIZER

Many contemporary guitar amplifiers brazenly marketed as "all tube" designs are anything but, they are full of semiconductors (solid-state components), if not in their audio stages, then certainly in their power supplies and effect modules and stages. A typical class of "tube" amplifiers features solid-state (usually based on integrated circuits) preamp stages driving a tube power stage. This makes them "hybrid" amps in every sense of the word.

We'll also look at a couple of vintage transistor amplifiers. Before you skip this section, think again. Some of these vintage cuties look and sound better (warmer, more euphonic, or "musical") than quite a few "pure" tube guitar amps of current manufacture!

KALAMAZOO MODEL 4

This amp is not only cute and well made, but it also sounds better than many tube combo amps, either vintage or modern. The 1966 catalog blurb does not specify the output power level, but soon we'll measure the actual maximum power on the test bench.

The cabinet is the funkiest I've seen; the upper slanted chamber that houses the electronics is permanently attached to the lower rectangular speaker case. Its top rear panel is an integral part of the cabinet and cannot be removed, adding to the strength and stiffness of the whole cabinet. The removable bottom cover allows access to the speaker and its replacement if needed.

The front panel and the whole chassis are removable from the amp's front once two hidden bolts are removed.

There is plenty of space on the control panel for add-ons. The amp does not have a headphone or external speaker output. Its tremolo effect can only be switched by hand; the switch is part of the tremolo speed potentiometer, just as the on-off power switch is mounted on the tone control pot.

The cabinet is large enough for a 12" speaker, although the stock 10" Alnico speaker sounds pretty darn good.

Kalamazoo Model 4

SPECS

- Year: 1966
- Solid state, PCB construction
- Tone, volume, tremolo (rate only)
- 10" 8Ω Alnico speaker
- 9 bipolar silicon transistors, 5 diodes
- Dimensions: 16.5"x13"x 7"
- Weight: 13 pounds

The internal wiring

The internal wiring and the thick, high-quality PCB are beautiful sights. The two output transistors are mounted directly on the steel chassis (1). The only faux-pass is that the on-off switch (3) is on the opposite side of the chassis, so a twisted cable (2) crosses its whole length and depth. A better option would be to install a dedicated power switch next to the fuse holder in position (4).

However, since there is no output jack for an external speaker or footswitch jack for turning the tremolo on and off, location (4) would be ideal for that purpose - two jacks can easily fit there.

Two 1,000μF/50V elkos (5) are much larger than today's replacements, so they should be changed over. One works as the first filtering cap in the power supply, the other as the output coupling capacitor.

The audio circuit - preamp stages

When a guitar jack is plugged into input 1, the normally closed auxiliary contact that grounds the input opens, placing the 22k resistor in series with the signal. The 68k resistor wired to the common point Y and the hot lead of input 2 remains grounded since nothing is plugged into input 2, thus creating a 22k-68k voltage divider. The output signal in point Y is 68/(22+68) = 68/90 = 0.76 or 76% of the input signal.

The situation is reversed with a guitar plugged into input two, and input one left unused. The 68k resistor is now in series with the signal, and the 22k one is grounded, so the 68k-22k voltage divider has an attenuation of 22/90 = 0.24, reducing the signal's amplitude to 24%. Thus, Input 1 is HI, and input 2 is LOW!

Notice a 1n5 input bypass cap between the base of Q1 and GND. Together with the 3k3 series resistor it forms a low pass filter with its upper -3dB frequency (amplitude drop to 0.71% of its midrange level) of $f_H = 1/(2\pi RC) = 32.2$ kHz. This is way above the upper limit of the audio spectrum (around 20kHz), so its purpose is to attenuate radio frequencies and other HF interference.

Before proceeding with the signal flow, let's glance over the whole preamp circuit and get a broad feel for its topology. There are two cascaded stages with one NPN transistor in each stage. Both collector loads are connected to the same positive DC supply, V2.

Thus, both operate as common emitter stages and use an identical base biasing method, a high-value resistor between collector (output of the stage) and base, the input to the stage.

Each stage has local NFB since AC signals are also brought back from the collector to the base. This reduces gain but widens the frequency range and stabilizes the transistors' operation; it also minimizes the chance of thermal runaway and transistor destruction.

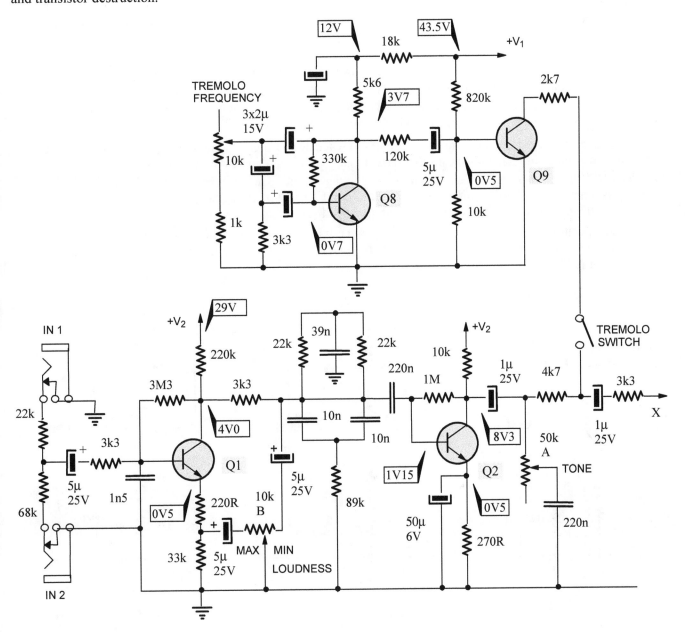

Twin T-filter

The filter stage between the two amplifying stages is called a Twin T-filter since components are arranged in the capital letter T. The two resistors/one capacitor in one filter are mirrored by two capacitors and one resistor in the other T-filter.

The analysis of this seemingly simple circuit is very complex and beyond the scope of this book. Usually, the values of the components are chosen so that the center rejection frequencies of the two filters coincide, but in this case, one filter has a CRF of 518Hz, the other 254Hz. The 518 Hz one is dominant since the overall CRF is just under 500Hz, as marked on this particular filter's attenuation-versus-frequency curve. At that frequency, the attenuation of the signal is maximal, around -17dB. The (-) sign indicates attenuation; a positive dB figure would indicate voltage gain.

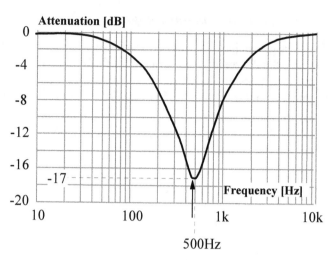

ABOVE: The attenuation characteristics of the Twin T filter in Kalamazoo Model 4 amplifier.

Loudness control

The Loudness control is a single pot, but its operation is far from simple. In the MAX position of the *Loudness* pot, its whole resistance of 10k is in series with the 5m bypass capacitor to GND. At 1kHz signal frequency, the reactance of that capacitor is around 32 ohms, which is way smaller than the 10k in series, so the voltage divider at the output of the 1st stage attenuates the signal down to 10/13.3 = 0.75 or 75% of its collector value. The other 5µF cap bypasses the 33k emitter resistor. Again, at 1kHz, its reactance of 32 ohms is way lower than the 33k it bypasses, so it can be considered a short circuit for AC signal. Thus, the AC gain is maximized.

In the MIN position, the whole 10k resistance of the *Loudness* pot is in series with the cathode bypass capacitor. Hence, a minimal bypassing action happens since the 10k is the same order of magnitude as the 33k emitter resistor. Thus, local negative feedback is introduced, which reduces voltage gain.

As the signal frequency drops into the bass region, the reactance of the cathode 5µF capacitor increases, and the bypass action reduces further. This means that bass frequencies are attenuated more than the treble range due to the increased NFB in the bass region (less AC signal bypassing action). The 32Ω reactance (at 1kHz) of the first 5µF capacitor is practically a short circuit across the output (point X), so for AC signals, the voltage in point X would be 32/3332 = 0.0096 or 0.96% of the collector output, practically zero.

In its middle position, the resistance of the linear tapered pot is equally split between the output and cathode bypass circuit. So, there is a lower output voltage compared to the MAX position and more cathode resistor bypass action compared to the MIN position.

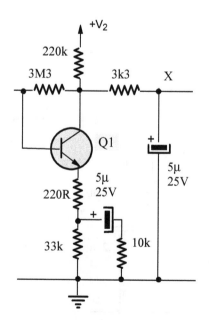

Loudness pot in MIN position (fully CCW)

Loudness pot in MIDDLE position

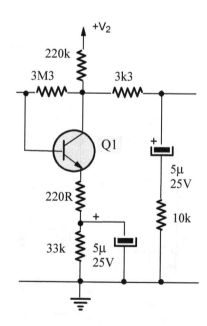

Loudness pot in MAX position (fully CW)

Notice that this Loudness control works in the opposite way to that of most loudness controls in hi-fi systems. Usually, as the volume control is turned CCW towards its min position, there is a need to boost the bass frequencies. This is due to the human ear's nonlinear sensitivity versus frequency characteristics (the so-called Fletcher-Munson curves). The bass-midrange-treble balance of the amp remains the same; all frequencies are reduced in proportion, but due to the lower power and thus reduced loudness levels (or air pressure levels from the speakers), the human ear perceives a tonal imbalance as if the bass was reduced more than the midrange and treble, so the sound is "thin" and lacks "body."

This amp's *Loudness* control reduces bass as the volume is reduced, so the treble is emphasized at lower volumes since the bass frequencies are cut more.

The power amplifier

Transistor technology has made tremendous advances since the mid-1960s when this amp was designed and built. However, the basic principles of transistor operation and the main building blocks and topologies have not changed much. The power amplifier of Kalamazoo Model 4 is a textbook example of a single-ended push-pull stage using quasi-complementary output transistors.

In the 1960s and 70s, it was hard to find fully complementary pairs of high power output transistors (NPN +PNP). Thus, somebody had an idea to use identical transistors in the output stage (both of PNP or NPN persuasion) and perform phase inversion in the preceding or driver stage, where it was not a problem to find matched but complementary transistor pairs (lower power transistors).

The top driver Q4 is an NPN type, but the bottom driver Q5 is its complementary PNP type transistor. Theoretically, in a quasi-complementary amp, both transistors in the driver stage (Q4 and Q5) would work in class B, as would the output transistors Q6 and Q7.

However, in practical designs, to minimize crossover distortion, both stages use a slightly elevated bias. 220Ω resistors achieve this in the bases of Q4 and Q5 and by a string of three diodes in their base circuit. Apart from providing approx. 1.8V (3x0.6V) voltage drop for biasing purposes, the three MR2064 diodes serve the same stabilizing purpose in the collector circuit of Q3, which works as a common emitter amplifier stage in class A.

Resistors marked with an * serve as temperature stabilizing elements, preventing or at least minimizing the possibility of thermal runaway. These 1Ω 1W resistors also serve as fuses. Since output power is $P=I^2/R$, the maximum current through these resistors is $I= \sqrt{(P/R)} = 1A_{RMS}$.

A much higher current can flow in peaks (for a short time), but should it persist, the resistor(s) will eventually overheat and burn out (go open circuit), just like a fuse.

Driver Q4 works as an emitter follower, connected between the base and emitter of Q6, and driver Q5 works as a CE stage. A common-emitter stage inverts the phase, but the emitter follower stage does not, so the required phase inversion is achieved.

Since both output transistors are an NPN type, their base driving signals must be equal but of opposite phase. The Q4 and Q5 driver stages both have a current gain approx. equal to the current gain β of their transistors.

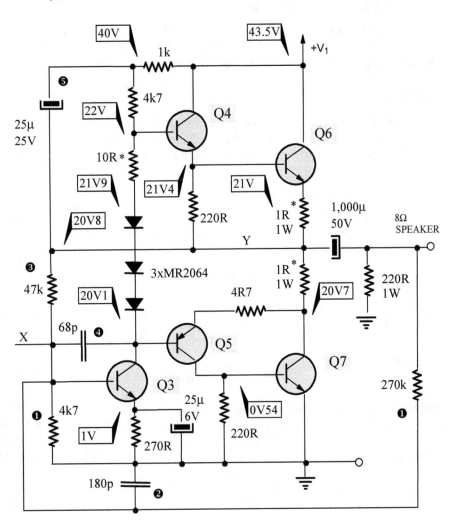

So, by choosing a complementary pair for Q4/Q5 (equal β) and making their emitter and collector resistors equal (220R), Q6 and Q7 receive driving signals of equal amplitude.

The Q4-Q6 arrangement of two (or more) transistors is the so-called Darlington compound or circuit, with one emitter follower driving another, so their cascaded current gains multiply. Thus, although EF provides no voltage gain, its current gain and thus power gain can be very high! The output impedance of this Darlington circuit is roughly equal to the load resistance R_L (220R here), while the input impedance is approx. equal to $\beta^2 * R_L$.

Q5-Q7 is not a Darlington circuit, simply a DC-coupled complementary pair of transistors. Since the asymmetrical power supply of approx. +44V is used, the output of the push-pull power stage (point Y) should be at a DC level exactly half that (+22V) to ensure equal positive and negative peaks of the signal. We measured 20.1V, slightly unbalanced, meaning the maximum output power will be somewhat reduced and the distortion slightly increased as one half-wave will be cut off sooner than the other.

This DC voltage at the output necessitates using a coupling capacitor (1,000μF) to prevent DC current from flowing through and destroying the loudspeaker. The lower -3dB frequency of the output high pass CR filter that cap forms with the 220R load resistor is $f_L=1/(2\pi RC) = 0.7$ Hz.

The global NFB from the output is taken via the 270k resistor back to the base of the Q3 CE stage (1). Notice a 180pF shunt capacitor to ground (2), thus forming an RC or low-pass filter with the upper -3dB frequency of $f_L=1/(2\pi RC)$ = 3,275 Hz. In series with the signal flow that would attenuate treble frequencies, but here it shunts the high-frequency feedback signals to GND. As signal frequency increases, the negative feedback is weaker and weaker, meaning this arrangement increases the treble content (treble boost).

The 47k resistor (3) provides NFB between the Q6-Q7 output rail (point Y) and the base of the Q3 driver. This ensures that point Y remains at one-half of the DC supply level $+V_1$, even if the DC power supply voltage fluctuates. This DC feedback means that the whole power amplifier is a unity-gain voltage follower for direct current and power supply fluctuations!

High-frequency oscillations can happen with certain combinations of source and load impedances. Transistors "transfer" resistance in both directions, unlike tubes where input circuits do not affect the output circuits. The 68pF capacitor (4) provides negative feedback between Q3's collector and base and thus provides an HF roll-off, reducing gain at high frequencies so oscillations cannot happen.

The 25μF capacitor (5) and 1k resistor (part of the base biasing pair 4k7 and 1k for Q4) form a decoupling circuit whose task is to filter out any objectionable hum caused by the AC ripple on the DC power supply line $+V_1$. However, the cap is not returned to GND but to point Y instead.

Topology

The whole power amp can be considered an operational amplifier with a high open-loop voltage gain (without negative feedback). This means that the voltage gain of the amplifier depends almost exclusively on the value of the series resistor R_S and the feedback resistor R_F: $A_V=-R_F/R_S=270k/(4k7+3k3)=33.8$.

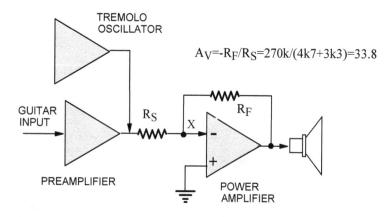

$A_V=-R_F/R_S=270k/(4k7+3k3)=33.8$

TREMOLO OSCILLATOR

GUITAR INPUT

PREAMPLIFIER

R_S

X

R_F

POWER AMPLIFIER

ABOVE: The topology of Kalamazoo Model 4

MEASURED RESULTS:
- -3dB BW: 27 Hz - 49 kHz (tone control in the mid position)
- P_{MAX}= 12 W (8Ω load, 1kHz signal)

Kalamazoo Model 4
TRANSISTOR & SUBSTITUTE CHART

- Q1: SPS4357 (PNP)
- Q2-Q8-Q9: SPS4035
- Q3: SPS4036
- Q4: SPS4037
- Q5: SPS44038
- Q6-Q7: SJ3408 (power transistors)

SUBSTITUTE CHART:
- SPS4035, SPS4036, SPS4037: NTE123AP, 2N3904 / BC547, PN2222, 2N4401 (NPN Si Transistor)
- SPS4038: NTE128, 2N3053 (NPN Si Transistor, 80V/1A)
- SPS4357: NTE159, 2N3906 (PNP Si Transistor)
- SJ3408: NTE175 (NPN Si Transistor 500V/3A)

JORDAN PERFORMER J100

Before I came across this early 1970s amp, I must admit that I had never even heard of Jordan Electronics, based in Pasadena, California. Schematics for their models J120 and J140 are available online, but they are totally different beasts from this model, J100. Luckily, there was a circuit diagram glued to the inside of its cabinet. That saved us hours of reverse-engineering the circuit from the actual amp.

J120 and J140 models use four transistors in the preamp stages (three in J100) and have two-knob tone controls (*Bass* and *Treble*). Their output stage uses a TS0059 integrated operational amplifier as a driver, followed by a complementary push-pull output stage. The supply voltage is +/-24V, compared to +/-17V in J100, so J120 and J140 are more powerful amps.

Notice the absence of a separate carrying handle. The back edge of the extruded aluminum chassis serves as a handle, a clever idea.

There is plenty of real estate for additional controls on top of the control panel and at the back. One option would be a Brite-Normal-Dark voicing switch, as implemented on J120 and J100 models.

There is no external speaker jack, headphone or line-out outputs, but these can be easily added should you like the sound of this amp.

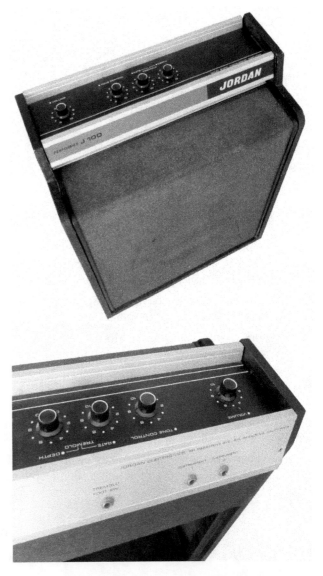

The internal layout

J100 uses all discrete components, soldered onto a large, well-laid printed circuit board (LEFT).

The copper side of the PCB only carries the two power supply filtering capacitors (1), 2,000 μF each.

The power transformer is mounted on the other (speaker) side of the particleboard shelf (2) that completely isolates and divides the speaker chamber from the top compartment.

The PCB is attached to the extruded aluminum chassis with no less than 13 self-tapered screws! Three of those attach the driver (3) and two output transistors (4) to the chassis via mica insulators (next page). That way, the chassis acts as a large heatsink.

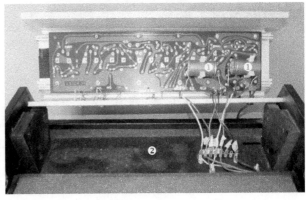

Lots of components were not installed (5), indicating that the same PCB was shared with another model, which had more features, including one more potentiometer (6).

The preamp consists of two stages, both with bipolar transistors. A common-emitter input stage is followed by a simple one-knob tone control (treble cut) and a compound stage with a PNP transistor working in a common-emitter mode, DC-coupled to an NPN transistor as an emitter follower.

A foot-switchable tremolo oscillator uses a PNP transistor as a phase-shift oscillator and a JFET transistor as an output switch.T

he stages are neatly arranged to follow the signal flow (left-to-right in the photo), making the circuit easy to follow and troubleshoot. All components are marked on the thick PCB.

Input stage   2nd and 3rd stages   Tremolo circuit   Power amplifier

The audio circuit - preamp stages

Before we proceed with the analysis of the preamp circuit, let me emphasize the primary purpose of this case study. It is not to become familiar with this relatively obscure and, at least from the circuit aspect (the chassis is mechanically well conceptualized and built), not exceptionally well-designed amp (more on that soon).

The aim is to gain at least a basic understanding of solid-state audio circuits and additional perspective by comparing them with tube circuits used in guitar amps.

Furthermore, while Kalamazoo Model 4 mainly used NPN transistors, a single power supply (+44V), and a quazi-complementary output stage, this amp uses mainly PNP transistors, dual power supplies (+/-17V), and a fully complementary output circuit.

The preamp circuit is powered from point Y, which is the -17V rail of the power amp. Notice the PNP transistor Q4, whose collector is not connected; its emitter is at the ground potential while its negative base is at -9V_{DC}. This means that the emitter-base PN junction is reversely polarized. This transistor does not work as any amplifying device but simply as a Zener diode, which, together with the 1k5 series resistor, regulates the -9V supply voltage.

Q6 is the heart of the tremolo oscillator, with 150k + 100k collector resistors. The output signal is taken from their junction (voltage divider) and fed to the 100k Depth potentiometer, whose wiper is capacitively coupled to the gate of a JFET transistor Q5.

The tremolo circuit and preamp stages of Jordan J100 amplifier

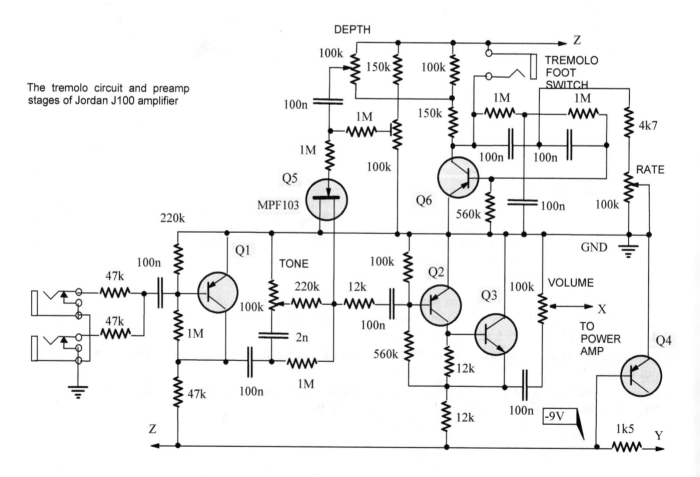

The gate is DC biased by the 100k trimmer potentiometer and the 150k resistor, so if your amp has this type of circuit (and quite a few brands used this JFET buffering method) and the tremolo does not work, before suspecting the JFET and/or replacing it, keep adjusting the trimmer until the JFET starts switching on and off at the tremolo frequency. This frequency or Rate is controllable from the front panel by the 100k potentiometer.

Notice that the tremolo effect is on by default unless you plug in a footswitch and short the collector of Q6 to GND and thus stop Q6 from oscillating.

Each of the two guitar inputs has its 47k series resistor and is capacitively coupled to the base of Q1, which works as a CE stage with a 47k collector load. The signal from the collector passes the simple one-pot Tone control (treble cut), the output of which is modulated by the treble signal through the Q5 mentioned above.

The second and third stages are DC coupled. A common-emitter (CE) stage (Q2) feeds an emitter follower Q3, in whose emitter leg is the load, a 100k Volume potentiometer.

The audio circuit - power stage

The output transistors Q15 & Q16 work as emitter followers, producing no voltage gain, but there is a power gain due to their high current gain. Since a symmetrical power supply is used, the output point is (at least in theory) at the GND potential, so there is no output coupling capacitor, which certainly has its sonic benefits. However, should one of the output transistors fail, a DC voltage would appear on the speaker and destroy it - the voice coil would burn out due to high DC current.

Notice the same bootstrapping circuit and positive feedback from the output through the 125µF elco back to the junction of the 82R and 150R resistors and, ultimately, the bases of the output transistors through the 1N4001 diode. This diode provides the bias to reduce the crossover distortion.

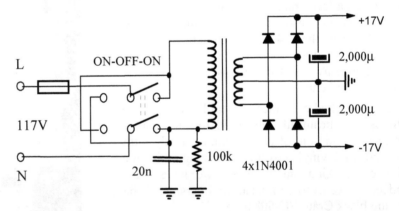

There are no thermal protection resistors in the emitter circuits of Q15/16. Q14 is an NPN driver whose base receives the signal from Q13, an emitter follower, which is in turn driven by the signal from the collector of Q12, one of the pair of transistors in the differential amplifier stage (Q11-Q12).

The power supply (ABOVE) and power amplifier (BELOW) of Jordan J100 guitar amplifier

Their common-collector load is the 100k resistor, whose upper-end DC voltage is reduced from +17V to +9V and stabilized by Q10, again connected as a Zener diode.

The global NFB is taken from the output and applied to the differential input stage's negative (inverting) input via the 560k resistor.

If we draw a block diagram of the amp, just as we did for Kalamazoo Model 4, we see that their topologies are almost identical.

The whole power amp can be considered an operational amplifier with a high open-loop voltage gain (without negative feedback). The voltage gain of the amplifier depends almost exclusively on the value of the series resistor R_S and the feedback resistor R_F:

$A_V = -R_F/R_S = 560k/18k = -31.1$ or 29.9 dB.

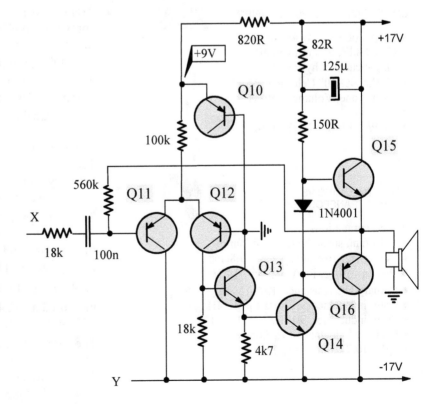

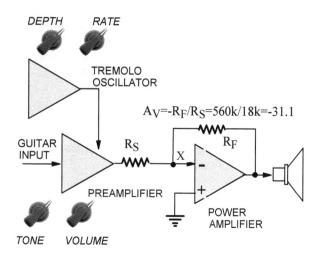

ABOVE: The topology of Jordan J100

$$A_V = -R_F/R_S = 560k/18k = -31.1$$

CRATE PALOMINO V8

We've had a few Crate power transformers, which were of decent construction and, of importance to us in Australia, had 240V primary taps. So, when this elegant-looking small combo amp appeared for sale on eBay USA for $150, we just had to get it onto our test bench.

The amp looked even nicer in real life than in the photos. The beautiful Trans-gold finished corner protectors and control panel contrast nicely with the creamy vinyl covering, making this the classiest looking small combo we have seen. Indeed, it seems to be a beautified version of the boring black Crate VC-508 amp.

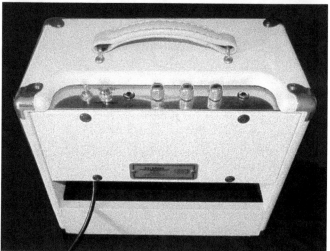

CRATE PALOMINO V8 AMPLIFIER   SPECS

- Output power: 5W @ 10% THD
- Gain 58dB, S/N ratio 66 dB
- Single channel hybrid amp
- 15Watt 4Ω 10" Celestion speaker
- Gain, Tone, Volume controls
- Line out
- 13"W x 12"H x 7-1/2"D, 18lbs

CRATE PALOMINO V8 OUTPUT TRANSFORMER:

- EI57 laminations
- a=19 mm, stack thickness S=20mm
- Center leg cross area A=aS = 3.8cm^2
- Power rating: P=A^2= 3.8^2 = 14W
- Primary DC resistance: R$_P$= 126 Ω
- Prim. inductance @120Hz L$_P$=13.4H
- Prim. inductance @1kHz L$_P$= 10.6H
- Leakage inductance L$_L$=5.6mH
- QF = 10.7/0.0056 = 1,911

Trans-gold is a widespread yellow powder-coated finish sprayed on nickel or chrome-plated parts. Once it is baked in an oven, the yellow powder becomes translucent and resembles gold plating.

The output transformer (both transformers were marked "Transtec") is rated at 14 Watts, an acceptable result for a 5-Watt rated amp. The low primary's DCR (126Ω) means it was wound with a thick wire and/or with a low number of turns. The primary inductance dropped from 13.4H @ 120Hz to 10.6H @1kHz, which would indicate that ordinary silicon steel laminations were used, not the more expensive and better performing GOSS material.

The leakage inductance at 1kHz was very low, only 5.6mH, indicating an excellent magnetic coupling between the primary and secondary windings, resulting in a high quality factor of Q=1,911.

With an 8Ω load, at 1 kHz, the primary impedance was measured as 5.24 kΩ, so the impedance ratio is 5,240/8 = 655, and the voltage and turns ratios are TR=VR=25.6.

The power transformer also uses EI57 laminations with S=30mm, so its power rating is P=A^2= 5.7^2= 32 Watts. The two tubes draw 0.8A and 0.6A, so the heater circuit requires 1.4*6.3 = 9 Watts.

The audio circuit draws around 60 mA in total, so the HV supply dissipates 0.06A*253V= 15W, plus say 2W for the op-amp chip and losses, for a total of 26 Watts. 26 out of 32 Watts can be considered a borderline case (since transformer losses need to be added to those 26 Watts of load power); this power transformer will get hot since it operates at the maximum capacity.

The PCB and topology issues

The under-the-chassis inspection shows the most common screw-up modern tube designers make - bundling all transformer output wires together (1). Also, notice the Line Out jack (2) very close to the power switch (3). Why is that important?

Because in this amp, the signal to the second tube stage and the output stage passes through the shorting contacts on the Line Out jack! The output transformer's primary is fed high voltage immediately after the rectifier (4). There is too much ripple at that point (not enough filtration).

The amp had a very pronounced hum (buzz) and also, as the Gain was increased, a pronounced high-frequency hiss.

No studio recording with this abomination, that's for sure. One or all three of these design errors could be the cause of the hum. The hiss is definitely generated in the two op-amp input stages.

Notice that the manufacturer did not use a DIL socket for some reason (to save a few cents) but soldered the input operational amplifier chip directly to the board (5). By the way, DIL stands for "Dual-In-Line" and refers to the fact that the eight pins are in dual (parallel) lines of four.

Replacing it with a superior-sounding chip (or one without hiss!) is now a tedious and frustrating desoldering exercise. Once you've done that, solder in an 8-pin DIL socket and then try various ICs to see if there is any difference, not just in sound but also in dynamic response and noise levels.

TL072 is a dual (two separate amplifiers in one chip) low-noise JFET-Input general-purpose operational amplifier. It has the maximum input voltage range of 30V (or ±15V).

The circuit diagram on the next page shows that the two 1N914 diodes connected between the positive input (pin 3) and the positive and negative voltage supply rails provide input overvoltage protection. When the amplitude of the input signal exceeds one of the supply voltages plus the forward voltage of a diode, the diode starts conducting and diverts the current to the power supply rather than into the op-amp input, which the excess current could damage.

The audio circuit - input stages

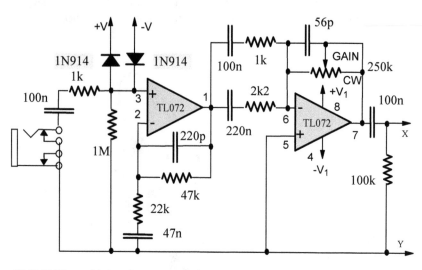

ABOVE: The solid state input stages of Crate Palomino V8 © St. Louis Music Inc.

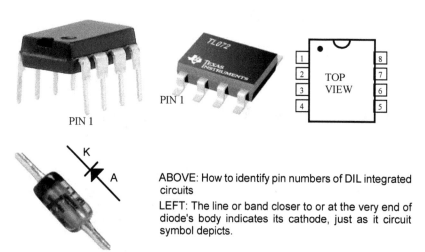

ABOVE: How to identify pin numbers of DIL integrated circuits

LEFT: The line or band closer to or at the very end of diode's body indicates its cathode, just as it circuit symbol depicts.

You may instinctively feel that there are two voltage amplification stages here, and you'd be correct in that assumption. But to understand how the circuit works, we need to sidetrack a bit and cover the basic circuits with operational amplifiers.

An op-amp has two inputs, a positive or non-inverting input and a negative or inverting input. Internally, op-amps consist of hundreds of components - diodes, resistors, bipolar and JFET transistors, even low capacitance capacitors, all etched on a single piece of silicon through the epitaxial process. Understanding their external operation isn't too mentally demanding despite their internal complexity.

In these simplifications and approximations, we assume that the open-loop gain of the op-amp is infinite, so the closed-loop voltage gain of these amplifying circuits will depend only on the external resistive components.

Closed-loop refers to the fact that all op-amp circuits use negative feedback from the output to one or both of the inputs. RF and CF are called feedback resistor and feedback capacitor and named with subscript "F"!

The voltage gain of the inverting amplifier is $A_0 = -R_F/R_S$, and the gain of the non-inverting amplifier is $A_0 = 1 + R_F/R_S$. The two linguistic terms have the same meaning as amplification stages with vacuum tubes; an inverting stage inverts the signal's phase, the non-inverting stage does not (the input and the output signal are "in phase," meaning there is no 180-degree phase shift between them).

The third example from the left is a voltage follower, as an ideal cathode follower, with a "gain" of one and without any phase inversion, so the output voltage "follows" the input voltage.

The circuit on the far right is equivalent to Palomino's 2nd stage. Disregarding the two capacitors for a moment (they don't change the midrange gain), the midrange voltage gain is $A_V = -R_F/R_S = -250/2.2 = 114$ times or $20*\log 114 = 41$dB.

This assumes that the "gain" pot is fully clockwise (marked CW on diagram), meaning the maximum volume. As that pot is turned CCW (towards "0"), more and more of the feedback resistor is shorted out and the gain of that stage and the whole amp drops.

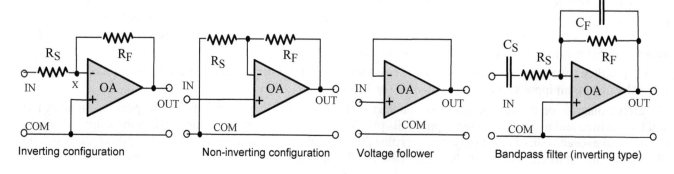

Inverting configuration   Non-inverting configuration   Voltage follower   Bandpass filter (inverting type)

The C_S and R_S form the low pass filter with R_F only in the NFB loop (no C_F). Likewise, the high pass filter is formed by C_F and R_F with R_S only at the input (no C_S). Combine the two, and you get a bandpass filter, whose A-f (amplitude versus frequency) curve is drawn on the left.

The midrange gain is A_0. It drops at the rate of 20dB per decade (ten times the ratio of frequencies). There are two frequencies of interest, the lower -3dB frequency f_L (determined by the series RC components) and the upper -3dB frequency f_U, determined by the parallel or feedback RC pair:

$f_L = 1/(2\pi R_S C_S)$, $f_U = 1/(2\pi R_S C_S)$

In Palomino's case, $f_L = 329$ Hz and $f_U = 1,137$ Hz. Notice that as R_F is lowered from maximum volume, f_U increases, so for instance when $R_F = 100k$, $f_U = 2,284$ Hz.

The gain of the stage drops to 1 or 0dB ($\log 1 = 0$) at frequencies f_{0L} and f_{0U}. They can be estimated as $f_{0L} = 1/(2\pi R_F C_S)$ and $f_{0U} = 1/(2\pi R_S C_F)$.

For the non-inverting 1st stage, the same situation applies. The series RC pair (22k and 47n) determine $f_L = 154$Hz and the parallel feedback RC pair (47k and 220pF) determine $f_U = 15,392$Hz.

If you calculate the two zero gain frequencies you will get $f_{0L} = 1/(2\pi R_F C_S) = 72$ Hz and $f_{0U} = 1/(2\pi R_S C_F) = 32,883$ Hz.

This input stage of the amp attenuates frequencies below the lowest guitar frequency of 80Hz and the ultrasonic and radio frequencies above 33 kHz.

Since there is that "1" in the gain formula $A_0 = 1 + R_F/R_S$, the voltage gain cannot drop below one, so the A-f curve looks slightly different, as illustrated on the right.

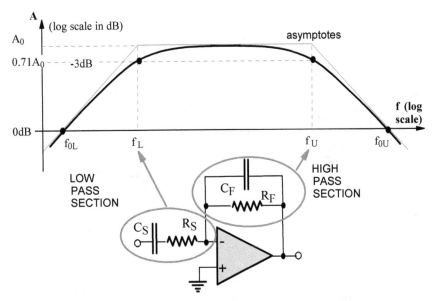

ABOVE: The amplitude-frequency characteristic of the inverting operational amplifier stage configured as a bandpass filter, as in Palomino V8's 2nd stage

BELOW: The amplitude-frequency characteristic of the non-inverting operational amplifier stage configured as a bandpass filter, as in Palomino V8's 1st stage

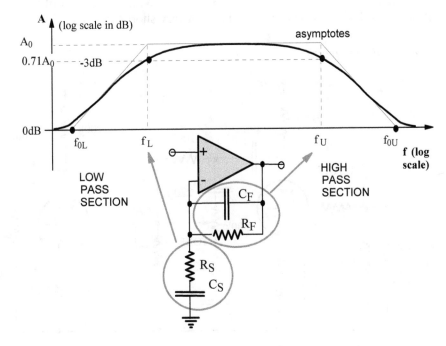

The audio circuit - tubed section

While the topology of the tube section seems standard, four issues stand out. First, the tone control circuit (1) is unusual; it does not operate in the usual fashion but as a midrange scoop. As with any tone control design, some guitar players will like it; others will loathe it. However, should you decide to keep it, notice that it is an integral part of the complex voltage divider (or the other way around?), formed by five resistors: 22k1, two times 221k, 100k, and the 10k "Level" pot (2).

At maximum volume, only the voltage across the pot is passed onto the second 12AX7 stage. The attenuation of this voltage divider is A = 10/(10+100+2x221+22.1) =1 0/574.1 = 0.0174, so only 1.74% of the signal is used! A gain of 1/0.0174 = 57.4 times would be needed to neutralize this attenuation, which is the voltage gain of one whole 12AX7 stage! So, Crate designers attenuated the signal 57.4 times and then included the hiss & hum-inducing input solid state section to bring the signal levels back to normal?!? This makes no sense to me.

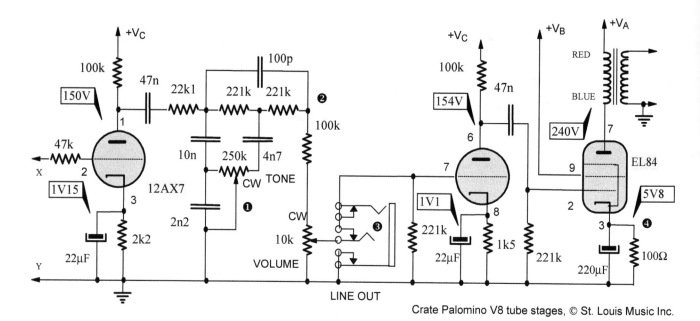

Crate Palomino V8 tube stages, © St. Louis Music Inc.

The possible explanation is that the line-out signal is taken from the wiper of the Volume potentiometer (3), so the signal had to be sufficiently attenuated. Notice that the signal loops through the two normally closed contacts of the line-out socket. Once a line-out jack is plugged in, the signal cannot reach pin 7 (the grid) of the 2nd stage, so the last two stages of the amp are thus inactive.

At idle, the cathode of EL84 (4) is at $5.8V_{DC}$, and since the cathode resistor is 100Ω, the cathode current is 58mA, which is way too high for EL84. With 56mA of anode current and 240-5.8=234 V between the anode and cathode, the output tube dissipates P=0.056*234 = 13.1 Watts. EL84 is only rated at 12 Watts, so another blunder from Palomino's designers. This tube will have a very short and stressful life.

The modified circuit

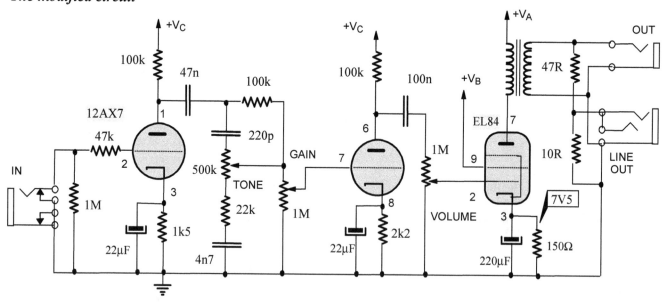

Since the original amp was performing so poorly, we sold the PCB and the tubes (which almost paid for the amp's cost) and rebuilt the circuit in a point-to-point fashion. The simplest circuit suggestion is illustrated above, but you can build any design you like.

The Line Out signal was taken from the speaker output through a simple voltage divider. Since $P=V^2R$, at the maximum power of P=5 Watts on R=8Ω load, the signal voltage is 6.32 Volts. To get the line level of 1 V at the LINE OUT, we need a voltage divider of around 6:1 ratio. With a 47Ω series resistor, the parallel resistor needs to have around five times lower resistance or 10Ω.

The power supply

The power supply is straightforward. The positive and negative voltages for the op-amp chip, +V and -V, are derived from the center-tapped heater secondary winding. Each side of the symmetrical supply has a voltage doubler using two 1N914 diodes and two 220µF capacitors.

As mentioned already, the HV power supply is inadequate. The anode voltage for the power stage isn't filtered enough. This is easily fixed by adding one elco, say 47 or 100µF, and a 100Ω resistor or a 3-10H choke, with 100mA minimum current capacity. Why this wasn't done in the factory is beyond comprehension!

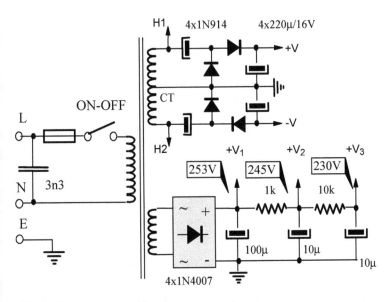

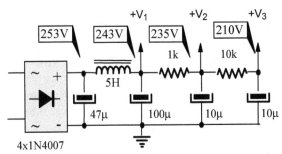

ABOVE: The proper way to filter the HV power supply with only two additional components

LEFT: The power supply of Crate Palomino V8 © St. Louis Music Inc.

The output stage

The output transformer's primary impedance was measured with an 8Ω load as 5.24kΩ, but then we noticed that the speaker in this amp is 4Ω. That means the output tube is working into an impedance of 2k6 (half of 5k24), which is way too low for EL84 pentode. In the original regime the tube operated above its maximum rated dissipation for more than half of the time, between points X and Y on the graph.

The new operating point will be at roughly the same anode voltage but higher bias, -7.5V instead of -5.8V. The anode current will decrease to around 46 mA (48 mA cathode current), so the power dissipation will be P=0.046*240 = 11.0 Watts, which is within the 12 Watt capability of the tube.

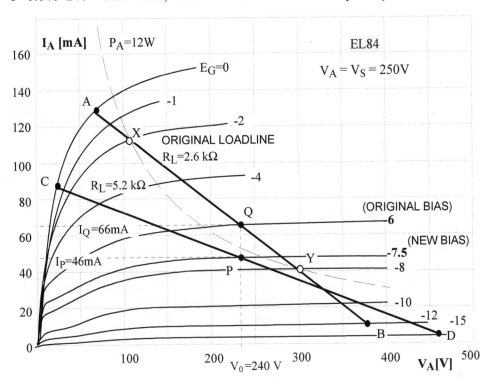

We have to either change the output transformer (to suit the 4Ω speaker) or replace the stock speaker with an 8Ω one.

The new cathode resistance needs to be $R_K = V_K/I_K = 7.5/0.048 = 156 \Omega$ and must dissipate $7.5*0.048 = 0.36$ Watts of power, so use a 150Ω, 2W resistor.

Notice how much longer the negative swing P-D is than the original Q-B, indicating much lower distortion of the output stage.

In conclusion, one of the best-looking amps but not so good in terms of its design, sound, noise, and hum (in its original state)!

FENDER VAPORIZER

Outside looks and features

Vaporizer was one of the "Pawn Shop Special" series combos, with a tube power stage and hybrid preamp stages. Apart from its space-age (meaning the 1950s!) look and funky colors such as Surf Green and Slate Blue, Vaporizer is a 12 Watt EL84 push-pull design, with spring reverb and two 10" speakers.

A local shop had two of these amps on sale for AU$450 (US$325 in 2016), which was apparently below the shop's cost, according to the salesman. The recommended retail price was AU$899.

The sticker inside shows that the amp was made in 2013, meaning the two amps languished on the shop floor for three years, which wasn't a good sign at all! Local amp buyers didn't like its looks, the tone, or both. Intrigued and needing case study amps for this book, we bought the blue one, hoping that we would be able to decipher why this model (and the whole Pawnshop series) had been a failure for Fender.

FENDER VAPORIZER
- Two inputs, Normal and Bright
- Controls: Volume, Tone and Reverb
- "Vaporizer Mode" (bypasses volume and tone controls)
- "Vaporizer" footswitch
- Spring Reverb (IC-driven)
- External speaker jack
- Two 10" 16Ω speakers (in parallel)
- Tubes: 2 x 12AX7, 2 x 6BQ5/EL84
- Output power: 12 Watts
- Dimensions: 24.75" (63 cm) wide, 17" (43 cm) high and 10.5" (27 cm) deep
- Weight: 37 lb. (16.8 kg)

ABOVE: You either love or hate Vaporizer's looks. Personally, I find it cute and funky, but I am a conservative man in his fifties, do younger guitar players like it is the real question!

The insides

Two inputs, volume, tone, and reverb control, make up a very Spartan control panel: no standby switch, no power reduction control of any kind, no presence or tremolo.

The Vaporizer is constructed on three printed circuit boards (next page). The smallest one (3) only carries a couple of capacitors and two 1/4" jacks for the speaker and "Vaporize" footswitch. All four tube sockets and associated components are on the larger board (4), while the top controls (pots and jacks) are on the main board (5), together with integrated circuits and the rest of the resistors and capacitors.

Of all contemporary amps featured in this book, Vaporizer is one of the least mod-friendly, meaning the most difficult to work with and modify.

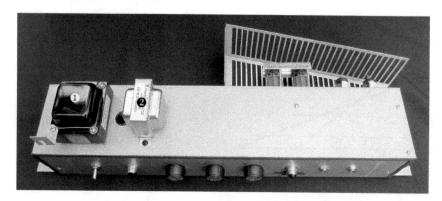

LEFT: The long and shallow chassis is held in place by only three thin bolts. However, both the mains (1) and output transformers (2) are so tiny and light that the mechanical support does not become an issue.

There is plenty of room on top of the chassis to install a choke should you wish to. The output transformer should be repositioned to the right with the choke installed in its place on the chassis, oriented with its main axis at 90 degrees to that of the output transformer.

There are four DIL (Dual-In-Line) integrated circuits and quite a few JFETs working as switches. These components are tiny and hard to reach. For that reason (and because it would take too much time for little benefit), we will not draw its complex circuit diagram; it is available online for download should you have nothing better to do with your life. We will discuss the block diagram instead.

As for the wiring, this must be the worst example of a lead dress we have ever seen in any piece of equipment (6). Since poor wiring practice is so widespread and the competition for the worst stuff-up is so stiff among amp makers, this speaks volumes. All the cables, high voltage, heater, signal are bundled together. Even worse, one wire is way too short, so there is a considerable strain on it (7).

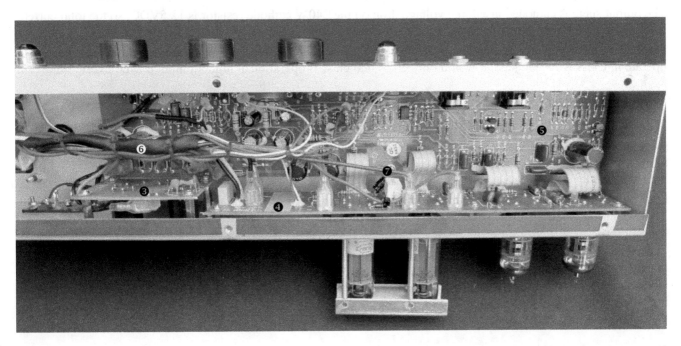

Topology

ABOVE: The inside view of Fender's Vaporizer amp.

BELOW: The simplified block diagram of Fender's Vaporizer amp. Not all JFETs are shown, neither is the dual 4560 comparator circuit that controls the operation of those semiconductor switches and reroutes the signal flow through the amp.

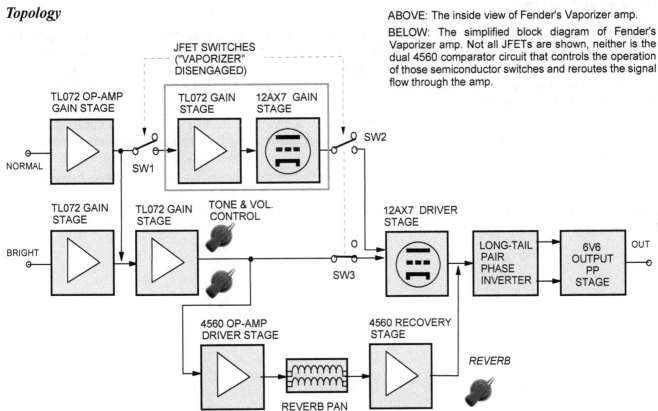

Both Normal and Bright channels feature an input stage with an operational amplifier, TL072 in this case. The voicing is achieved through different RC circuits in op amps' feedback networks. The two outputs join together and feed another gain stage with TL072 op-amp. The tone and volume control potentiometers are at that stage's output.

With "Vaporizer" off, the switches SW1 and SW2 are open, and SW3 is closed (as shown), so the signal goes through the two op-amp stages, volume and tone controls, 12AX7 tube stage, long-tail phase inverter, and feeds the grids of the EL84 output tubes.

The spring reverb is driven by a single op-amp gain stage using a 4560 integrated circuit; a similar gain recovery stage is at its output.

In "Vaporizer" mode, switches 1 and 2 are closed, and SW3 is open, so instead of the TL072 gain stage with volume and tone controls feeding the 12AX7 driver stage, the signal goes through another TL072 stage and an additional 12AX7 tube stage. These two stages (in the light gray frame) are only active in the "Vaporizer" mode.

The output stage is cathode biased; the voltage drop on the 240W cathode resistor is 12.5VDC, so the total current is12.5/240=0.052A, or 26mA per tube, which is a pretty "cold" biasing, closer to class B operation.

Some EL84 push-pull output stages work at 35-40 mA per tube, closer to class A operation, meaning the stage would operate in Class A for longer and would only cross into class B operation at higher output power levels.

Output transformer

Even at first glance, the output transformer seems tiny. After some strategical use of a ruler and calculator, indeed, the power rating of its core is under 14 Watts, meaning less clean power at higher power levels.

Like its brethren of similar power rating, the Excelsior amp, Vaporizer's output transformer foreshadows plenty of early distortion.

The DC resistance of the primary winding's halves is almost identical, a well-balanced design. Although the primary inductance is on the low side, the extremely low leakage inductance increases the quality factor to a very high 2,000+ figure!

FENDER VAPORIZER OUTPUT TRANSFORMER:

- EI56, a=18.7mm, S=20mm
- $A = aS = 3.7 cm^2$
- $P = A^2 = 3.7^2 = 13.7$ W
- Primary inductance L_P= 9.5H@1kHz,
- Leakage inductance L_L=4.7 mH
- Quality factor: $QF = 9.5/0.0047 = 2,021$
- Impedance ratio: 772
- Primary impedance with 8Ω speaker: 6.2kΩ

Power up and tone test

There was no residual hum when the amp was powered up, but a few glaring problems quickly became apparent. First, turning the Tone control clockwise boosted the treble but also introduced a significant high-frequency hiss.

Second, turning the reverb know clockwise had hardly any effect but added low-frequency hum so much so that the amp became unusable! Only at the maximum setting of the potentiometer was there any noticeable reverb effect! So, not just incredibly noisy but also a very weak reverb.

Finally, as soon as the "Vaporizer" mode is activated (via the footswitch), the hiss increases to unbearable levels. Even without the hiss, the idea of bypassing the volume control and cranking the amp to its maximum power level is questionable.

This is one seriously flawed design and an amp that tarnishes Fender's reputation. It should have never been released in this form.

As with Crate palomino V8, the only option was to sell its printed circuit boards off and build a completely different, hardwired amp in its funky cabinet.

OTHER TUBE AMPLIFIER BOOKS BY IGOR S. POPOVICH

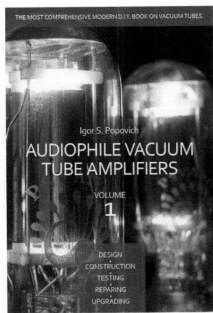

Audiophile Vacuum Tube Amplifiers, Vol. 1:

- BASIC ELECTRONIC CIRCUIT THEORY
- ELECTRONIC COMPONENTS
- AUDIO FREQUENCY AMPLIFIERS
- PHYSICAL FUNDAMENTALS OF VACUUM TUBE OPERATION
- VOLTAGE AMPLIFICATION WITH TRIODES - THE COMMON CATHODE STAGE
- OTHER VOLTAGE AMPLIFICATION STAGES WITH TRIODES
- TETRODES AND PENTODES AS VOLTAGE AMPLIFIERS
- FREQUENCY RESPONSE OF VACUUM TUBE AMPLIFIERS
- IMPEDANCE-COUPLED STAGES AND INTERSTAGE TRANSFORMERS
- NEGATIVE FEEDBACK
- TONE CONTROLS, ACTIVE CROSSOVERS AND OTHER CIRCUITS
- PRACTICAL LINE-LEVEL PREAMPLIFIER DESIGNS
- PHONO PREAMPLIFIERS
- SINGLE-ENDED TRIODE OUTPUT STAGE
- PRACTICAL SINGLE-ENDED TRIODE AMPLIFIER DESIGNS
- PRACTICAL SINGLE-ENDED PSEUDO-TRIODE DESIGNS
- SINGLE-ENDED PENTODE AND ULTRALINEAR OUTPUT STAGES

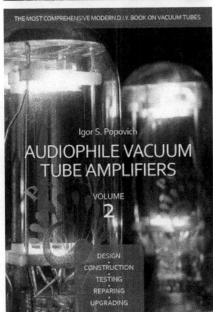

Audiophile Vacuum Tube Amplifiers, Vol. 2:

- PRACTICAL SINGLE-ENDED PENTODE AND ULTRALINEAR DESIGNS
- PUSH-PULL OUTPUT STAGES
- PRACTICAL PUSH-PULL AMPLIFIER DESIGNS
- BALANCED, BRIDGE AND OTL (OUTPUT TRANSFORMERLESS) AMPLIFIERS
- THE DESIGN PROCESS
- FUNDAMENTALS OF MAGNETIC CIRCUITS AND TRANSFORMERS
- MAINS TRANSFORMERS AND FILTERING CHOKES
- POWER SUPPLIES FOR TUBE AMPLIFIERS
- AUDIO TRANSFORMERS
- TROUBLESHOOTING AND REPAIRING TUBE AMPLIFIERS
- UPGRADING & IMPROVING TUBE AMPLIFIERS
- SOUND CONSTRUCTION PRACTICES
- AUDIO TESTS & MEASUREMENTS
- TESTING & MATCHING VACUUM TUBES

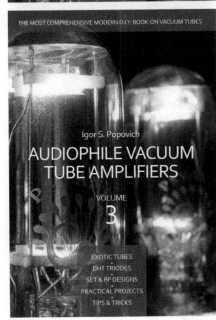

Audiophile Vacuum Tube Amplifiers, Vol. 3:

- THE FRONT-END: SUPERIOR INPUT & DRIVER STAGES
- FROM SHOCKING TO SUBLIME: LESSONS FROM COMMERCIAL LINE STAGES
- DIY LINE-LEVEL PREAMPLIFIERS: $10,000 SOUND ON $500-$1,000 BUDGET
- THE STARS OF THE AUDION ERA: ANCIENT TUBES IN MODERN AMPS
- CHEAP & CHEERFUL: PREAMP & DRIVER TUBES FOR AUDIO EXPLORERS
- SLEEPING GIANTS: OUTPUT TUBES FOR THOSE WHO WANT TO BE DIFFERENT
- THE QUEEN OF HEARTS: SINGLE-ENDED AMPLIFIERS WITH 300B TRIODES
- TRIODES, PENTODES AND BEAM TUBES: MORE SINGLE-ENDED DESIGNS
- BIG BOTTLES: SET AMPLIFIERS WITH HIGH VOLTAGE TRANSMITTING TUBES
- THE WAY IT USED TO BE: VINTAGE PUSH-PULL AMPLIFIERS
- NEW? IMPROVED? MODERN PUSH-PULL AMPLIFIER DESIGNS
- CUTE, CLEVER OR CONTROVERSIAL? INTERESTING IDEAS FROM TUBE AUDIO'S PAST AND PRESENT
- THRIFTY TIPS & TRICKS: TIME & MONEY SAVING IDEAS
- OUTPUT AND INTERSTAGE TRANSFORMERS: FROM COMMERCIAL BENCHMARKS TO YOUR OWN DESIGNS
- MEASUREMENTS VERSUS LISTENING AND OTHER AUDIO DESIGN DILEMMAS

INDEX

Printed in the USA
CPSIA information can be obtained
at www.ICGtesting.com
LVHW082328230324
775346LV00004B/589